INTERPRETING LAND RECORDS

Other titles by Donald A. Wilson

Deed Descriptions I Have Known But Could Have Done Without
Easements and Reversions
Brown's Boundary Control & Legal Principles (with Walter G. Robillard)
Evidence and Procedures for Boundary Location (with Walter G. Robillard)

INTERPRETING LAND RECORDS

DONALD A. WILSON, BSF, MSF, LLS, PLS, RPF

LAND BOUNDARY CONSULTANT, NEWFIELDS, NEW HAMPSHIRE

JOHN WILEY & SONS, INC.

Published by John Wiley & Sons, Inc., Hoboken, New Jersey
Published simultaneously in Canada

For general information on our other products and services or for technical support, please contact our
Customer Care Department within the United States at (800) 762-2974, outside the United States at
(317) 572-3993 or fax (317) 572-4002.

Wiley also publishes its books in a variety of electronic formats. Some content that appears in print may
not be available in electronic books. For more information about Wiley products, visit our web site at
www.wiley.com.

Library of Congress Cataloging-in-Publication Data:

Wilson, Donald A., 1941-
    Interpreting land records / Donald A. Wilson
        p. cm.
    Includes bibliographical references and index.
    ISBN-13: 978-0-471-71543-6 (cloth)
    ISBN-10: 0-471-71543-3 (cloth)
    1. Surveying.   2. Land tenure.   I. Title.
    TA549.W55 2006
    526.9--dc22

                                        2005033918

Printed in the United States of America

10 9 8 7 6 5 4 3 2

CONTENTS

FOREWORD

The land surveyors' work can be divided into two categories: retracement of established boundaries, and performing work preliminary to the establishment of new boundaries. *Interpreting Land Records* is primarily directed toward the retracement surveyor and persons studying to become land surveyors. In addition to land surveyors, lawyers, judges, title abstractors, and others in the real estate field, will also find much of value.

Retracement, by its very nature, is often the most difficult task of the surveyor. It is more than measuring objects found on the ground. New Hampshire surveyor E. N. Roberts wrote some thirty years ago, that one must first determine what is to be measured, and then determine where. The first step is to recover the records. The next step is to determine the legal meaning of the words in those records. *Interpreting Land Records* thoroughly details everything you need to know to start the work.

The next step is to apply the words in the documents to conditions on the ground. This must be done without subtracting from the rights of others. The land surveyor is, in other words, asked to give an opinion as to where the actual on-the-ground locations of certain boundaries are. That opinion should be defendable in a court of law. As a surveyor you must always ask yourself: Would the judge approve? Would the jury agree? *Interpreting Land Records* thoroughly covers the law regarding ambiguous words, conflicts between words in the documents, and conflicts between those words and items outside the documents, as pertains to the location of property boundaries.

Disagreements between surveyors are rooted in the following failures: the failure to recover the necessary documents, the failure to correctly read the document, the failure to apply the correct principle of boundary line law, and the failure to correlate the words in the documents to conditions on the ground. Most, if not all, of those disagreements would not occur if the surveyors engaged in the retracement of land boundaries had a working knowledge equal to what is given herein. My regret is that this treatise was not available forty years ago.

The surveyor establishing markers on the ground for new division lines, and writing words to be included in legal documents that will be used to create new property lines—by understanding retracement—will be able to produce a product that will be free of the ambiguities too often found in present and past records. Such ambiguities are costly to investigate, lead to discord among neighbors, hinder the management and use of real estate, upset the peace of the community, and often result in litigation that is costly and often never-ending.

Read and understand this book—go and sin no more.

GEORGE F. BUTTS, L.S.
Chittenden, Vermont

Note to the Reader

This work has been produced from the insights of many people dealing with a host of situations. It is not intended as advice, legal or otherwise, but was assembled with survey-related issues in mind. The author's goal was to bring together support, explanations and illustrations of those elements that make up land description, whether written or unwritten. The book is intended for *guidance* in resolving ordinary problems and explaining difficult, misleading, or ambiguous records as an aid in leading to resolution of conflicts, in and out of the record.

Historical research and investigation requires creativity, patience and skill, and this book should be considered a marker on the path towards competency. . . .

ACKNOWLEDGMENTS

I wish to express appreciation to Charles C. Scott, author of *Photographic Evidence*, Volumes I, II, and III, which I have read many times and referred numerous persons to for information, for his granting permission to use selected examples for the Chapter 11 document examination. He helped illustrate several techniques, demonstrating that a picture truly is "worth a thousand words."

The Eastman Kodak Company of Rochester, New York, a leader in technological advancements of photographic media and techniques, provided materials and literature for the solution of individual problems. Their gracious permission to use some of their fine examples is greatly appreciated. Appreciation is also expressed to the Sanborn Mapping Company for the use of illustrations from fire insurance maps in Chapter 9.

The illustrations in Chapters 9 and 11 provided by the aforementioned have greatly enhanced the book by providing real-life examples of what was discussed in the text. Without them, readers would have been left to visualize for themselves.

I wish to thank Francois "Bud" Uzes for his permission to copy the example of the surveyor's chain used in Chapter 2. Bud is an historian and collector of antique surveying equipment, and his publications contain excellent illustrations of early surveying tools.

I want to especially thank two close friends and colleagues who took time from their busy schedules to give me support, encouragement, and much needed literary input, criticism, and correction. Both Mike Shores of Cartographic Associates in Littleton, New Hampshire, and George Butts, surveying consultant and former staff surveyor for the Green Mountain National Forest in Vermont, were of great assistance. Mike reworked part of the chapters on maps and photographs, while George proofread the entire manuscript and offered helpful comments throughout. These gentlemen approached this project as if it were their own, and fine-tuned those things that an author sometimes gets too close to and overlooks.

My wife, Christine, deserves an expression of appreciation for her patience and tolerance in helping me overcome enough of the idiosyncrasies of a new computer system to get this manuscript up and running. Her knowledge and experience assisted me in moving from a typewriter to a keyboard without a major crisis, although she stood by me through a few minor ones.

And finally to my publisher, John Wiley & Sons, thank you for believing in this book and including it into your growing list of titles for land surveying and land investigation. Hopefully it will serve as a natural complement to other works in this series.

DONALD A. WILSON
Land Boundary Consultant

Our ignorance of history makes us libel our own times.

People have always been like this.

—Gustave Flaubert (1821-1880)

CHAPTER 1

INTRODUCTION TO LAND RECORDS

It is common experience that descriptions of property
in rural communities are somewhat indefinite, and even
inaccurate, and courts deal leniently with such cases in
seeking to ascertain the intention of the parties.

The antiquated doctrine that a document must be construed
solely within its four corners, no matter how puzzling
the problem, is no longer the law of this state.

—People v. Call (1927)
129 Misc. 862, 223 N.Y. Supp. 257

The term "land records" means different things to different people. Strictly defined, it means *records* dealing with, or relating to, *land*. In its narrow sense, people immediately think of deeds. But in its broadest form, it encompasses what I like to think of as anything, in any form of record, that tells something about the land. Land itself includes the space above it and the soil below it, and records include anything and everything that preserves part of the story—wherever found and in whatever form.

With that definition in place, this book deals with the interpretation thereof—reading and understanding the story told by the record. Again, depending on who is asked, and whether they are laypeople or professionals such as surveyors, lawyers, or title examiners, a host of definitions comes to one's attention. Land records tell something of the boundaries, of the ownership, of what the property used to look like, or what improvements existed at some point in time. But interpretation goes much further than that. There are more court decisions relating to real property—boundaries and ownership—than almost anything else. And interpretation extends beyond mere reading and understanding. What if the record is lost? What if it is of such poor condition that it cannot be read? What do certain words and symbols signify? And, ultimately, what is the author or creator trying to tell us, in his or her way, that is entirely different from your approach or mine?

Each record is somewhat of a work of art. Some are originals and some are copies. Some copies are good copies, while others are poor or incorrect. Some are legitimate, whereas others are fakes or forgeries. Some are readily available, but others are lost forever. And some are humorous, interesting, and pleasant to look at, while others lead to discouragement and despair. Some are simple, but some are intricate, complex, or confusing. Nonetheless, they all have meaning and demand consideration.

The average person, particularly one not well trained in research, when confronted with the term *land record*, will immediately think of a deed. And some will think only in terms of current deeds, while others will think strictly in terms of the *public* record. These approaches are superficial and often lead to incomplete, or worse, incorrect, conclusions and resolutions of problems and conflicts. It will readily be seen that many of the standards for researching land records go much farther than that. And the supreme guidance—the law—has clearly stated for ages that just the deed, or merely the public record, is rarely sufficient.

However, like many things, more research and more information sometimes lead to more confusion and additional problems. Interpretation then requires more expertise, but frequently the answer lies therein, often clearly reciting an explanation or offering a solution not available with limited information.

Perhaps Brian Clarke stated it best:[1]

The critical difference between the expert at anything, and the non-expert, is not information, but understanding.

The non-expert fails most of the time because his success depends upon meeting conditions which coincide with a fixed, and usually limited, range of mentally-catalogued techniques; whereas the expert, because of his fundamental understanding of what he is trying to achieve, thinks more in terms of how and why, than of what; and thus is able to devise specific techniques in response to the demands of specific conditions. Through understanding, he achieves a kind of infinite flexibility.

The truth is that all books really can do is act as a catalyst, by providing enough basic information to fire the interest. They cannot, on behalf of the reader, translate this fire-side knowledge into better returns. We can learn only so much by proxy, at second hand; and really to improve one's performance requires *commitment* on the part of each individual, and *effort*. No one else can do the work for us; and if we rely on books and the written word, the task will be over and gone before, book in hand, tools in the other, we have gotten half way down the index in an effort to identify the cause of all the interest. It is essential, therefore, that anyone who hopes to improve his performance on a basis of more thought, is willing to put in the other work without which his aspirations will never be fulfilled.

[1] Clarke, Brian, *The Pursuit of Stillwater Trout.* London: Adam and Charles Black. 1975.

That is what this book is about: it is an aid for the interpretation of the record relating to some aspect of the land, whether the boundary, the title, its access, its size, its location, or some other characteristic. Hopefully, proper interpretation of the available record will lead to correct understanding of what it says.

CHAPTER 2

GEOMETRY OF THE DESCRIPTION

> By metes in strictness may be understood the exact length of each
> line, and the exact quantity of land in square feet, rods, or acres.
> . . . Metes result from bounds; and where the latter are
> definitely fixed, there can be no question about the former.
>
> —*Buck v. Hardy*, 6 Me. 162 (1829)

Land descriptions consist of geometric figures, however crude or complex they may be. These figures are defined by elements, or calls,[1] which form some combination of the following:

- Monuments
- Direction
- Distance
- Quantity

The use of a combination of calls outlines, or defines, the land parcel being identified as the subject matter of the instrument.

---

[1] A reference to, or statement of, an object, course, distance, or other matter of description, in a survey or grant, requiring or calling for a corresponding object, etc. on the land. *King v. Watkins*, C.C.Va., 98 F. 913, 922, reversed on other grounds, *Watkins v. King*, 118 F. 524, 55 C.C.A. 290.

Monuments. Monuments are of two kinds, natural or artificial. A *natural monument* is defined as one that has been placed by nature, while an *artificial monument* is one that has been placed by the hand of man.[2] Examples of natural monuments commonly found in land descriptions are bodies of water, marshes, rocks, trees, and the like. Artificial monuments may be anything man-made, or placed by man, such as buildings, streets, fences, posts, pins, pipes, stakes, stone piles, stones, and similar objects. See Figure 2.1.

Directions. Directions are sometimes termed *courses*[3] but may also occur in other forms, or be the result of angles. Historically, a course would be some direction based on the cardinal points of the compass, such as north, south, east, or west, or northerly, northeasterly, and the like. Or they may be more refined by being expressed in angular degrees, such as North 30° East, or in equivalent *azimuth*, 30°.

A *bearing* is defined as the direction from north or south to the east or west, expressed in degrees. An *azimuth* is the *clockwise* angle, in degrees, from a defined meridian, which is usually North. The meridian may be South, however, or may be anything the user wishes to define it as, so long as it is defined, and the angles from it are measured clockwise.

Figure 2.1   A stake and stones. One of the most commonly called-for monuments in early land descriptions.

---

[2] *Parran v. Wilson*, 154 A. 449, 160 Md. 604 (1931). *Timme v. Squires*, 225 N.W. 825, 199 Wis. 178 (1929).

[3] A "course," as used with reference to boundaries, is the direction of a line run with a compass or transit and with reference to a meridian. See 12 Am Jur 2d, Boundaries, § 10; *M'Iver v. Walker*, 9. Cranch (US) 173, 3 L Ed 694 (1815).

Angles may be found in a variety of forms. For equivalents and conversions, refer to Table 2.1. Measured with some device, or in early descriptions often estimated, angles may be clockwise or counterclockwise, may be right or left, or may be deflection angles. See Figure 2.2.

A *deflection angle* is one that is measured either right or left from the prolongation of a line, usually the line of travel. See Figure 2.3.

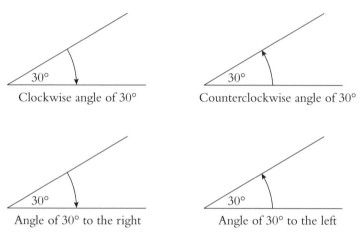

Clockwise angle of 30°   Counterclockwise angle of 30°

Angle of 30° to the right   Angle of 30° to the left

Figure 2.2   Angles clockwise, counterclockwise, right, and left.

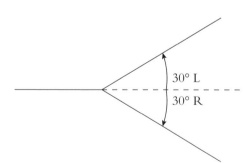

Figure 2.3   Deflection angles.

Angular Measure

Sexagesimal System

1 minute	=	60 seconds (")
1 degree	=	60 minutes (')

1 side	=	30 degrees (°)
1 right angle	=	90 degrees (°)
1 circle	=	4 right angles = 360 degrees

Metric System

1 centesimal minute	=	100 centesimal seconds
1 grade	=	100 centesimal minutes
1 quadrant	=	100 grades
1 circle	=	4 quadrants

Other Units

1 radian	=	57.295,78 degrees
	=	63.661,98 grades
	=	0.159,154,9 circles or $1/2\pi$ circles
π	=	3.141,592,65

Table 2.1   Angular measure and conversions.

Meridians. A meridian is technically defined as a great circle on the surface of the earth, passing through the poles at any given place. It is a line used as a reference from which to measure angles of direction.

There are several meridians that may be used as a reference, the most common in land descriptions being *true north* and *magnetic north*. South (true or magnetic) may be used, a selected line in a survey may be used, an assumed meridian may suffice, or the reference may be one of several other definitions.

Astronomic north, also known as geographic north and true north, is based on the geographic location of the North Pole. Astronomic north implies that the line was determined through astronomic observations such as on the Sun or the North Star, while true north, may be a line computed, or determined through methods other than from astronomic observation. *Grid north* is the north defined by a superimposed grid, such as Mercator's Projection. *Geodetic North* is north based on a meridian line defined by some representation of the geoid or ellipsoid shape of the earth and is approximately that of astronomic North. See Figure 2.4.

Bearings are measured by quadrants, Northeast or Southeast and do not exceed 90°. This is shown in Figure 2.5.

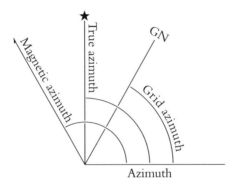

Figure 2.4 Illustration of the relationship of several North meridians.

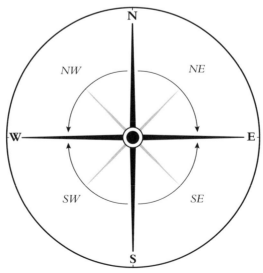

Figure 2.5 Measurement of bearings by quadrants.

Azimuths, always measured clockwise, range from 0° at the defined meridian all the way around, full circle to 360° back to the defined meridian.

The measurement of angles, azimuths, and bearings is illustrated in Figure 2.6.

Bearings are courses stated as a number of degrees *east or west from the north or south meridian,* as illustrated in Figure 2.7.

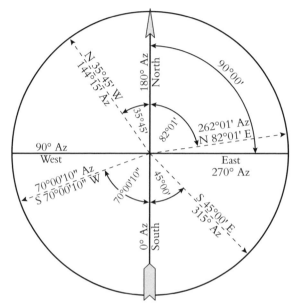

Figure 2.6 Directions of lines: angles, azimuths, and bearings. From Brown, Robillard & Wilson, 1986.

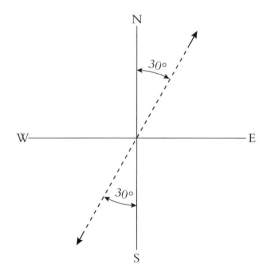

Figure 2.7 Measurements of bearings.

The represented line would be designated as running North 30 degrees East, usually termed N 30° E, or South 30° West, depending on which direction the observer or reader was facing. Thus, in actuality, a straight line has two directions, 180 degrees opposite each other.

In early surveys it was not uncommon when a line was close to the east-west meridian, to express its direction in reverse. For an example, see Figure 2.8.

When this is encountered in the records, the danger is in merely treating it as a "scrivener's error" and reversing the letters. Doing so in this case would result in a direction of North 10° East, whereas the direction is really North 80° East (90° – 10° = 80°). Swapping the letters has introduced a 70° error into the description.

When a surveyor, or a scrivener, stated E 10° N, he meant just that, a line with a direction of 10° north of the east meridian. Today, such direction would be the complement of the angle, or N 80° E.

In the earlier surveys made in the United States, it seems to have been the practice to run the lines on the ground according to the magnetic meridian and not according to the true meridian. In localities where this is assumed to have been the case, the courts, in interpreting descriptions of boundaries in deeds and patents based on such surveys, may recognize a presumption that the lines were run according to the magnetic meridian.[4]

However, the opposite can, on occasion, be shown. That is, the bearings run were true, or something else.

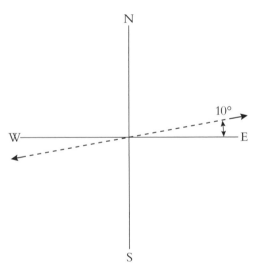

Figure 2.8   Bearing cited as East 10° North.

---

[4] *Taylor v. Fomby*, 22 So. 910, 116 Ala 621 (1897); *Bryan v. Beckley*, 16 Ky (Litt. Sel. Cas.) 91(1809).

Magnetic Declination. The angular difference between the true meridian and the magnetic meridian, expressed in degrees, is known as *magnetic declination*. See Figure 2.9.

True north, or astronomic north, also known as the North Pole, is a stationary point at the top of the earth where the meridians of longitude converge. It is fixed in position and does not change, whereas the magnetic pole, or north reference point, is continually changing its position. Therefore, the angular difference between true and magnetic north also changes, and this is monitored over time. Tables of this variation, published by state or by position of latitude and longitude, may be purchased from:

U.S. Department of Commerce

National Oceanic and Atmospheric Administration

Environmental Data Service

National Geophysical and Solar-Terrestrial Data Center

Boulder, Colorado 80302

www.ngdc.noaa.gov

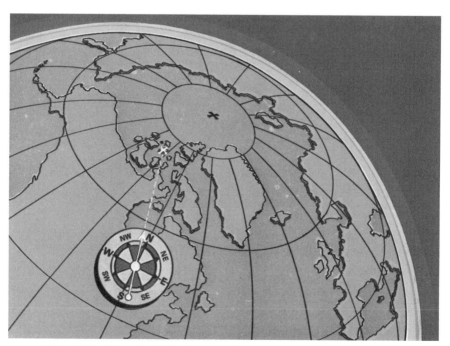

Figure 2.9   Magnetic declination caused by the fact that the magnetic pole is located almost 1,000 miles from the true North Pole. From *To Measure The Earth,* a reprint from *OUR PUBLIC LANDS,* Bureau of Land Management.

Table 2.2 provides an example of this.

Values of Magnetic Declination

LAT	44	44	44	44	44	44	44	44	44	LAT
LONG	69	70	71	72	73	74	75	76	77	LONG
1750	10 00W	09 31W	09 04W	08 37W	08 18W	07 58W	07 29W	06 59W	06 14W	1750
1760	09 44	09 11	08 39	08 07	07 42	07 17	06 47	06 16	05 30	1760
1770	09 36	08 59	08 22	07 45	07 15	06 44	06 10	05 35	04 48	1770
1780	09 41	08 59	08 16	07 32	06 54	06 16	05 39	05 01	04 13	1780
1790	09 53	09 06	08 18	07 30	06 47	06 03	05 22	04 39	03 49	1790
1800	10 17	09 26	08 33	07 39	06 50	06 01	05 15	04 27	03 35	1800
1810	10 48	09 53	08 55	07 57	07 03	06 09	05 19	04 27	03 33	1810
1820	11 26	10 28	09 27	08 25	07 27	06 29	05 35	04 40	03 43	1820
1830	12 09	11 09	10 05	09 01	08 00	06 59	06 02	05 04	04 05	1830
1840	12 56	11 54	10 49	09 43	08 41	07 38	06 38	05 37	04 36	1840
1850	13 44	12 41	11 34	10 27	09 23	08 19	07 17	06 14	05 12	1850
1860	14 29	13 26	12 19	11 11	10 06	09 00	07 57	06 53	05 50	1860
1870	14 59	13 57	12 52	11 46	10 43	09 39	08 37	07 33	06 30	1870
1880	15 25	14 27	13 26	12 24	11 24	10 24	09 25	08 24	07 21	1880
1890	15 43	14 47	13 48	12 48	11 50	10 52	09 55	08 57	07 57	1890
1900	16 06	15 13	14 17	13 20	12 24	11 27	10 31	09 34	08 35	1900
1905	16 32	15 39	14 43	13 46	12 50	11 52	10 55	09 56	08 57	1905
1910	17 00	16 07	15 12	14 15	13 18	12 20	11 21	10 22	09 21	1910
1915	17 27	16 34	15 40	14 43	13 46	12 47	11 48	10 48	09 47	1915
1920	17 46	16 54	15 59	15 03	14 05	13 06	12 07	11 06	10 05	1920
1925	18 09	17 17	16 23	15 27	14 29	13 30	12 31	11 31	10 29	1925
1930	18 25	17 34	16 40	15 45	14 47	13 48	12 49	11 48	10 47	1930
1935	18 38	17 46	16 52	15 57	14 59	14 01	13 01	12 00	10 58	1935
1940	18 43	17 51	16 56	16 00	15 01	14 02	13 02	12 00	10 57	1940
1945	18 47	17 54	16 59	16 01	15 02	14 02	13 01	11 58	10 55	1945
1950	18 41	17 47	16 51	15 53	14 54	13 53	12 52	11 49	10 46	1950
1955	18 38	17 44	16 49	15 51	14 52	13 51	12 50	11 47	10 44	1955
1960	18 36	17 43	16 48	15 52	14 53	13 53	12 52	11 49	10 47	1960
1965	18 27	17 36	16 44	15 49	14 52	13 53	12 54	11 53	10 51	1965
1970	18 19	17 31	16 40	15 46	14 51	13 54	12 57	11 57	10 57	1970
1975	18 13	17 27	16 39	15 49	14 56	14 03	13 08	12 11	11 13	1975
1980	18 11W	17 29W	16 44W	15 57W	15 08W	14 18W	13 26W	12 32W	11 37W	1980
1985	18 01W	17 23W	16 42W	15 59W	15 14W	14 27W	13 38W	12 47W	11 54W	1985
1990	18 06W	17 31W	16 53W	16 13W	15 31W	14 46W	14 00W	13 11W	12 20W	1990

Table 2.2   Table of declination values. Prepared by National Geophysical Data Center, NESDIS. NOAA 08/12/1987.

The variation on a daily basis, called the *daily variation*, is slight and, for all intents and purposes here, can be ignored. *Annual change*, amounting to several seconds, is also relatively small, but an accumulation of changes over a several-year period must be taken into account in order to retrace older descriptions. The field notes in Figure 2.10, recorded in 1875, reveal a variation between a deed description and an actual survey.

While past performance of the movement of the magnetic pole is monitored and can be shown, predictions of what the future differences will be is not possible. Knowing the past performance, conversion from one system to another (usually from the date of the deed, to present time) may be accomplished through the use of the table. To do so, we must know the years, and the location by latitude and longitude (scaled from a U.S.G.S. topographic map is sufficient).

Use of Table 2.2 to convert a past magnetic bearing to the current magnetic bearing is illustrated in the following problem:

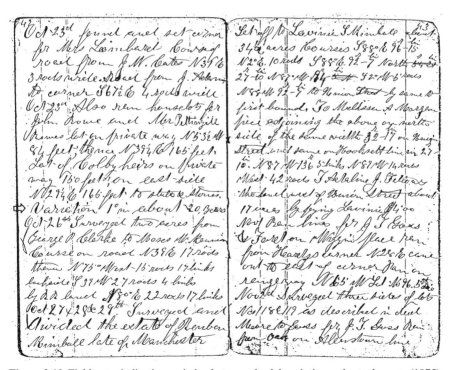

Figure 2.10 Field notes indicating variation between deed description and actual survey (1875).

PROBLEM. AN 1850 DEED DESCRIPTION IS:

Determining the directions of the lines today would be done as follows:

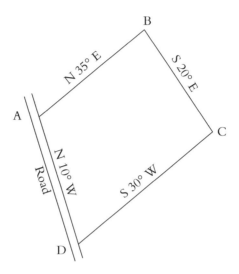

In 1850, the declination from the chart was found to be 10° West, while today the declination is found to be 20° West.

The true bearing of line A-B is found:

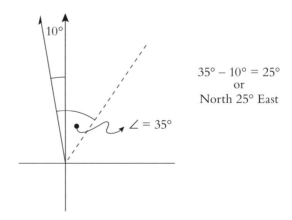

$$35° - 10° = 25°$$
or
North 25° East

Today the magnetic bearing of line A-B is:

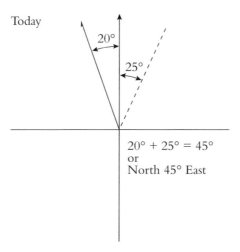

If the reader will make the computations for lines B-C, C-D and D-A, the resulting magnetic directions for the parcel will appear as follows:

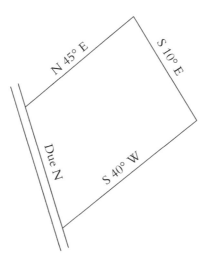

Always remember, *the physical location of the property lines do not change*, nor do the angles between them. The only thing that changes is the reference meridian, magnetic north.

To check for correctness, compute the angles between the lines:

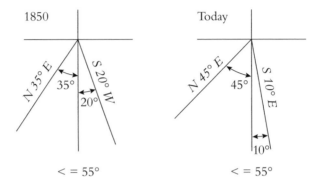

It is crucial to be as near as possible with such conversions in order to be able to find the precise location of the point, or the evidence marking the point.

An error of 1° will result in a horizontal error of 92 feet at the end of a mile (see Figure 2.11).

Taking a line 500 feet in length (about 1/10 of a mile) and an error (possibly due to not making the conversion) of 5°, the resulting error at the point in question would be about 50 feet. For example:

$$9.2 \times 5 = 46 \text{ feet}$$

In heavy brush, swamp, or where there is a heavy cover of leaves and debris after a hundred or more years, the precise point may not be found.

Numerous other errors may have affected the given bearings as well, so it is important to refine the work as much as possible. Local attraction, as noted in Figure 2.12, the inherent characteristics of the observer's compass, the original observer's techniques, and other factors may all serve to induce errors. Some of these errors may be compensating and some may be accumulating, depending on the type of error and circumstances of the situation.

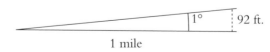

Figure 2.11 Linear error incurred as a result of angular error.

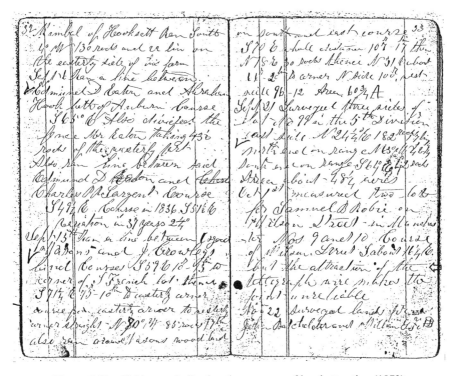

Figure 2.12   Field notes indicating the presence of local attraction (1873).

As an example, the following real situation was encountered, and it will be seen what adjustments were made to achieve success. An actual 1832 deed description reads:

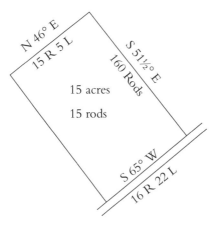

The declination in 1832 was 8° W; the declination in 1982 was 15° W. Conversion in 1982 resulted in the following description:

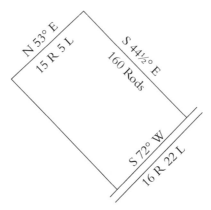

At the site, the observed bearing on the road was N 71° E.

It was ultimately found that there was a difference in observations between 1832 and 1982 of 6°, or 2° less than the conversion for declination gave. The actual conversion did not quite line up with the evidence found, so the bearings were "field adjusted" based on the known line, and the following results were obtained:

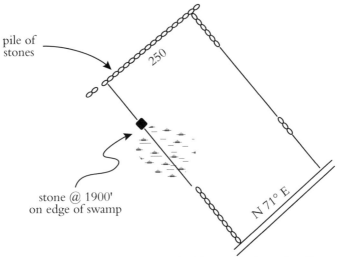

In surveying the public lands of the United States, the boundary lines are required to be run according to the true meridian. Courses and distances in private lands, however, in the older states at least, are usually run according to the magnetic meridian, and are presumed to have been according to it unless something appears to show that a different mode is intended in the conveyance. In other states it is presumed that lines were run according to the true meridian.[5]

---

[5] 43 USC § 751; *Riley v. Griffin,* 16 Ga 141 (1854); Doe ex dem. *Taylor v. Roe,* 11 N.C. (4 Hawks) 116 (1825), *M'Iver v. Walker,* 9 Cranch (U.S.) 173, 3 L. Ed. 694 (1815), *Wells v. Jackson Iron Mfg. Co.,* 47 N.H. 235 (1866), *Reed v. Tacoma Bldg. & Sav. Assoc.,* 26 P. 252, 2 Wash. 198 (1891).

It is part of the common law in many states that courses in deeds of private lands are to be run according to the magnetic meridian when no other is specifically designated. *Wells v. Jackson Mfg. Co.,* 47 N.H. 235 (1866).

DETERMINATION OF TRUE OR MAGNETIC BEARING.

The question frequently arises as to whether the bearing in the description is a true bearing or a magnetic one. Sometimes a clue is given, such as "directions by the needle," or if the variation (declination) is shown on the plan, whether it had been determined and possibly considered; also, the North Arrow might be shown with a full head (true north) or a half-head (magnetic north, with the half-point on the left or right side, indicating the declination as west or east). Figure 2.13 provides an example.

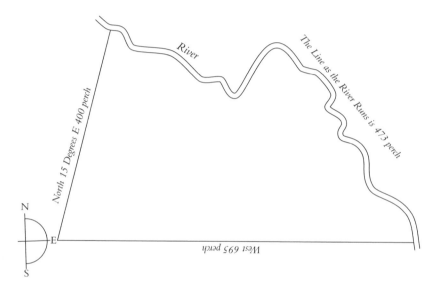

By virtue of A Grant made by the Great & General Court, to yᵉ town of Medford, I the Subscriber have Surveyed and Laid out, (with the Assistance of L John Goffe and mʳ Ephraim Bushnall Chane-men), one thousand Acres of Land in the following manner (Viz) bounded Southerly by a tract of Land Laid out to the Grantees of yᵉ town-Ship Called by the Name of Old Harryᶜˢ town Westerly by Province Land, Notherly and Easterly by Pescataquogg River, the lines beginning att a pitch pine tree on the bak of Sᵈ River (about two miles west of Merrimack River) Markt M F Then Running Due West by yᵉ Nedle with a line of Markt trees: 693 perch, then turning No 15 Degrees E to a Maple tree Standing on the bank of the Aforesᵈ Pescataquogg River Markt M F 400 perch, then turning and Running with sᵈ Pescataquogg River untill it Coms to yᵉ pitch pine first Mentioned which plan is protracted by a Scale of 80 poles or perch, to one Inch June the 16th 1736

By Me   Caleb Brooks Surveyʳ

In Surveying this farm there was Given one Chain in fifty for Broken Land and Sagg of Chain

Figure 2.13   Example of a survey "run by the needle." From *Mass. Maps and Plans*, Vol. 10, p. 13.

While it is true that, if nothing else appears, the calls in a deed must be followed as of the date thereof, when it clearly appears on the face of the deed, or when the evidence shows, that a line as established on a prior date was adopted and copied in the deed according to the courses and distances thereof, it is necessary to take into consideration the variations of a magnetic needle in locating the same.[6] Since the constant and predictable variation of magnetic north from true north is a scientific fact commonly known, a court may take judicial notice of such fact.[7]

The courts have faced and ruled on the question of true or magnetic bearing for some time.

In weighing evidence of recent surveys of ancient lines, consideration must be given to variations of needle in determining magnetic course.[8]

In 1866, the New Hampshire court stated in a leading case, "the courses in a deed are to be run according to the magnetic meridian, unless something appears to show that a different mode is intended."[9] Also, the Indiana court ruled in 1903 that, "where land was conveyed by metes and bounds, the position of a corner not fixed by measurement will be determined by courses and distances according to the magnetic meridian."[10]

In the Public Land System, the Missouri court stated, "a call was, 'thence south, 42 chains, to a stake on said line between said [government] sections, and thence to the beginning.' No stake could be found. Held, that the line should be run due south to the section line, making proper allowances for the variation of the magnetic needle from the true meridian."[11]

And, "Inasmuch as both true and magnetic meridians have been adopted in survey of public lands, fixed, determinate judicial construction cannot be given to words 'west' or 'due west.'"[12]

Considering the court's determination of the meaning of the word "due," a recent Rhode Island decision stated, "it will not be presumed that either magnetic or true course was intended by the grantor of a deed describing the course as 'due' north or south; rather, it will be left in each case to the trier of facts to determine from the evidence presented to him the course to be followed."[13]

In that case, the court went on to say that "the fact that a plat prepared by predecessors in title to owners of one property involved in boundary dispute showed the boundary line set out on a 'true' bearing did not require a conclusion that the term

[6] *Greer v. Hayes*, 5 S.E.2d 169, 216 N.C. 396 (1939).

[7] *Parker v. T.O. Sutton & Sons*, 384 S.W.2d 433 (Tex.Civ.App., 1964).

[8] *Milliken v. Buswell*, 313 A.2d 111 (Me., 1973).

[9] *Wells v. Jackson Iron Mfg. Co.*, 47 N.H. 235, 90 Am. Dec. 575 (1866).

[10] *Ayers v. Huddleston*, 66 N.E. 60, 30 Ind. App. 242 (1903).

[11] *Hoffman v. Riehl*, 27 Mo. 554 (1858).

[12] *McKinney v. McKinney*, 8 Ohio St. 423 (1858).

[13] *Martin v. Tucker*, 300 A.2d 480 (R.I., 1973).

'due south' contained in the descriptive portion of a deed executed in 1906 meant 'true' south rather than 'magnetic' south where there was no evidence that the plat had ever been recorded and it was drawn up subsequent to the 1906 deed and showed the boundary line meeting a stone bound which was situated on the shore and distinguished by a drill hole not mentioned in the 1906 deed."[14]

There are other types of directions found in the records. Two of the most common are *compass points* and *general directions*.

Compass Points. Some early descriptions, particularly those near the seacoast, have directions represented by words such as north-northwest, east-southeast, and the like. These are, or are equivalent to, points of the compass, and must be converted to degrees of azimuth, and perhaps, then to bearings. See Table 2.3.

A quarter point is 2° 48' 45". The quarter points proceed thus:

North - N 1/4 E - N 1/2 E - N 3/4 E

N. by E. - N. by E 1/4 E - N. by E 1/2 E

and so on, around the compass.[15]

General Directions. Sometimes, only general directions are given, such as "northwardly," "northeasterly," "generally westerly," and the like. This may mean that the direction was merely estimated, although it could be that it was based on a compass observation and the surveyor was faced with difficult sighting conditions or only made his observation in a general manner. The courts have frequently been faced with the interpretation of such observations.

In a Maine case the court stated, "in identifying descriptions in deeds, the words north or northerly, east or easterly, does not always indicate a direction that is due north, east, the use of south or southerly, west or westerly, south or west."[16]

The Kentucky court stated that "northwardly" was not synonymous with "north."[17] Later, the Kentucky court ruled that the term "south course" in a deed did not mean due south, but meant a southwardly course.[18]

The Illinois court refined its thinking even more, and said "merely because a deed gives a call for north does not require the line to be run on true north or astronomical north."[19]

---

[14] Ibid.

[15] For a complete table showing all the quarter points and their values in angular measure, see *American Practical Navigator* by Nathaniel Bowditch, U.S. Hydrographic Office.

[16] *Milliken v. Buswell*, 313 A.2d 111(Me., 1973).

[17] *Craig v. Hawkins' Heirs*, 4 Ky. 53, 1 Bibb 53 (Ky., 1808).

[18] *Martt v. McBrayer*, 166 S.W.2d 823, 292 Ky. 479 (Ky., 1943).

[19] *Dorsey v. Ryan*, 66 Ill. Dec. 263, 442 N.E.2d 689, 110 Ill. App.3d 577(1982).

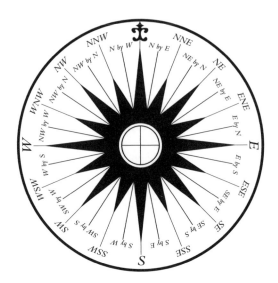

Conversion of Compass Points to Degrees

	Points	Angular Measure			Points	Angular Measure	
		°	'			°	'
North to East				South to West			
North	0	0	0	South	16	180	0
N by E	1	11	15	S by W	17	191	15
NNE	2	22	30	SSW	18	202	30
NE by N	3	33	45	SW by S	19	213	45
NE	4	45	0	SW	20	225	0
NE by E	5	56	15	SW by W	21	236	15
ENE	6	67	30	WSW	22	247	30
E by N	7	78	45	W by S	23	258	45
East to South				West to North			
East	8	90	0	West	24	270	0
E by S	9	101	15	W by N	25	281	15
ESE	10	112	30	WNW	26	292	30
SE by E	11	123	45	NW by W	27	303	45
SE	12	135	0	NW	28	315	0
SE by S	13	146	15	NW by N	29	326	15
SSE	14	157	30	NNW	30	337	30
S by E	15	168	45	N by W	31	348	45
				North	32	360	0

Table 2.3   Conversion of compass points to degrees.

The Rhode Island court summed up the thinking as follows: "Terms such as 'south' or 'due south,' when they appear in the descriptive portion of a deed, cannot be given inflexible judicial construction, their meaning must depend upon and be controlled by extraneous facts."[20]

Distances. Of primary importance are those distances or lengths between boundary points, or property corners. The length of frontage on a road or body of water is often of great importance for subdivision and zoning purposes, as well as for the assessment of value.

Today, distances are usually measured or expressed in feet, tenths of feet, and hundredths of feet, although some measurements are expressed in units of the metric system. As with any language, problems do not usually occur with the use of the language or the system itself, but are encountered when translations are made or when conversions from one to the other are undertaken. Therefore, it is of prime importance to understand the basis of the measurement. This will become extremely critical in discussion of measurements and units from an earlier time. See Tables 2.4 to 2.6.

English System		
1 inch	=	1,000 miles
1 hand	=	4 inches
1 foot	=	12 inches
1 yard	=	3 feet
1 furlong	=	660 feet
		220 yards
		1/8 statute mile
1 statute mile	=	63,360 inches
		5,280 feet
		1,760 yards
		8 furlongs

Table 2.4   English system and conversions.

---

[20] *Martin v. Tucker*, 300 A.2d 480, 111 R.I. 192 (1973), see 70 A.L.R.3d 1215.

1 nautical mile	=	6,076.1155 feet
	=	1.1508 statute mile
	=	2,027.00 yards
	=	1.85325 kilometers
	=	1' or 1/600 of a great circle of the earth
1 league	=	15,840 feet
		5,280 yards
		3 statute miles
1 fathom	=	6 feet
1 U.S. Survey foot	=	1200/3937 meters
1 U.S. Standard foot	=	0.3048 meters

Table 2.5   English system and conversions.

Metric System			
1 millimeter	=	1000 microns	
1 centimeter	=	10 millimeters	= 0.393 inches
1 decimeter	=	10 centimeters	= 3.937 inches
1 meter	=	10 decimeters	= 39.37 inches
			3.281 feet
			1.094 yards
	1 decameter	=	10 meters
	1 hectometer	=	10 decameters
	1 kilometer	=	10 hectometers
			1000 meters
			3280.833 feet
		=	0.62137 statute miles
		=	0.540 nautical miles

Table 2.6   Metric system and conversions.

Distances, or lengths, in early documents were expressed in different units. When engineering surveying became popular, and when most surveying was taught by engineering schools, the chain was frequently used. Chains are of different lengths and graduated in various units, but commonly used was the engineer's chain, which was generally 100 feet in length, and graduated into 100 links of 1.00 foot each. Longer tapes may be found, in lengths of 200, 300, and even 500 feet. Lengths were expressed in feet, tenths and hundredths, for example:

$$523.78'$$

There are many other types of chain, of different lengths and graduations. Chains will be found for measurements in meters, varas, and so on. Early chains were made up of individual links with markers at intervals. Most of the early surveying in the United States was done with an instrument of this nature. See Figure 2.14.

Rathborne's Chain. Before 1620, in England, *Rathborne's chain* was used to measure distances. Aaron Rathborne was a surveyor who published several books on the subject and developed, among other devices, a decimal chain. This was commonly used before the invention of the *Gunter's chain* and may have been used for some time after. The Rathborne chain was commonly two statute poles of 16 1/2 feet each, in length, but was sometimes three. Poles were further divided into *primes* of 19 2/3 inches, and *seconds* of 10 per prime, so that there were 100 divisions of each pole or perch, of 49/50 of an inch each, known as *seconds*.

As Rathborne described it,[21] the chain is divided into three *termes*, "*Vnites*," "*Primes*," and "*Seconds*," whose characters are these 0.1.2. So that you would express 26.Vnites, 4. Primes, and 5. Seconds, they are thus to be written 26°. 4_.5_. or together thus, 2645. or briefly thus, 2645'.

Figure 2.14 Example of a surveying chain.

[21] Richeson, *English Land Measuring to 1800: Instruments and Practice.*

Gunter's Chain. The *Gunter's chain*, commonly called the "surveyor's chain," was used to do most of the measuring in the English colonies and the eastern part of this country. It was developed by an English mathematician, Edmund Gunter, in 1620. While the most commonly used chain was 66 feet in length, containing 100 links of 0.66 feet or 7.92 inches, it is not uncommon to find chains of 33 feet in length containing 50 links. In fact, the standard measuring device for the U.S. Rectangular System was set to be the "two-pole chain," which was 33 feet in length.

The chain was also graduated in rods of 16 1/2 feet each, there being 4 rods in a full chain of 66 feet. Also known as a pole, or a perch, the rod was the standard unit of measurement for many years and is prevalent in descriptions in the eastern states. Distances were ordinarily expressed in rods and links, as illustrated in Figure 2.15.

Occasionally, links were further expressed in decimals, but this was rare. Depending on the geographical area, any of the three units, rods, poles or perches, may be commonly encountered.

In England, the rod was set by statute because of its easy conversion to acreage. The statute rod or perch was 16 1/2 feet. However, since quality of land was of concern as well as quantity, allowance was made by several methods when lands were laid out, or divided. For a time in England the length of the rod varied, from 12 feet to 22 feet.

In many places the 12-foot rod was used for arable lands, an 18-foot rod was used for meadows and a 20- or 22-foot rod was used for rock land and woodland. These rod measures were referred to as *customary rods*, and the acres derived from these measures were called *customary acres* in order to differentiate them from the later *statute acres* measured with the statute rod. These customary rods and acres were used throughout the eighteenth century and possibly to some extent as late as the nineteenth century in surveying in England.[22]

Since much of the New World was settled by English, and much surveying was based on English methods, it is essential to have an appreciation for the methods and units employed. As an example, some of the areas settled by Scotch-Irish were laid out with a rod 18 feet in length. Other definitions of length are found in other areas, and it is pretty much a matter of making the determination on a case by case basis. The important thing to keep in mind is that when the unit *rod* is encountered, it may not always be 16.5 feet in length. See Table 2.7.

16 rods 11 links

tree   stone

Figure 2.15   Distance between two objects expressed in rods and links.

---

[22] Ibid.

Early English Units		
1 link	=	7.92 inches
		0.66 feet
1 rod	=	25 links
		16½ feet (usually)
		5.50 yards
1 chain	=	100 links
		66 feet
		4 rods
1 statute mile	=	80 chains
		320 rods

Table 2.7   Early English units and conversions.

Not every area was settled and surveyed by the English, so several other units of measurement may be found. It is critical to have an appreciation of how an area may have been settled, and by whom; what type of measurement was used; and what was customary, or statutory, at the time.

In French-settled areas, the *arpent* was a unit of measurement. Like the rod, however, the arpent was a unit of area as well as distance. Like many units of measurement, different values will be found in different areas. It is necessary to determine the correct definition for the area being considered. Even in a particular area, variations may be found, and it is important to investigate the local custom. See Table 2.8.

The size of the arpent depends on its origin and on local custom.[23]

In Spanish-settled areas, the unit of measurement was the *vara*.[24] Care must be taken in conversion of the vara, because it has different definitions in different locations. All are approximately 33 inches; however, inducing a slight error may make a significant difference in a long line.

The measurement of length used in grants by either the Spanish or the Mexican government is the "vara" of Mexico. In Arizona, California, and New Mexico it is considered to be 33 inches,[25] in Florida to 33.372 inches, and in Texas to be 33.3333 inches or 36 varas = 100 feet. Its exact equivalent would appear to be 33.38676 inches.

---

[23] Webster's Dictionary states that the *arpent*, as a linear measure, is equal to about 11.5 rods, used locally in Canada.

[24] For a compiled list of the different definitions, consult Wattles, *Land Survey Descriptions*. See also *Patton on Titles*, Chapter 4.

[25] Actually, 32.99312 inches.

French Units		
1 arpent	=	192.50 feet (Arkansas & Missouri)
		58.674 meters
1 arpent	=	191.994 feet (Mississippi, Louisiana, Alabama, & Florida)
1 perch	=	29.166 links (Gunter's chain)
10 perches	=	1 arpent lineal
		2.9166 chains
27 arpents	=	approx. 80 chains or one mile
84 arpents	=	approx. 245 chains or one league
In Canada:		
1 pied or Paris foot	=	1.065.75 feet
	=	12.789 inches
	=	32.484,06 centimeters
1 arpent	=	180 pieds
	=	191.835 feet
	=	58.471,31 meters

Table 2.8   French units and conversions.

Early surveys in Texas were made under Spanish and Mexican laws and regulations. Texas was an independent republic from 1836 until 1845, when it became part of the United States. While a province of Mexico, land measures were established by decrees. Lineal measure was to be by a vara of "three geometrical feet" (33⅓ inches), a mile of one thousand varas, and a league of five thousand varas. See Figure 2.9.

To reduce varas to feet, multiply by 100 and then divide by 36.

A *cana* is a Spanish measure of length varying (in different localities) from about five to seven feet.

Pace. Some distances are "paced," or walked, counting the number of steps taken from one point to another. Knowing the length of the pace, one can convert to a distance in feet, or other units. While not very precise, pacing will suffice in many instances, and distances expressed in paces are found in the records.

Spanish or Mexican Units		
1 cordel	=	50 varas
1 league	=	100 cordels
		5,000 varas
1 vara	=	33⅓ inches
36 varas	=	100 feet
108 varas	=	100 yards
1,000 varas	=	2,777 feet, 9⅓ inches
5,000 varas	=	1 league
		2 miles, 3,328 feet, 10⅔ inches

Table 2.9   Spanish or Mexican units and conversions.

Webster's Dictionary defines a pace as being from 30 to 40 inches and the *military pace* as 30 inches, or 36 inches in double time. The *Roman pace*, measured as being from the heel of one foot to the heel of the same foot in the next stride, was 5 Roman feet, or 58.1 inches, now known as a *geometric pace* of about 5 feet.

Confusion sometimes arises as to whether a pace is one step or two. Surveying texts usually define a pace as being equivalent to one step, while a stride is equivalent to two paces, or the distance from where the heel of one foot leaves the ground to the point where the same heel strikes the ground again. For the average person, the pace may vary between 2½ and 3 feet. If a lot of pacing is to be done, it is advisable to "calibrate" the step, or compare the number of steps against a known distance. In situations where it is important to know the length of someone else's pace, it is essential to do the same thing, compare the number of paces with a known distance within the particular description being analyzed, in order to determine a standard.

Few definitions are found in court decisions, but there is an occasional one:

Vt. 1964. "Pace," the unit of measurement employed by deed, was not precise, but did not render the deed so ambiguous as to justify resort to extraneous evidence and could be treated as approximately three feet.[26]

[26] *Haklits v. Oldenburg*, 201 A.2d 690, 124 Vt. 199 (1964).

Slope Measurement vs. Horizontal. Another consideration in the interpretation of distance is whether the measurement was made horizontally or along the slope. Recent measurements are almost always horizontal distances, although they may have been taken on a slope and reduced to horizontal distance. Earlier measurements were sometimes expressed as taken without any corrections being made. In extremely steep terrain, a significant error can result in long lines. See Figure 2.16.

In some states, such as North Carolina, Kentucky, and Tennessee, slope measurements were common, but they may be encountered anywhere. It is essential to determine what kind of measurement was made because of the magnitude of the error that may be introduced.

In some cases, particularly those involving large tracts of mountain lands, it has been held or recognized that the correct method of surveying the boundaries of such lands is by using horizontal measurement rather than surface measurement. Additionally, it has been held, pursuant to this view, that this is a legal rule, which local custom as to surface measurement cannot change. But in other cases the view has been taken that surface measure was proper where such method was the custom of the locality or was dictated by circumstances.[27]

In considering whether distances were measured in a horizontal manner or along the slope, courts have found the following:

In *Justice v. McCoy*,[28] "where the proper method of measuring the depth of a city lot which fronted on one street and extended part of the way up a hillside toward a parallel street was involved, and the evidence tended strongly to support the theory that the depth of the lot should be measured along the surface of the hillside, the court, in holding that such method of measurement was proper, said that it found authority for the proposition that a surface measurement was proper where it was a custom of the locality, or where it was dictated by the circumstances of the case."

In a North Carolina case,[29] the court recognized that in making surveys of mountainous regions the early surveyors, at least, had universally adopted surface measurement. Since this has been found to be true in several states, it may often be considered whenever mountainous terrain is a consideration.

Straight Line Presumption. Where a line is described as running from one point to another, it is presumed, unless a different line is described in the instrument, or marked on the ground, to be a straight line, so that by ascertaining the points at the angles of a parcel of land the boundary lines can at once be determined. The rule

---

[27] *Cody v. England*, 19 S.E.2d 10; 221 N.C. 40 (1942); *Bleidorn v. Pilot Mountain Coal & Min. Co.*, 15 S.W. 737, 89 Tenn. 16 (1890); *Justice v. McCoy*, 332 S.W.2d 846 (Ky, 1960); 80 ALR2d 1206.

In measuring up and down mountainsides or over other steep acclivities or depressions to constitute a legal survey, the chain must be leveled so as to approximate, to a reasonable extent, a horizontal measurement and not along or upon the surface; this is a legal rule which local custom as to surface measurement cannot change, *McEwen v. Den*, 24 How (U.S.) 242, 16 L Ed 672 (1860).

[28] 332 S.W.2d 846 (Ky., 1960), see 80 ALR2d 1206.

[29] *Duncan v. Hall*, 23 S.E. 362, 117 N.C. 443 (1895).

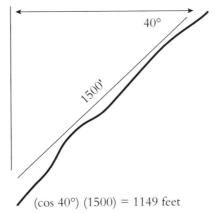

(cos 40°) (1500) = 1149 feet

Figure 2.16   Reduction of slope distance to horizontal.

of surveying, as well as of law, is to reach the point of destination by the line of shortest distance, and lines should never be deflected, except in order to conform to the intention of the parties.[30]

In *Jenks v. Morgan*,[31] the court held that a line given in a deed to be run from one landmark to another was to be a straight one unless a different line was described in the deed.

Where two points are determined, the line between them should be a direct one, and should not vary to conform with a line of marked trees not mentioned in the deed.[32]

In *Cowan v. Fauntleroy*,[33] the court said, "If, in making a survey a line shall have been marked only part of its distance, it would seem that the boundary of the survey, for the residue of the distance, should be ascertained by running a line as nearly direct as practicable from the termination of the marked line to the corner called for."

In *Hagan v. Campbell*,[34] the Alabama court said, "where a line is described as running towards one of the cardinal points, it must run directly in that course, unless it is controlled by some object."

Additionally, the North Carolina court stated in 1857, "where an ascertained object is called for in the description of land, it must be reached by a straight line without regard to an erroneous description by course and distance."[35]

In *Taylor v. Watkins*,[36] it was held that a line to be run up a creek for a distance of 1,400 poles was to be determined by a straight line, and not by the meanders of the stream.

---

[30] *Halstead v. Aliff*, 89 S.E. 721, 78 W.Va. 480 (1916), *Bartlett Land, etc., Co. v. Saunders*, (N.H.), 103 U.S. 316, 26 L.Ed. 546 (1889), *Leonard v. Smith*, 36 So. 101, 111 La. 1008 (1904), *Platt v. Jones et al*, 43 Cal. 219 (1872).

[31] 6 Gray 448 (Mass., 1856).

[32] Den ex dem. *Wynne v. Alexander*, 29 N.C. (7 Ired. L.) 237, 47 Am. Dec. 326 (1847).

[33] 2 Bibb 261 (Ky., 1810).

[34] 8 Port. 9, 33 Am. Dec. 267 (1838).

[35] *Campbell v. Branch*, 49 N.C. (4 Jones L.) 313 (1857).

[36] 7 J.J. Marsh. 363 (Ky., 1832).

Arbitrary Corrections. Early surveyors frequently took into account factors such as hilly terrain, poor-quality land, wet areas, boulders, and land that presented measurement difficulties by allowing for errors in their final results. Ordinarily, it is necessary to review the surveyor's notes or the final report to know what, if any, corrections were made. Sometimes allowances were made by the chain, sometimes by the mile. See Figures 2.17 to 2.19.

On the question of "allowance" for uneven terrain, in the case of *Bryan v. Beckley*,[37] an action involving the location of lost corners and line of a survey made in 1774, the court recognized that allowance should be made for the unevenness of the ground over which the lines ran, and said that reasonable accuracy could never be attained in completing the business without good instruments, careful chain carriers, and such allowance, or other safeguard, for unevenness of surface as would be equivalent to horizontal admeasurement, upon which the art and rules of surveying were founded.

May ye 21st 1738. Then Finished the Surveying and Laying out of a Township of ye Contents of Six miles Square. To Satisfie a Grant of ye Great and General Court of ye Province of ye Massachusets Bay made ye 16th Day of January 1737 on the Petition of Samuel Haywood and others and their Assotiates; Lying on the Easterly Sid of a Great Hill Called Manadnock Hill between Said Hill and a Township Laidout to ye Inhabitants of Salam and others who Servid in ye Expedition to Canada anno 1690 and Lyeth on the Southerly branch of Contokock River near the hed there of said Branch Runing throughit. It began att a Black Burch tree ye South East Corner and from thence it Ran West Six Miles and Sixty Eight Rods by a line of marked trees to a Spruse tree marked for ye South west Corner, from thence it Ran

north by a line of markd trees six Miles and Sixty Rods to a Stake and Piller of Stons ye northwest Corner and from thence it Ran East by a line of Markd trees Six Miles and Sixty Eight Rods to a Stake and heep of Stons the Northeast Corner and from thence Straight to where it began Six Miles and Sixty Rods. the Lines above said Contains ye Contents of six miles Square with ye alowance of one Chane in thirty for Sagg of Chane and fifty acres for apond

☙ Joseph Wilder Jun Surveyr

Figure 2.17 Return of a 1738 survey, which allowed one chain in thirty for sag of chain and 50 acres for a pond.

---

[37] 16 Ky (Litt Sel Cas) 91, 12 Am. Dec. 276 (1809).

June the 12, 1736— Then Capt Josiah Willard Survey^r &
George Macfarland & James Johnson Chainmen were sworn to
Deal faithfully & Impartially In Surveying & measuring The
Township Granted by the Gen^ll Court to Capt Joseph Silvester &
Company &c—

Before me   Saml Thaxter Js peace

June ye 19th 1736.

Then finished the Laying out a Tract of Land on the Easterly
Side of Arlington Granted by the Great and General Court held
at Boston In June 1735— for a Plantation or Township of the
Contents of Six Miles Square and is Granted to Capt Joseph Sil-
vister and his Company and is bounded thus beginning at a piller
of Stones Erected for the South Easterly corner of Arlington from
thence runing East Six miles partly on a Town Ship Lately Laid
out and partly on province Land— to a Hemlock tree marked
with J. S. from thence Running North Six miles & fifty Six
Perches to a maple tree marked with J. S— from thence West 44
degres North Seven hundred and thirteen Perch to an heap of
Stones on the Lower ashewelot line and from thence South forty
three deg. West on the ashewelot line one thousand one hundred
and ninety three perch to a piller of Stones then West Eighteen
deg. North Six hundred & fourty Perch to a piller of stones then
Running South on Arlington line five miles and one hundred and
fifty one Rods to a piller of Stones the first mentioned Bounds

there's allowed about one Rod in thirty for uneven land and Swag
of Chain, also there is allowed one hundred acres for a farm all
ready Granted to Coll. Josiah Willard with five hundred acres for
ponds—

                              ℘ Josiah Willard Surveyor

a scale of 310 perch to an inch

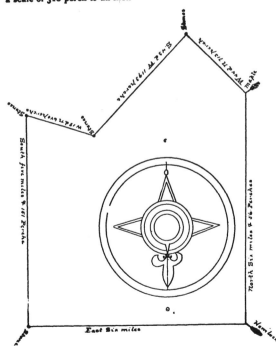

Figure 2.18 Return of survey and plan, allowing one rod in thirty for uneven land
and swag of chain, plus 100 acres for a previous grant and 500 acres for ponds.

Mass : ss. April 20 1739
Then John Stevens as Surveyor, Jeremiah Hall and Seth Heaton,
as Chainmen made Solemn Oath that in Surveying the Lands included in the Plat of the Lower Ashuelot that Lyes within the
Township of Earlington so Called, and in Surveying and Laying
Out an Equivalent therefor they would Severally Act according to
their Best Skill and Judgement therein

Before   Thomas Berry J Pacis

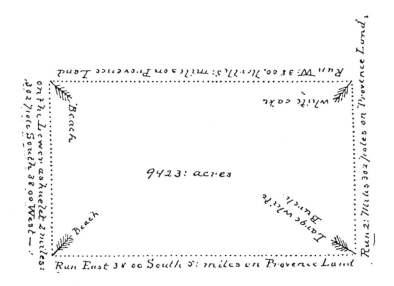

This Plot Contains Nine thousand four hundred and Twenty
three acres of Land and was Laid out by the order of Thomas
Berry Esq[r] and others In Considerration of Three thousand and
one hundred and forty one acres of Land Taken out of the Lower
Ashuelet by Earlington line and Bounded as folows—(viz)—Be-
gining at A Beach Tree marked In the line of Said Lower
Ashuelet about Three miles and forty Pole from the North East
Corner and Run East: 38 Degrees South: 5 miles To a Large
white Burch and from thence Run North 38 East Two Miles Three
hundred and Two pole To a White Oake Corner from thence Run
West 38, Degrees North five miles To the line of ashuelet afore
Said To a Beach Tree a bout Sixty Pole from the Corner of Said
Town Then on Said Town line To the Corner whear it first Begun with the Common Allowance of aboute Twelve Chains To a
mile for Sagg of Chain—

Surveyed Aprile 1739 ⅊   John Stevens Surveyer

Figure 2.19   Return of survey and plan allowing 12 chains to
the mile for sag of chain. Note two types of bearing expression.

In *Beckley v. Bryan*,[38] an earlier appeal of the foregoing case, the court also recognized that in running the lines of the survey a proper allowance should be made for the unevenness of the ground over which the lines ran.

Quantity. Land area is derived from computation of the courses and distances observed. Different surveyors had different methods for area determination, and techniques changed over time. One key to description and interpretation is that when area is expressed in precise, or refined, units, such as acres and rods, it is based on some type of measurement and probably on a survey.

Units. Modern surveys are expressed in acres or square feet and occasionally hectares or square meters. See Tables 2.10 and 2.11.

		English System
1 square foot	=	144 square inches
1 square yard	=	1,296 square inches
	=	9 square feet
1 acre	=	43,560 square feet
	=	4,840 square yards
1 square mile	=	640 acres
	=	3,097,600 square yards
	=	27,878,400 square feet
1 square inch	=	6.452 square centimeters
1 square foot	=	929 square centimeters
1 square yard	=	0.8361 square meters
1 acre	=	4047 square meters
	=	40.4687 ares
	=	0.4047 hectares

Table 2.10   English system and conversions.

---

[38] 2 Ky (Sneed) 91 (1801).

Metric System

1 square centimeter	=	100 square millimeters
	=	0.155 square inches
1 square meter	=	10,000 square centimeters
	=	10.75 square feet
	=	1.196 square yard
	=	0.0002471 acres
1 square decameter	=	100 square meters
1 hectare	=	100 ares
	=	10,000 square meters
	=	2.471,054 acres
1 square kilometer	=	100 hectares
	=	1,000,000 square meters
	=	0.386,102,2 square miles

Table 2.11   Metric system and conversions.

Earlier surveys were expressed in acres or square rods, or arpents and other units where areas were settled and surveyed by other groups. See Table 2.12.

Roods. For a time, in some areas, the English term "rood" was used. A rood is a measurement of area equivalent to 40 square rods, or 1/4 acre.

1 acre = 4 roods = 160 square rods
1 rood = 40 square rods = 1/4 acre

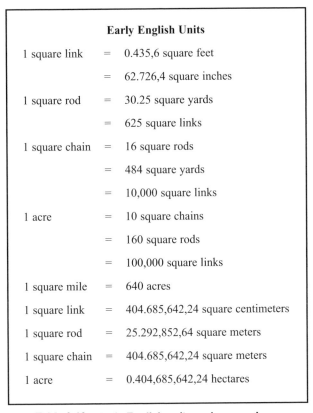

Early English Units

1 square link	=	0.435,6 square feet
	=	62.726,4 square inches
1 square rod	=	30.25 square yards
	=	625 square links
1 square chain	=	16 square rods
	=	484 square yards
	=	10,000 square links
1 acre	=	10 square chains
	=	160 square rods
	=	100,000 square links
1 square mile	=	640 acres
1 square link	=	404.685,642,24 square centimeters
1 square rod	=	25.292,852,64 square meters
1 square chain	=	404.685,642,24 square meters
1 acre	=	0.404,685,642,24 hectares

Table 2.12   Early English units and conversions.

Beware the change in description where some scrivener or conveyancer, in copying an earlier description, thought there was a misspelling and changed the word *rood* to *rod*, not realizing that the area was being altered by a factor of one or more 1/4 acres. See Table 2.13.

An *are* is a surface measure in the French law, in the form of a square, equal to 1076.441 square feet.

French Units		
1 arpent	=	0.8507 acres (Arkansas and Missouri)
	=	0.845 acres (Louisiana)
1 arpent	=	0.84725 acres (Louisiana, Mississippi, Alabama, and northwestern Florida when no other definition is known).
1 arpent	=	1 acre
1 league square	=	1,056 arpents
	=	6,002.5 acres
In Canada:		
1 arpent[39]	=	32,400 square pieds or Paris feet
	=	36,800.67 square feet
	=	0.844,827,1 acres
	=	0.341,889,4 hectares

Table 2.13   French units and conversions.

Spanish or Mexican Units		
1 acre	=	5,645.376 varas
1 labor	=	1,000,000 square varas
	=	177⅙ acres
1 sitio	=	1 league square
	=	25,000,000 square varas
	=	4,428.402 acres

To reduce square varas to acres, multiply by 177, delete the six figures on the right, and the remaining figures determine the acres (approximately).

Table 2.14   Spanish or Mexican units and conversions.

[39] Webster's Dictionary states that the *arpent* is an old French land measure of varying value, esp. one equal to 0.84 acres still common in parts of Canada.

The Mexican land area measures are as follows:

> *Sitio de granada mayor*, a square 5,000 x 5,000 varas
> *Sitio de granada menor*, a square 3333⅓ x 3333⅓ varas
> *Criadero de granada mayor*, a square 2,500 x 2,500 varas
> *Criadero de granada menor*, a square 1666⅔ x 1666⅔ varas
> *Caballeria*, a rectangle 1,104 x 552 varas
> *Media caballeria*, a square, 552 x 552 varas
> *Cuarto caballeria*, a rectangle, 552 x 276 varas
> *Sala para casa*, a square, 50 x 50 varas
> *Fundo legal para pueblos*, a square, 1,200 x 1,200 varas

Local Units. Because there is significant variation in measurement units depending on local custom, one should always be aware that such may exist, particularly in those areas, and during those times, when no measurement standard was in effect.

Some areas, some projects, and some industries have adopted their own systems. For instance, in Manchester, New Hampshire, when the Amoskeag Manufacturing Company laid out lots for their employees, they used what was known as the *Amoskeag Foot*.

The *Amoskeag Foot* was actually 1.01 feet long so that in laying out a block of ten lots, each 100 feet wide, there would be an extra 10 feet, or 1 foot per hundred "for good measure." Frequently, today, the searcher is on the lookout for descriptions carrying the phrase, "U.S. Standard Measure," indicating that the Amoskeag Foot was not used.

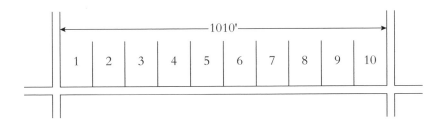

Various other cities have a particular unit of conversion, which is not necessarily consistent throughout. For example, in Philadelphia, the constant for a block must be determined before measuring a lot in that particular block. That constant may or may not be the same for an adjacent block. See Figure 2.20.

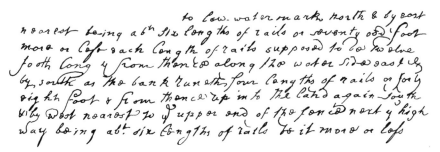

Figure 2.20 Deed description employing "fence lengths" as unit of measurement.

The *Builder's Acre* is a common term in the construction industry. It is equivalent to 40,000 square feet instead of the standard 43,560, because of its ease of computation. A *Builder's Acre* is equivalent to a square that is 200 feet on a side, rather than 208.71033 feet.

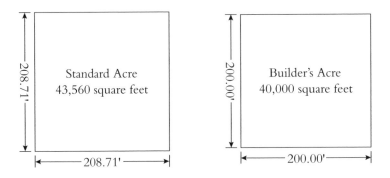

On Block Island, in Rhode Island, there is commonly found what is known as the *Block Island Acre*. Like the Builder's Acre, it contains 40,000 square feet.

Overrun in Area. Because allowances were made for "swag" of chain, difficult conditions, and poor-quality land, large original lots are usually larger than the area expressed. For example, 100-acre lots are frequently 106 to 112 acres in size. Also, measurement with an 18-foot rod and conversion by 160 rods per acre will give an overrun. This fact becomes extremely critical when original lots are divided by 1/2s and 1/4s.

A frequent occurrence is the conveyance of 1/2 of a 100-acre lot, being 50 acres, when in fact the lot is more than 100 acres. So, each 1/2 would also be more than 50 acres.

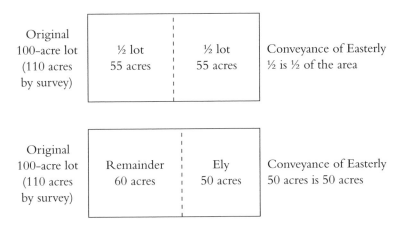

| Original 100–acre lot (110 acres by survey) | ½ lot 55 acres | ½ lot 55 acres | Conveyance of Easterly ½ is ½ of the area |

| Original 100-acre lot (110 acres by survey) | Remainder 60 acres | Ely 50 acres | Conveyance of Easterly 50 acres is 50 acres |

CIRCLES AND CURVES

The geometry of the circle is as follows (see Figure 2.21):

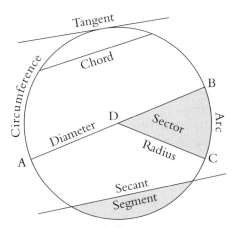

Figure 2.21 Curve geometry.

A *circle* is a plane figure bounded by a curved line called a *circumference*, every point of which is equally distant from the point within called the *center*.

The *circumference* is the bounding line of the circle. An *arc* is any part of a circumference, and is designated in the diagram as B-C.1/360 of any circumference is a *degree* of the circumference. If the space about a point is divided into 360 equal parts or angles, by straight lines meeting at the point, each angle is an angle of 1 degree.

The *diameter* is a line segment drawn through the center and terminated both ways by the circumference. It is designated in the diagram as A-B.

A *chord* is any line segment terminated both ways by the circumference.

Segments are the parts into which a chord or a secant divides the circle.

The *radius* is a line segment from the center to the circumference. It is designated in the diagram as D-C.

A *secant* is a line cutting the circumference in two points.

A *tangent* is a line that touches the circumference at only one point, however far the line is prolonged.

Curve Elements. The terminology of the curve is expressed in the following diagram with formulas for the determination of the various elements. Frequently it is important to know the curve distance for frontage, but other elements are expressed in the description. Determination of additional elements is a simple matter through the use of the formulas. See Figures 2.22 and 2.23.

Highway Curves vs. Railroad Curves. Highway curves and railroad curves are based on different definitions, therefore the same formulas do not apply to both. The computations for a railroad curve are based on a chord distance of 100 feet, whereas a highway curve is based on an arc length of 100 feet. See Figures 2.24 and 2.25.

Stationing. Route surveys and locations, such as highways, railroads, pipelines, power line rights of way, and the like, are surveyed and designated by stations. Ordinarily, stations are 100.00 feet apart along the route (centerline, sideline, or survey line) and are numbered consecutively from the beginning, or 0 point, to the end of the route, or end of survey. One hundred foot station increments are numbered as shown in Figure 2.26.

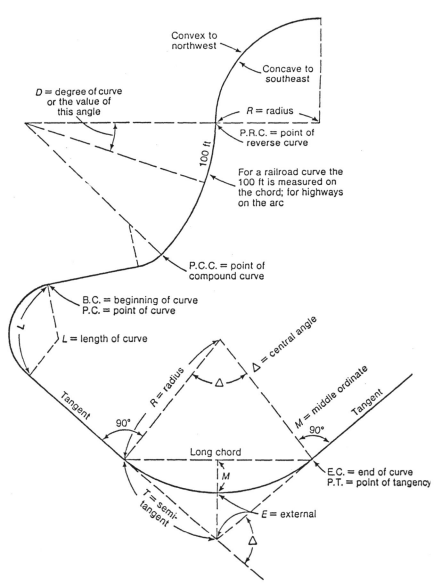

Figure 2.22 Curve elements. From Brown, Robillard & Wilson, 1995.

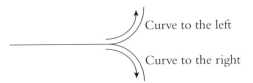

Figure 2.23 Curve geometry.

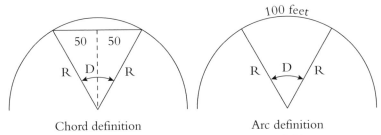

Chord definition   Arc definition

Figure 2.24 Railroad and highway curve definitions.

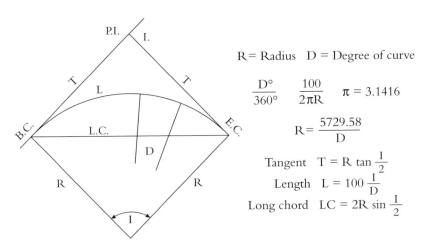

R = Radius   D = Degree of curve

$$\frac{D°}{360°} \quad \frac{100}{2\pi R} \quad \pi = 3.1416$$

$$R = \frac{5729.58}{D}$$

Tangent $T = R \tan \frac{I}{2}$

Length $L = 100 \frac{I}{D}$

Long chord $LC = 2R \sin \frac{I}{2}$

Figure 2.25 Highway curve elements.

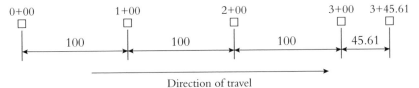

Direction of travel

Figure 2.26 A distance of 345.61 feet expressed in stations.

When working with documents concerning routes, or rights of way, it is often necessary to make the conversion from stations to distances in feet. Also, descriptions of transfer and taking from a private entity to the right of way authority (such as highway, railroad, power line, pipe line, and the like) are often related to stationing.

THE UNITED STATES RECTANGULAR SYSTEM

While there are exceptions in a few isolated locations, especially in Ohio, where the system began, there is a basic design to the U.S. Rectangular System. It comprises all of the states except the original colonies, Texas, and Hawaii. There are also small areas in Kentucky and Tennessee that are part of this scheme. Without going into the details of the origin and evolution of the system, and the various exceptions and special cases thereof, a few paragraphs will describe its geometry.[40]

Public lands of the United States are first surveyed into rectangular tracts, by running parallel lines north and south and crossing them at right angles with other parallel lines, so as to form rectangles 6 miles square, called townships. The first north and south line is a true meridian, and is called a *principal* or *prime meridian*. The north and south lines parallel to it are called *range lines*. The six-mile strips bounded by these lines are called *ranges*. The first east and west line is called a *base line*. The other east and west lines drawn parallel with it are called *township lines*. The six-mile strips bounded by these lines are called *townships*. The rectangles formed by the two sets of lines are also called townships. They are the largest subdivisions of the survey. They are frequently used as the unit of a political subdivision and given a name, but as a unit in the government survey their only designation is by the numbers of the township strip and of the range strip in which located—for example, township 30 north (from the base line), of range 10 west of the _____ meridian; or, township 30 south, of range 10 east of the _____ meridian.

The townships are subdivided into thirty-six tracts, each a mile square, called *sections*, and containing, as near as may be, 640 acres each. They are numbered consecutively, commencing with Section 1 at the northeast corner, and proceeding west to Section 6; then in the next tier, numbering east from Section 7 to Section 12; and so on back and forth until Section 36 is reached in the southeast corner. See Figure 2.27.

Section corners are marked with monuments. Halfway between these, monuments or stakes are set to mark what are known as "*quarter corners*." Imaginary lines drawn from the latter divide the section into four "quarter sections" of approximately 160 acres each. In this way, two of the outside corners of each "quarter" are marked by stakes or otherwise as "quarter corners," and one coincides with and is marked as a section corner. The inside corner, being the center of the section, is not marked, but may be determined by locating the intersection of the half-section lines run from

---

[40] For other aspects, consult Cazier, Pattison, Peters, Sherman, Stewart, and White (see references at end of chapter).

opposite quarter corners. The quarters were not subdivided by the government survey, but rules for their subdivision were established by the General Land Office.[41]

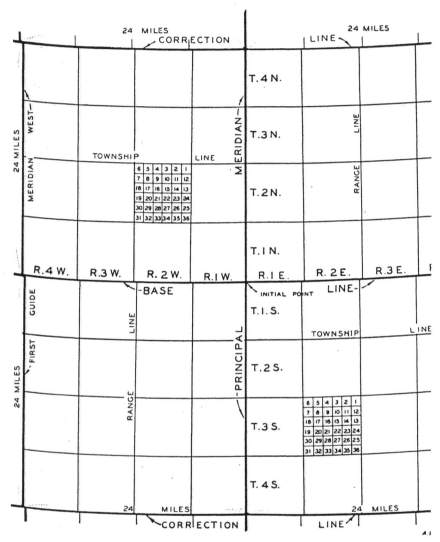

Figure 2.27 Scheme of the U.S. Rectangular Survey System. From *Manual of Survey Instructions, 1973.*

[41] Patton, § 116.

Where reservations, navigable lakes or streams, and the like intercept the surveys, they produce fractional divisions, known as *fractional sections*, *fractional quarters*, and so on. The divisions of a fractional section are numbered, and called *lots*. The sinuosities of a shore are marked, usually a short distance back from it, by a surveyed line, called a *meander line*. Meander corners are set at all points where the subdivision lines of a section intersect the meander line.[42]

The standard area for a section of this survey is 640 acres. Nevertheless, because of the curvature of the earth's surface, it is impossible for all townships to be exactly square. The necessary deficiencies or excesses are apportioned.[43] See Figures 2.28 and 2.29.

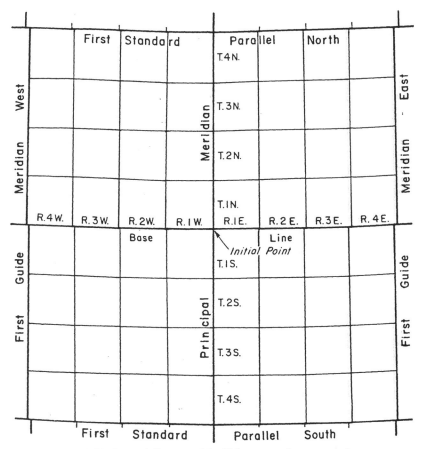

Figure 2.28 General diagram of the U.S. rectangular system of surveys. From *Maps For America*, U.S.G.S.

[42] Ibid.

[43] Ibid.

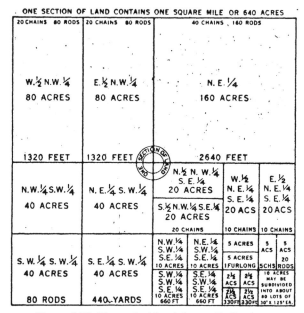

Figure 2.29 Theoretical breakdown of a Section.

FOR FURTHER REFERENCE

Bedini, Silvio A. *Early American Scientific Instruments and Their Makers*. Rancho Cordova, CA: Landmark Enterprises. 1986.

Brinker, Russell C., and Roy Minnick, Eds. *The Surveying Handbook*. New York: Van Nostrand Reinhold Company. 1987.

Cazier, Lola. *Surveys and Surveyors of the Public Domain, 1785–1975*. Washington: U.S.G.P.O. 1978.

Chermside, Herbert B., Jr. *Boundaries: Description in deed as relating to magnetic or true meridian*. 70 A.L.R. 3d 1220.

Estopinal, Stephen V. *A Guide to Understanding Land Surveys*. Eau Claire: Professional Education Systems, Inc. 1989.

Henry, H. H. Boundaries: *Measurement in horizontal line or along surface or contour*. 80 A.L.R. 2d 1208.

Howe, H. Herbert, and L. Hurwitz. *Magnetic Surveys*. U.S. Department of Commerce, Coast & Geodetic Survey Serial No. 718. Washington: U.S.G.P.O., 1964.

J.Q.L. *Distance as determined by straight line or other method*. 54 A.L.R. 781.

Love, John. *Geodaesia*. London: Printed for John Taylor. Several editions.

Pattison, William D. *Beginnings of the American Rectangular Survey System, 1784–1800*.

The University of Chicago Department of Geography Research Paper No. 50. Chicago. 1964.

Patton, Rufford G., and Carroll G. Patton. *Patton on Land Titles*, second edition, Volume I. St. Paul: West Publishing Co. 1957.

Peters, William E. *Ohio Lands and Their History*, third edition, reprinted 1979. New York: Arno Press, Inc. 1930.

Richeson, A. W. *English Land Measuring to 1800: Instruments and Practices*. Cambridge: The Society for the History of Technology and The M.I.T. Press. 1966.

Sherman, C. E. *Original Ohio Land Subdivisions*. State of Ohio, Department of Natural Resources, Division of Geological Survey, Columbus. 1925, reprinted 1982.

Stewart, Lowell O. *Public Land Surveys: History, Instructions, Methods*. Ames: Collegiate Press, Inc. 1935, limited 1975 reprinting by The Meyers Printing Co., Minneapolis.

Wattles, William C. *Land Survey Descriptions*. Revised and published by Gurdon H. Wattles, Orange, CA. several editions.

White, C. Albert. *A History of the Rectangular Survey System*. Washington: USGPO.

REFERENCES

Brown, Curtis M., Walter G. Robillard, and Donald A. Wilson. *Brown's Boundary Control and Legal Principles*, fourth edition. New York: John Wiley & Sons, Inc. 1995.

Richeson, A. W. *English Land Measuring to 1800*. Cambridge, MA: The M.I.T. Press and The Society for the History of Technology. 1966. Chapter 3

CHAPTER 3

RECORDS RESEARCH: TITLE SEARCH OR DEED SEARCH

Ancient deeds are to be construed as written in
light of the then use of properties conveyed and adjacent
land, and cannot be cut down by vagueness in subsequent
conveyances.

—*Harvey v. Inhabitants of Town of Sandwich,*
152 N.E. 625, 256 Mass. 379 (1926).

Lawyers are not surveyors.

—Abraham Moskow
14 Boston Bar Bulletin 237 (1943)

Although often believed to be the same thing, title searching and boundary, or deed, searching are very different. For each, much of the search is done in the same places, many of the same records are reviewed, but the goals are very different. The acceptable standards for the two types of searches are not alike, and the results of the searches are entirely different. True, in each close scrutiny is given to the description, but for different reasons. For a rule of thumb:

> The title search is performed to determine what the parcel is, who owns it and what encumbers it; while the deed search is performed to find out where the parcel is, what its boundaries are, and how large it is.

Title searching, or more properly, *title examination*, is a look at the *quality of title* concerning (usually) one parcel of land, although it may involve more than just one parcel, if they are in the same immediate area, in the same ownership, or in same title chain. The examiner's goal is to determine who has what interests or rights in the parcel and if or how it may be encumbered. All of these aspects are to be reported after having researched and examined all of the documents concerning the land *and its owners* to date, over a period of time set by the title examination standards. The period of search is usually 35 years, 40 years, 60 years, or whatever, or to a warranty deed or similar base of title that affords some guarantee that the title is good, or defendable.

The examiner's report will usually contain a chain of title and copies, or abstracts, of all documents affecting the title, and a list of all outstanding interests and encumbrances. See Figure 3.1.

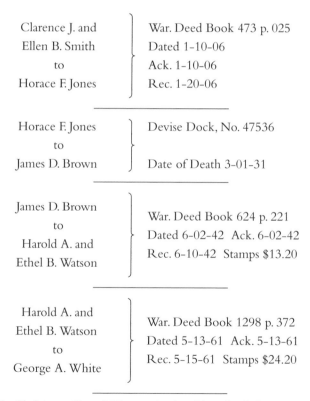

Figure 3.1 Skeleton outline of title examination. More detailed reports may include copies of the descriptions, and other pertinent information. Results vary from place to place and from time to time. Older *abstracts* were much more detailed.

Title examinations, abstracts, reports, opinions, and title insurance policies are generally subject to the following items:

- Items outside the period of search
- Items not on the public record
- Items that would be disclosed by an accurate survey

Items Outside the Period of Search. There are frequently better, more complete property descriptions appearing in the earlier records that are often outside the period of search. In addition, the practice for many years has been to merely copy the old description into the new deed when the sale takes place, without updating the description. This is particularly true in some of the older states, and the older the original description the more of a potential problem there is. In some areas, current descriptions are the same as they were in the eighteenth century and do not accurately reflect the description of the property. Abutters have long since changed, and often abutting roads have changed, encumbrances have been created, and sometimes even the property itself has changed.

There are many things wrong with this practice, but two of the biggest problems are transcription errors and additional information appearing in older records but not carried forth into later records. In working with older records, partially because they are difficult to read and because terminology, word interpretation, and even meanings of words change, transcription errors occur in abundance. As an example, a landowner opted to stake out his own deed description, which read:

> Bounded as follows: Southerly by the road, 200 feet;
> easterly by land of Smith, 157 feet; north by land
> of Jones, 208 feet; and westerly by land of Brown,
> 175 feet.

Had this description been investigated, the records would have shown that none of the corners were ever called for in any description, so the points found may or may not be correctly marking the true corners. Brown's land had been surveyed about 20 years previously, and a plan of record depicted Brown's line marked by two stone bounds 157.00 feet apart. The back one was not found because it was flush with the ground so that it would not interfere with mowing, and grass had grown over it. In addition, Brown's description has always called the line 157 feet long, and earlier descriptions of the subject tract also called for the line to be 157 feet in length. At some time, the numbers had been transposed.

Examining the site, the owner found the following conditions shown in Figure 3.2.

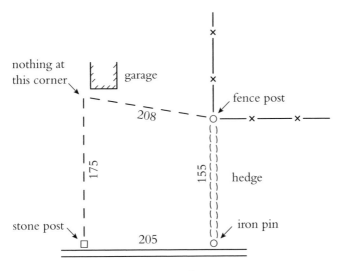

Figure 3.2   Sample property.

The owner accepted the physical corners found, and set a new corner, as close as he could, where the one was missing. See Figure 3.3.

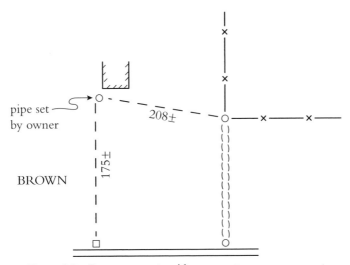

Figure 3.3   Former property with owner-set corner monument.

Not only has this owner created a title problem and an overlap of two properties, but now the garage, which is fairly new and was built according to the description of the northerly abutting parcel, is in violation of the setback requirement according to the line as staked. One can easily see how such situations become worse as time passes, owners change, and memories fade. This situation is definitely a potential lawsuit, or at the very least, an almost guaranteed loss of a sale when a mortgage inspection is done on the northerly lot for lending purposes.

The other type of common problem arises when easements and other rights created in early descriptions are dropped from the descriptions and are not reflected in later deeds. Many of these easements and rights "run with the land" and are as much in effect today as when they were created.

Consider the recent example of a seventy-five-unit trailer park that has been designed, been surveyed, and had the title examined. A small burial ground was found on the property and, being undevelopable, was not considered. The surveyor had done the necessary deed research but had not investigated the graveyard, which should always be done. It was not felt necessary for the survey and was well outside the scope of the title examination. It was depicted on the survey as shown in Figure 3.4.

Later investigation uncovered the following 1864 description:

> . . . a certain tract or parcel of land situated in said town and known as the bury-ing ground, or yard of the late Isaac Smith deceased, situated on the homestead of sd deceased, inclosed on all sides by a stone wall, together with a half a rod of land in width on the outside of said yard, carrying the breadth of a half a rod in width on each and every side of the aforesaid yard, joining said yard wall, with the privilege of passing and repassing at any and all times in the usual passways, from the road to said burying ground.

The plan was amended and the lots redesigned.

Family burial plots are always a potential problem area in the records. Others may have the right to burial, and at the very least, family members and perhaps others have a right to visit the lot. In some states, certain activities cannot take place within a certain distance of a graveyard.[1]

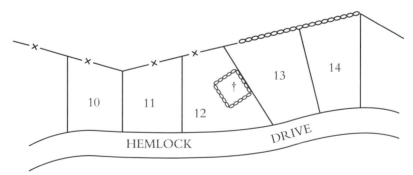

Figure 3.4   Trailer park lots designed around old burial ground.

[1] Where a deed reserved "one-half acre where the graveyard is now situate," reference was made to grave-yard to identify location of land reserved, the graveyard must be taken as the common center of the tract reserved. *Honey v. Gambriel*, 135 N.E. 25, 203 Ill. 74 (1922).

Similar potentially severe problems that frequently do not present themselves in routine title examinations are rights of way, water and spring rights, roads laid out but never built, outstanding mill rights and privileges, and a host of other exceptions and reservations, even outsales not reflected in later conveyances.

Items Not of Record. There are a number of documents relevant to or affecting a property that never appear on the public record. Usually, except in special circumstances, they are not part of or considered by a title examination. However, they may be discovered in a records investigation.

Sometimes deeds are just not recorded. See Figure 3.5.

Neither an examination of the grantee indexes for Samuel Smith nor the grantor indexes (reviewed to date in case of a late recording) for William Brown produced this deed. On an outside chance that William Brown died and left the property to Samuel Smith, Brown's probate was checked. Brown's will recited that his son Charles was to receive all of the real estate, and the land in question did not appear in his inventory. Sources such as court records and other obscure possibilities were checked. This conveyance is believed to have never been placed on record, although there is a record of its existence. We can continue with the search and the chain of title, but cannot account for the document.

In one instance, it was necessary to locate such a deed. Research indicated that a 5-acre parcel was divided in an estate as shown in Figure 3.6.

The chain of title to parcel 1 was complete, and its description clear (see Figure 3.7).

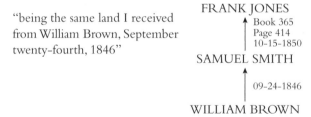

Figure 3.5 Sample chain of title.

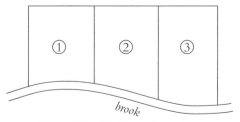

Figure 3.6 Division of an estate.

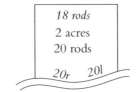

Figure 3.7   Diagram of parcel division in Figure 3.6.

This parcel had been surveyed, and stone bounds were found at all four corners.

The interest was in the other two parcels, particularly in lot 3. While the chain of title for parcel 2 was clear to date, there was no description for the lot and the beginning deed was not on record. Nothing was found for parcel 3. The title appeared as shown in Figure 3.8.

Examination of the Estate of Stephen N. Chase resulted in a Return from a License to Sell Real Estate. This report stated that the following sales were made:

David Gove, woodland	$132.00
Frank Brown, woodland	$ 85.00
George A. Weare and	
Roby Morrill, woodland	$105.94

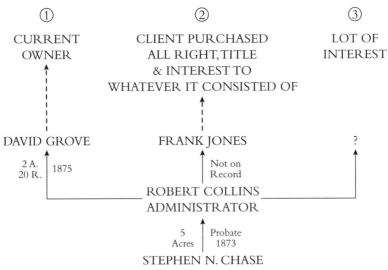

Figure 3.8   Chain of title for above example.

Checking George A. Weare and Roby Morrill in the grantor indexes to date did not reveal a sale of the parcel in question. The parcel was not specifically mentioned in either probate file, but without conveying it, the assumption was made that they died with their interests. Heirs can be traced and found, but at this point there was no description for the property, or even an idea of its size.

The next *three years* were spent tracking Weare and Morrill heirs until, one day the original deed was found in an attic trunk belonging to a distant relative:

More often it is a plan or a sketch that is not recorded. Some landowners will have a survey done but not place it on record because they don't want others to know about their land, particularly the tax assessor. Early surveys were almost never recorded except in the case of probate proceedings when dower and partition were part of the settlement of the estate.[2] Until fairly recently copying facilities did not exist, so any plans were one-of-a-kind, and each had to be hand drawn separately. The usual practice in early surveying was to complete a survey, compute it (sometimes at the site), make a drawing, and present it to the landowner. A record of the survey may have been kept in the form of field notes from which another drawing could be made at any time. Sometimes the computations and plan were on the same sheet of paper. See Figure 3.9.

---

[2] That the old map had not been placed on public record did not affect its admissibility. *Hamilton v. Town of Warrior*, 112 So. 136, 215 Ala. 670 (1927).

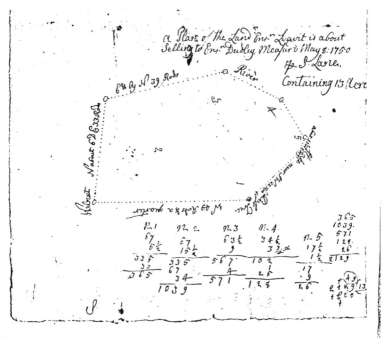

Figure 3.9 A plan of 13 acres made in 1750 by Samuel Lane showing boundaries and computations.

Occasionally, a plan reference will be found in an early deed and, even though the plan itself is not of record, it might be located when the date and surveyor are known:[3] See Figure 3.10.

More often, the surveyor is not mentioned, but knowing the geographical area and who might have practiced there at that time, sometimes a plan can be uncovered (see Figure 3.11):

> Metes and bounds deed descriptions in 1802, 1790, and 1755 were found by tracing the subject and abutting tracts back in time. The following plan was found in the collection of New Hampshire surveyor Samuel Lane, which was reviewed because it was known that he surveyed in this area around that time period.

---

[3] Generally, maps or plats of surveys, maps or plats attached to deeds or grants or referred to therein, maps that have been recognized as correct by former owners of the land, ancient maps and plats, and maps referred to in the evidence, are admissible when verified or authenticated. *Daniel v. Finn, Garrett & Holcomb*, 119 S.E. 307, 156 Ga. 310 (1923); *McDaniel v. Leuer*, 230 S.W. 633 (Mo., 1921); *Nichols v. Turney*, 15 Conn. 101 (1842), *Gates v. McCormick*, 97 S.E. 626, 176 N.C. 640 (1918); *Kearce v. Maloy*, 142 S.E. 271, 166 Ga. 89 (1928), *Cashion v. Meredith*, 64 S.W.2d 670, 333 Mo. 970 (1933), *Christian v. Bulbeck*, 90 S.E. 661, 120 Va. 74 (1916).

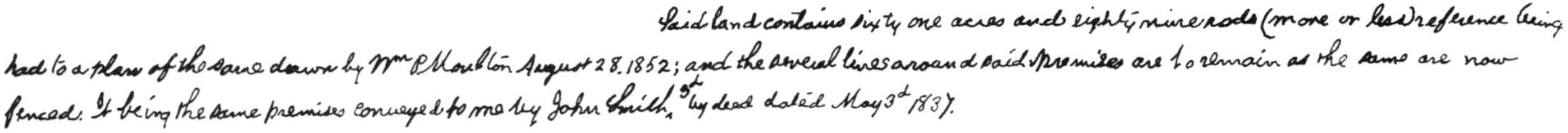
Said land contains sixty one acres and eighty nine rods (more or less) reference being had to a plan of the same drawn by Wm P Moulton August 28. 1852; and the several lines around said premises are to remain as the same are now fenced. It being the same premises conveyed to me by John Smith 3d by deed dated May 3d 1837.

Figure 3.10 Copy of a portion of an 1852 deed description.

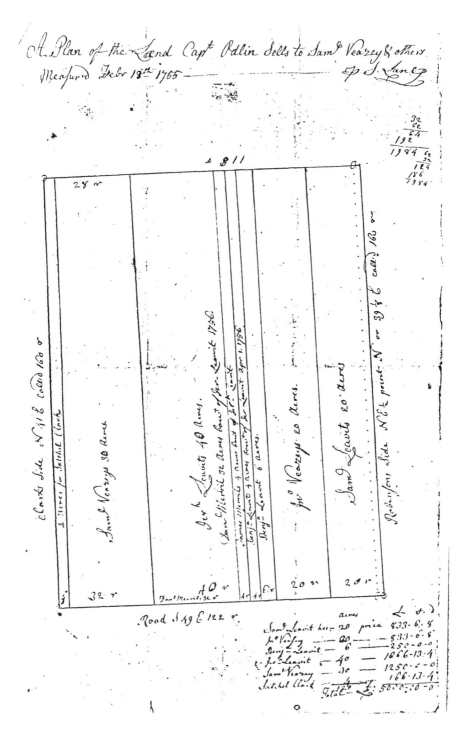

Figure 3.11 1755 plan made by Samuel Lane.

The foregoing example demonstrates why it is vitally important for surveyors' records to be preserved in libraries, historical societies, and archives, or even in other surveyors' offices. Sometimes these old plans can be invaluable, and occasionally they are the *only* source of information on a property. In some areas, most of the land, or at least the original, controlling, boundaries, was surveyed at least once in the past. It remains a matter of becoming aware of, then locating, the remaining records of those surveys.

Other unrecorded, relevant documents include agreements between adjoining landowners, sketches, notations in family bibles, and similar material.

Public Records Other Than the Court House. Usually a title examination is confined to the records at the court house or similar repository of deed and probate information. Many, *many* important records that have a significant effect on the property are not found at the court house but are a matter of public record elsewhere.[4] These include, but are by no means limited to, the following:

- International boundary information.
- State and county boundary descriptions and surveys[5] (see Figure 3.12).
- Town line descriptions, perambulations, and surveys.[6]
- Highway and other road layouts and surveys[7] (see Figure 3.13).
- Vital statistics—state, town and church records of births, deaths, marriages, adoptions, and so on.

---

[4] Generally, relevant documentary evidence is admissible, whereas irrelevant and immaterial evidence is not.

For the purpose of identifying the premises sought to be described, reference may be had to other conveyances, plats, or records, well known in the neighborhood, or on file in public offices. *Pittsburgh, C., C. & St. L. Ry. Co. v. Beck*, 53 N.E. 439, 152 Ind. 421 (1899).

Proper aids for identifying land described in deeds include judgments, certificates, patents, deeds, leases, and maps. *Sorsby v. State*, 624 S.W.2d 227 (1981).

[5] Courts are bound to take cognizance of the boundaries in fact claimed by the state, and will exercise jurisdiction accordingly. *State v. Dunwell*, 3 R.I. 127 (1855).

The court will take judicial notice of the lines of counties and the towns enclosed therein. *Ham v. Ham*, 39 Me. 263 (1855); *State v. Jackson*, 39 Me. 291 (1855).

[6] Perambulations of the line between two towns, made by the selectmen of such towns, are competent evidence of the true line in a suit between individuals owning land on opposite sides of the line. *Adams v. Stanyan*, 24 N.H. 405, 4 Fost. 405; (1852). *Lawrence v. Haynes*, 20 Am. Dec. 5511, 5 N.H. 33 (1829).

On a dispute between private parties as to the division line between their private lands, which was also the boundary line between two towns, a "field book" of one of them is admissible as an "ancient record," when it came into possession of the town clerk a "large number of years ago,"and deeds of land in the town, executed over 45 years previous to the trial, referred to it, and it was in the town clerk's custody at the time of the trial. *Aldrich v. Griffith*, 29 A. 376, 66 Vt. 390 (1893).

[7] Where a deed describes property as being bounded on a public highway, the return of the surveyors laying out such highway is competent evidence to establish the boundary, without showing their appointment or whether the highway was legally laid out. *Haring v. Van Houten*, 22 N.J. Law, 61 (N.J. Sup., 1849).

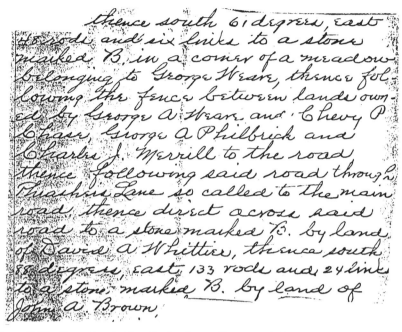

Figure 3.12 Example of a New England town line perambulation containing boundary information in addition to the description of the town boundary.

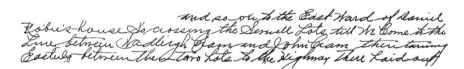

Figure 3.13 Example of a town road layout containing boundary information in addition to the description of the road.

- Tax records.
- Court records, including not only decisions and testimony but also exhibits that may not be found elsewhere.[8] These may be the source of descriptions of and actual condemnations. See Figure 3.14.

---

[8] In establishment of ancient boundaries, judgments of trial court were held admissible, notwithstanding parties were not the same. *Blaffer v. State*, 31 S.W.2d 172 (Tex. Civ. App., 1930).

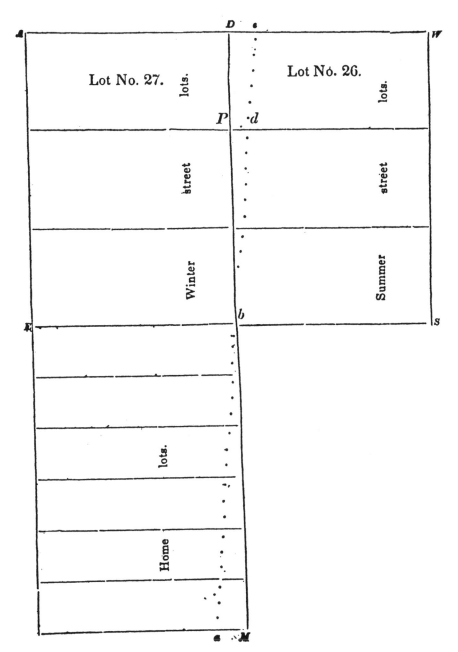

Figure 3.14   Example of a New Hampshire Supreme Court decision that determined the location of a boundary line.

- Records of state, county, and municipal lands.[9]
- Records of state, county, and municipal easements.[10]
- Public utility easements and surveys (see Figure 3.15).

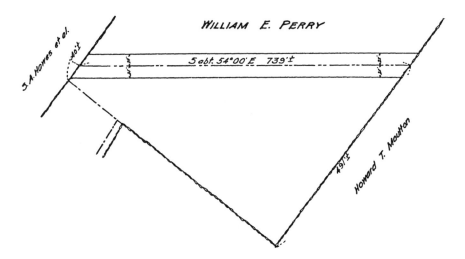

Figure 3.15   Plan showing electric company right of way across land of William E. Perry.

- Federal lands and surveys, including field notes.
- Records from other jurisdictions. Political boundaries change, but records sometimes remain within the original jurisdiction. This is particularly true of county and town records.

---

[9] In trespass to determine the location of a boundary line between certain lots, an old plat of the town obtained from a volume of state papers, showing the division of the lots, etc., was admissible, in the absence of objection as to its authority. *Twombly v. Lord*, 66 A. 486, 74 N.H. 211 (1907).

Ancient surveys of a city, showing streets and lots appearing on the county records, are presumed to have been recorded by authority, though not formally certified for record. *City of Lexington v. Hoskins*, 50 So. 561, 96 Miss. 163 (1900).

[10] In boundary dispute, copy of profile of street sewer showing boundary line and fence, dated October, 1892, and produced by city engineer's office, held competent. *Hews v. Troiani*, 179 N.E. 622, 278 Mass. 224 (1932).

As an example, historic boundary and title research was done concerning land in the Town of Bartlett, New Hampshire. Records were found in the following locations:

Bartlett granted in 1765, 1769
Initially in Rockingham County
part of Grafton County, established 1773
then part of Coos County, established 1805
Carroll County established in 1840
Bartlett annexed to Carroll County 1853, where it remains today

It was necessary to research the records for this land in the records of four counties.

• Railroad records—lands and surveys (see Figure 3.16)

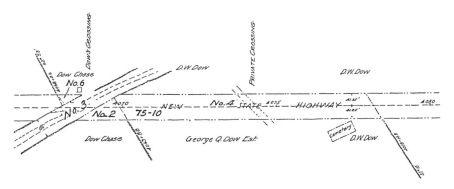

Figure 3.16   Example of a section of railroad right of way depicting boundary and title information.

Genealogical Information. While usually not a matter of any public records, except in the case of vital statistics, genealogical data can be most helpful, especially in overcoming gaps in chains of title.

A recent example was a title that could be traced back to a will of Mary E. Giddings in 1910, which contained an "18 acre woodlot," but further search failed to uncover earlier records. Through the use of calls in descriptions of abutting tracts, it was determined that Zebulon Wiggin had owned this tract and had sold it to a Mary E. Scammon in 1857. It then was a question of whether these are two persons or one who married and changed her name. The answer was found in the family genealogy. See Figure 3.17.

84 John Lyford (*Thomas*,[30] *Thomas*,[11] *Thomas*,[2] *Francis*[1]) born 1 Mar., 1777; died 1803; published 30 Aug., 1799, to Anna Hilton of Kingston (born 4 Sept., 1776), daughter of Andrew Hilton of Exeter (son of Benjamin Hilton) and Jemimah Prescott (born 23 Oct.,

1742). Letters of administration on the estate were granted 23 Jan., 1804. The widow married for her second husband, Kinsley Lyford[71], and she died 7 Mar., 1865, in Exeter.

Children, all born in Exeter:

198. A CHILD.
199. JAMES, b. 1800; d. at sea while still a young man.
200. ANN, b. 4 June, 1803; d. 4 Mar., 1857, in Stratham; m. 6 June, 1828, Ira Scammon of Stratham. He was b. 11 June, 1803, and d. 14 Jan., 1852. Their dau. Mary Ellen, b. 1 Mar., 1837; m. 22 Oct., 1882, John Colcord Giddings.

Figure 3.17   Excerpt from *Francis Lyford, of Boston, and Exeter, and some of his descendants.*

This information was actually helpful in two ways. In addition to the above, the tract in question was made up of two parcels, one of which was sold to "Ellen" Giddings. It could only be assumed that the "E." in Mary E. Giddings stood for Ellen, but this information confirmed it.

Important information may often be found in published genealogies, genealogical data in published town histories, or in personal family records.

Items Disclosed by an Accurate Survey. There is no way that the records can accurately reflect everything physical on the property. Descriptions and drawings can be made and cover a large percentage of things describing the land or affecting the title, but they cannot be 100 percent inclusive.

Ideally, every land transfer would be accompanied by an up-to-date title examination, a records investigation, and a survey. When surveys are done for title insurance purposes, parties are getting closer to this goal. The surveyor is able to inspect the property, discover, and report many things relevant to the title of the property that will not be reflected in the records. Burial grounds not mentioned in the documents, encroachments, use by others (potentially becoming a prescriptive right of way), abutting roads existing outside the formal right of way, erosion or accretion to shore-lines, and so on, are all things that should be noted on a survey but would be impossible to discover in an examination of the record title. Thus, it is one of the reasons for the standard exception in a certificate or a title policy.

A typical set of title examination standards will probably contain, among its many requirements, at least the following items:

TITLE EXAMINATION STANDARDS

ARTICLE I. PURPOSE

1.1. The objective of the title examiner is to determine whether or not the title in question is satisfactory of record. Objections to the title should be made only when the defect or defects could reasonably be expected to expose the prospective owner, tenant, or lienor to adverse claims. The following Title Standards express the practice considered reasonable by members of the _____ Bar Association with respect to some recurring problems, of title examiners.

ARTICLE II. PERIOD OF TITLE SEARCH

2.1. A title examination shall cover a minimum Period of thirty-five (35) years commencing with a deed with warranty covenants.

ARTICLE III. AFFIDAVITS AND RECITALS

ARTICLE IV. NAMES

4-1. The doctrine of Idem Sonans (namely, that if two names, as commonly pronounced in the English language, are sounded alike, a variance in their spelling is immaterial; and even a slight difference in their pronunciation is unimportant if the attentive ear finds difficulty in distinguishing between the two names when pronounced, and although spelled differently, they are to be regarded as the same) should be applied broadly, and the fact of identity of the party presumed in spite of variations in the spelling of the same.

ARTICLE V. DEEDS

ARTICLE VI. MORTGAGES

ARTICLE VII. ESTATES

ARTICLE VIII. CORPORATIONS

ARTICLE IX. LIENS

ARTICLE X. ABSTRACTOR'S CERTIFICATE

10-1.   The Certificate of the abstractor should state that a careful examination of the records in the Registry of Deeds and the Registry of Probate (and of the Superior Court, when applicable) has been made insofar as they relate to the title to the land in question, and that the abstract contains all conveyances and other instruments of record properly indexed affecting this title for the period covered by the examination.

N.B. The following suggested abstractor's certificate is not to be interpreted as an opinion on the title. It is merely a certification as to the actual record of matters that affect the title. An attorney may review the information given in the abstract to which this certificate is attached, and relating that information to his interpretation of the law, may be able to give an opinion on the title.

(Suggested Abstractor's Certificate)

Date and hour _____ "The undersigned certifies that the records at the _____   _____ County Registries of Deeds and Probate, and at the _____ County Superior Court, when applicable, have been carefully searched, and the following (abstract) (title report) sets forth all matters pertaining to the title to the premises in caption which were properly indexed therein from _____, 19 _ to the above day

and hour. Unless otherwise noted herein, all
conveyances were properly signed, sealed, witnessed
and acknowledged and dower, curtesy and homestead
were properly released. The (abstract) (title report)
was prepared for the sole use and benefit of
_____with the understanding
that the undersigned reserves and retains the sole
right to reproduce the same. Inquiry should be made
as to the existence of possible liens, rights and encumbrances which may not appear of record in the
records searched.
This (abstract) (title report) does not constitute a
guaranty or opinion of title.
(Description of Property Searched)"

Boundary research, on the other hand, is far more extensive. Most states have rather comprehensive standards including, as a minimum, requirements such as:

1. The subject parcel will be searched back to its origin of description.

Unless this is done, the searcher cannot know if the present description is accurate, for it may have been copied incorrectly some time in the past. Additionally, earlier descriptions may have contained information since dropped and not appearing in later descriptions. The original description reflects the intentions of the parties and establishes senior/junior rights, and, if it contains survey information, the origin may provide the date necessary for declination computation.[11]

2. At least the immediate abutters will be searched.

Additional information is frequently found in abutting deed descriptions. Common boundary lines must be compared for agreement, and to ensure that overlaps and gaps

---

[11] Generally, any deed or grant having a tendency to identify and to fix a disputed boundary is admissible. Deeds of adjacent lands are admissible if relevant. *Delaware Securities Corporation v. Kahn*, 175 So. 779 (Fla., 1937), *Marvil Package Co. v. Ginther*, 140 A. 95, 154 Md. 213 (1928), *Barnes v. Callaway*, 269 S.W. 1085 (Tex.Civ.App., 1924), *Hargrove v. Harris*, 189 S.E. 307, 167 Va. 320 (1937). In determining boundary between properties conveyed by partition deeds, deed to adjoining lot made as part of the same transaction could be considered, *Smith v. Harbaugh*, 112 So. 914, 216 Ala. 202 (1927), *Roberts v. Atlanta Cemetery Ass'n*, 91 S.E. 675, 146 Ga. 490 (1917). In a contest arising from conflicting state land grants, the calls in one grant tending to locate the prior grant are competent evidence in aiding the location of the prior grant. *State v. Cooper*, 53 S.W. 391 (Tenn. Ch., 1899).

do not occur within the descriptions or, if they do, that they are discovered and identified. Without researching abutting parcels, their surveys are often not detected.[12]

3.   Easements of record and visible signs of possession and easements will be reported.

The surveyor is the "eyes of the title examiner," and numerous things affecting the title may be apparent on the ground, but not within the scope of the title exam or known to the examiner.

4.   Other encroachments, encumbrances, and any potential effect on the title will be located and/or reported.

Wells, springs, graveyards, trails, cutting of trees, mowing of grass, and an infinite number of other uses by others that are incompatible to the true owner may have an ultimate effect on the title. All should be considered and reported.

Great care must be taken in using or relying on either of these searches, not for what they contain but for what their limitations are. Any search is only as good as (1) the standards under which it was carried out, and (2) the expertise of the individual doing the search. Even knowing this, the user must keep in mind at all times that documents not of record are often not available, but they may appear at some time in the future. Unrecorded documents are sometimes not even suspected, since they are not always referenced in other, subsequent documents. Pertinent materials may be forever lost or may have been destroyed.

Too often a surveyor's client will request that the surveyor dispense with researching the property because there has already been a title search done. What the client does not realize is that only a very small amount of the relevant, available information and evidence has been reviewed, and even that may not be reliable. Only more extensive searching will demonstrate if it is or is not. Also, a client will sometimes request a "title search" from a surveyor. If a title search is truly wanted, the client should be speaking with either a title examiner or a title attorney. On the other hand, if the client wants to know about his boundaries, or know about his neighbor's lot, and any easements relating to his ownership, then he should be requesting "deed research," not a title search.

A review of survey research standards around the United States revealed that, among others, minimum requirements common to many states were these:[13]

---

[12] Ibid.

[13] Wilson, 1987. *The Land Surveyor and His Relationship To Minimum Standards For Record Research.* Paper delivered at the ACSM-ASPRS Fall Convention Reconvene, Honolulu, Hawaii.

A subject tract shall be researched back at least to its origin.

All abutting tracts shall be researched as far back as practicable.

Tracts other than the subject and abutting tracts shall be researched whenever necessary.

All public and private rights of way or easements that are of record or apparent, adjoining or crossing the land surveyed shall be identified.

Cemeteries and burial grounds shall be shown along with their records, when available.

Lots, townships and ranges shall be identified.

Streets, roads, easements of record and/or usage shall be investigated.

Where applicable, certificates of title, section corner ties, subdivision plats, county surveyors' records, and section line information shall be investigated.

Proper research resources shall include, but not be limited to:

> Records of previous surveys
> Registry of deeds and registry of probate
> Records of highways
> Railroad and utility line records and plans
> Tax assessment maps and records
> Topographic maps
> Local histories
> Aerial photographs
> Genealogies
> Court records

In all cases, reference must be made to the source of information used in making the survey, such as the recorded deed description or other conveyance, a recorded or unrecorded plat, or other claim of right.

Procedure. In doing a search, usually the first event is an examination of the current deed, or other relevant document, and its description. To fully understand the subject matter of the conveyance, or the document, the description should be studied and a drawing (a freehand sketch is sufficient) made of the parcel(s). The drawing should be made somewhat to scale and thoroughly labeled, such as shown in Figure 3.18 a and b.

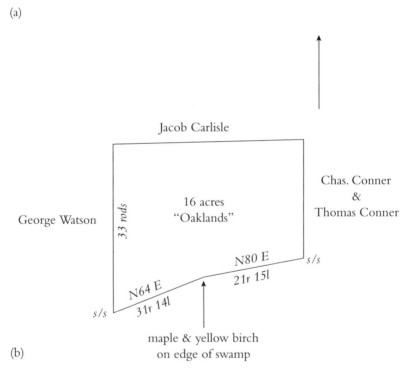

A certain tract or parcel of land situated in Exeter in said County and State, in the "Oaklands" (so called), containing sixteen acres more or less, and bounded and described as follows: Easterly by land of Chas. Conner and Thomas Conner. Northerly by land of Jacob Carlisle; Westerly thirty three (33) rods by land of George Watson to a stake and stones, thence running North sixty four degrees East (N. 64° E.) to a maple and a yellow birch tree on the edge of the swamp thirty one (31) rods and fourteen (14) links, thence running North eighty degrees East (N. 80° E.) twenty one (21) rods and fifteen (15) links to a stake and stones the same being the bound begun at.

(a)

Jacob Carlisle

33 rods

George Watson

16 acres
"Oaklands"

Chas. Conner
&
Thomas Conner

N80 E

21r 15l

s/s

N64 E

31r 14l

s/s

maple & yellow birch
on edge of swamp

(b)

Figure 3.18   Example of a simple deed description and its accompanying sketch.

A description should be read a *minimum* of three times. The first reading will give one an idea of the parcel, and an evaluation can be made as to whether it is the correct parcel. The second reading should be done item by item as the sketch is drawn. The third reading should be a check and verification of the sketch, making sure that *everything was put on the sketch that appeared in the description and only what was in the description was placed in the sketch.* The sketch should be as accurate as possible in representing the writing. If there are two interpretations, two drawings should be made, one for each.

When drawing a description, write down exactly what it says, even if you don't agree, and even if you know it is erroneous. That is what the description says. You did not write it, you are merely reading it and drawing what you read. Sometimes, it is later determined that the description was not in error after all. With additional information, you may find that you were the one in error, and the description was correct all along.

Running vs. Bounding Description. There are two types of description that may be encountered. The *running description* basically "runs" around the property in either a clockwise or counterclockwise manner, reciting information such as directions, distances, monuments, and abutters as they are encountered. In a closed figure, the ending point will be the same as the point of beginning and the figure will "close" verbally. A *bounding description* is one in which the scrivener (or description writer), and the reader, is on the inside of the parcel looking out and describing its bounds.

For example, the following descriptions apply to Figure 3.19.
A running description would read somewhat as follows:

Beginning at the southwest corner of the parcel being described,

thence running in a northerly direction by land of Smith, 100 feet;

thence easterly by land of White 150 feet; **thence** southerly by Jones,

100 feet; **thence** westerly by the highway, 150 feet to the point of

beginning.

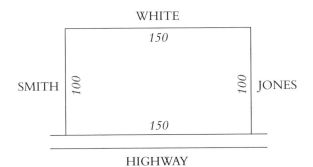

Figure 3.19   Sample tract for description.

A bounding description would read as follows:

> **Bounded** Westerly by land of Smith, 100 feet; northerly by
> White, 150 feet; easterly by Jones, 100 feet and southerly by the
> highway, 150 feet.

One can quickly see that if one type of description is interpreted as the other, the parcel will be sketched at 90° to what it should be.

This may seem very basic, but a word of caution is in store. There are descriptions where it is impossible to tell whether they are running or bounding descriptions unless they are drawn out and compared with known features, such as the directions of roads. The following description and accompanying sketch illustrate this type of problem (see Figure 3.20).

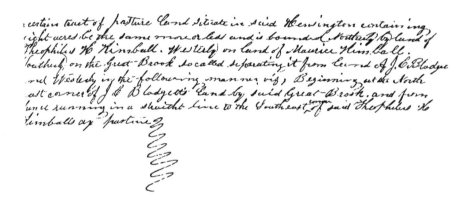

Which is it, running or bounding?

Taking the time to draw it, it would appear as shown in Figure 3.20.

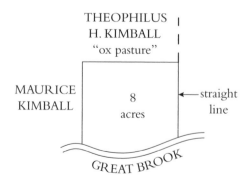

Figure 3.20 Sketch of foregoing written description.

And this is actually the worst one could expect, a combination of both running and bounding descriptions. This confusing entity can easily give the reader a false sense of security. The best advice is to not take chances. *Always draw the description.* And compare it, if only mentally, to known surroundings, features, and orientation. By following this advice, mistakes will be minimized. Mistakes may still be made, but they will be minimized.

Tracing the Chain of Title Back.   There are a number of reasons for tracing the description back to its origin.

- To check for transcription errors. Without comparing the present description with the original, there is no way of knowing if the present one is accurate. It could easily be in error merely from the transposition of numbers in copying.
- To determine the base year for declination computation. Without the earliest date attainable, conversion of magnetic bearings is, at best, only a guess. Even an error of just 50 years could mean a difference of several degrees, resulting in a failure to find an existing monument, or setting a new one in the wrong place.
- To determine the order of conveyancing and senior rights. Without knowing the precise date that a parcel was created, it is impossible to know how it fits in the order of conveyancing, and what priority to give it in proportioning, or allowing for, errors in measurement.
- To gain insight as to the intention of the parties at the time of the original description. Names, conditions, laws, and customs change. While this will be addressed in detail later, Figure 3.21 will suffice to illustrate what can happen if the wrong date is selected.

There are significant differences between the top (1919 edition) topographic map and the bottom (1957 edition) one.

A. Marchs Pond in 1919 is one large pond, whereas in 1957 Marchs Pond is the easterly, smaller, portion of the pond, while Chalk Pond is the westerly, larger, portion of the same pond. Descriptions on Marchs Pond may be construed to be in more than one location.

B. Club Pond exists in 1957, but only a stream exists in 1919. Descriptions calling for a stream in 1919 may now be under water.

C. A cemetery shown on the 1957 edition does not appear on the 1919 edition, indicating that it *may not* have existed in 1919.

D. The Boston & Maine Railroad tracks were in existence in 1919, but were gone by 1957.

E. The Merrymeeting River is a mere thread in 1919, but a wide, flooded river in 1957. Again, some ownerships may be partly, or wholly, under water.

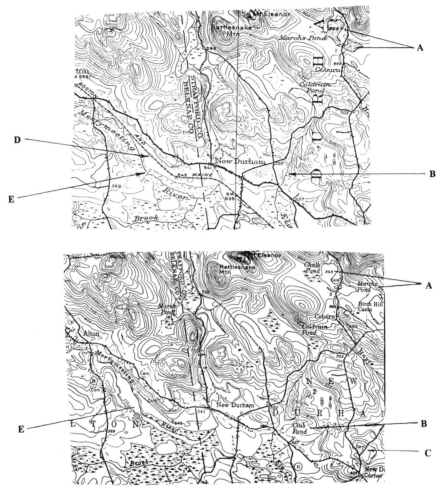

Figure 3.21 The top edition was published in 1919. The bottom one of showing the same area was published in 1957 and indicates changes during the interim period.

The next procedure is to trace the deeds and their descriptions back through the chain of title. Frequently, deeds will recite their source of title or the previous conveyance, in which case it is a simple matter of looking up the referenced document, comparing the descriptions, redrawing if there are any differences, and moving on.

When no reference appears in the deed, it is necessary to use the grantee indexes. Look up the name of the grantor (seller) in the "deed in hand" as a grantee (buyer) to determine what he or she bought and from whom. You may find that this person bought many parcels, and you may have to look up all of them. You may also find that this grantee assembled several parcels to make up what was sold when he or she became the grantor. When this is found, you will be faced with tracing several more titles and fitting those parcels together. This is essential to make certain that no pieces are missing, or that comparison can be made with the most recent description. If not, then whatever differences occur must be accounted for, whether they are out-sales, additions, valid changes in the description, or mistakes. A sample chain of title is shown in Figure 3.22.

After all parcels have been traced to their origin of description and assembled so that a picture of the subject parcel is available, one requirement remains. It is now necessary to make an assessment of whether there are *particular boundaries* affecting the parcel in question. If so, these may need to be examined as separate enti-ties, traced to *their* origin, or given special consideration. As an example, consider Figure 3.23.

In Figure 3.23, the subject parcel is composed of parcels 1, 2, 3, and 4, which define the perimeter boundary. The ownership may be affected, however, by addi-tional boundaries, which must also be examined.

Tracing Abutters. In boundary research, unlike title examination, it is necessary to trace all of the abutting parcels. There are several good reasons for this:

- To ensure agreement and compatibility of adjoining tracts. If the parcels do not fit together properly there may be problems with gaps or overlaps. If so, they have to be identified and reconciled.
- To understand the adjoiners' titles and descriptions in order to identify differ-ences in line descriptions by direction, length, or monumentation.
- To identify seniority of conveyancing when they have the same source of title.

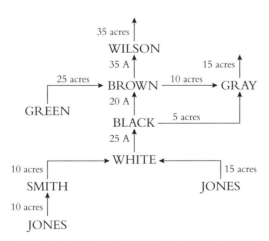

Figure 3.22 Skeleton chain of title showing purchases and sales.

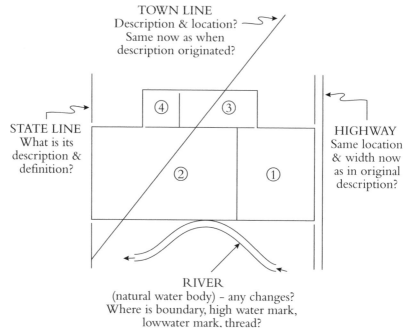

TOWN LINE
Description & location?
Same now as when
description originated?

STATE LINE
What is its
description &
definition?

HIGHWAY
Same location
& width now
as in original
description?

RIVER
(natural water body) – any changes?
Where is boundary, high water mark,
lowwater mark, thread?

Figure 3.23   Sample tract made up of four individual parcels and affected by
boundaries other than the perimeter or the interior lines of the four parcels.

- To obtain additional or, sometimes, better description elements appearing in abutting descriptions that don't appear in the descriptions of the subject parcel(s).[14]
- To locate and take advantage of surveys of abutting parcels. All of them may not be
  found without searching these abutting parcels, especially if they are not of record.

---

[14] Evidence of courses and distances in prior deeds or maps of the same and adjoining lands, and of other evidence aliunde competent to prove any actual fact, is competent to confirm or control courses and distances in a deed, on a question of boundary. *Opdyke v. Stephens*, 28 N.J. Law, 83 (N.J. Sup., 1859).

Where the exact location of the land claimed by plaintiff was uncertain, deeds of adjacent land were admissible to locate plaintiff's land as it formerly existed. *Raughtigan v. Norwich Nickel & Brass Co.*, 85 A. 517, 86 Conn. 281 (1912).

In a question of boundary of a parcel of land, ancient deeds of adjoining patents are admissible. *Townsend v. Johnson*, 3 N.J. Law, 706 (1810).

Where a course is evidently omitted in a deed to land sold according to a plan of lots the missing course may be shown from the description in a deed to the adjoining lot. *Zerbey v. Allan*, 64 A. 587, 215 Pa. 383 (1906).

To establish the location of demanded premises, grants of adjacent lands between strangers are admissible evidence. *Hathaway v. Spooner*, 26 Mass. 23, 9 Pick. 23 (1829); *Sparhawk v. Bullard*, 42 Mass. 95, 1 Metc. 95 (1840). See *Morris v. Callanan*, 105 Mass. 132 (1870).

Recitals in ancient deeds are admissible to prove the position of a line from the location of which the bound in dispute can be determined. *Sparhawk v. Bullard*, 42 Mass. 95, 1 Metc. 95 (1840); *Morris v. Callanan*, 105 Mass. 129; *Hathaway v. Evans*, 113 Mass. 264; Mass. 1879 (1873); *Drury v. Midland R. Co.*,127 Mass. 571 (1879).

Researching Groups of Lots. Sometimes it is necessary to research and establish complete chains of title for entire blocks, or groups of lots.[15] It is first necessary to determine if all of the lots were created in a simultaneous manner, as in modern subdivisions, or if the lots were created individually, in a sequential manner. Occasionally, it is a combination of both. See Figure 3.24.

It is important to make such a determination in order to deal with excess and deficiency of distance and areas when encountered. If lots were created simultaneously, excess or deficiency is distributed proportionately or evenly throughout. In a sequential situation, distances and/or areas are allotted in the manner by which they were created. The first lot is given what is called for, the second is given whatever comes next, and so on until the last lot is reached and it can only be given whatever is left, whether there is more or less than what is called for or described. Sometimes, depending on the description and the circumstances, small unconveyed or unaccounted-for parcels are discovered.

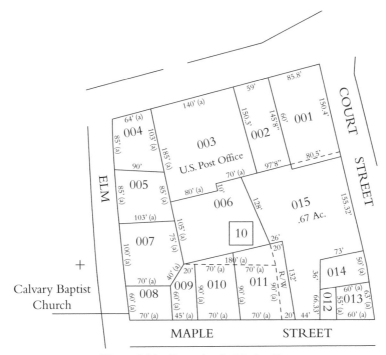

Figure 3.24   Example of a block of lots.

---

[15] Civ. App. 1938. Surveys constituting a block are not to be treated as separate and individual surveys, nor can each tract be located independently of the rest, by its own individual lines or calls or courses and distances, but surveys are to be located together as one block or one large tract. *Beck v. Gulf Production Co.*, 113 S.W.2d 258 (Civ.App., 1838).

See also *Duval County Ranch Co. v. Rogers*, 150 S.W.2d 880 (Tex., 1941).

See also 97 A.L.R. 1227. *Rights as between grantees in severalty of lots or parts of same tract, where actual measurements vary from those given in the deeds or indicated on the map or plat.*

The other important result of researching blocks or groups of lots is learning what additional nearby evidence may be available to assist in the location of a parcel or its survey. In addition, surveys of nearby lots may also be of some assistance.

As an example, in laying out a lot, measurements have to begin from somewhere, and any monumentation found that is not original must be verified for location. Original monumentation must be verified as to location as well to ensure that it has not been disturbed since it was originally set. See Figure 3.25.

Some markers are merely "plunked in" by owners, as shown in Figure 3.25. Other possible corner markers, after investigation, may turn out to be things like mail box posts, curb stones, horseshoe pins, pins to tie the dog to, and the like. *All found markers should be questioned.* If not original, and in its original location, a marker *must be verified* as to origin and significance.

In-Depth Research Gives a Broader Picture.   Researching the subject tract and its abutters, plus related tracts when necessary, will result in a much broader view of how the conveyancing took place and what the record really says about ownership(s). There are four categories that need to be examined in any research investigation.

The order of conveyancing (date of conveyance/recordation)

The boundaries of the source parcel

Any particular boundaries that may affect the subject tract

Changes in related titles, such as abutting roads and the like

Order of Conveyancing.   As previously shown, this is necessary for several reasons. Excess and deficiency must be dealt with when encountered, and senior-junior rights, boundaries, and surveys must be considered.

It is not enough to merely stake out 100 feet of frontage just because the description calls for 100 feet on the street. For an example, see Figure 3.26.

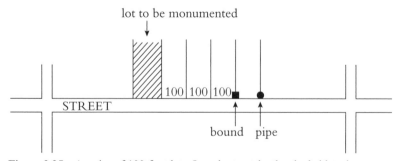

Figure 3.25   A series of 100-foot lots. In order to stake the shaded lot, the nearest marker is three lots away. It measures 298.5 feet and is 104.8 feet from the next marker. Which one can be relied on to measure from? Both? Or neither?

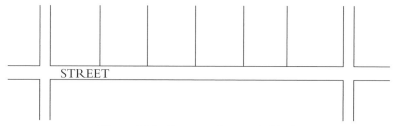

Figure 3.26   Frontage of a sample block.

The deeds describe six 100-foot lots from the time of their creation, but they were created sequentially from a parcel *thought to be* 600 feet wide, when in reality it is only 593 feet wide.

If the lots were created from one end to the other, the last lot would bear the deficiency, as shown in Figure 3.27.

However, if the lots were not sold in that order, but at random, almost anything can happen, as shown in Figure 3.28.

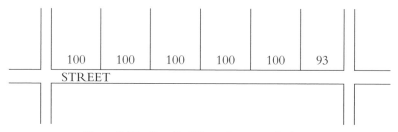

Figure 3.27   Result of foregoing example due to sequential conveyancing from one end.

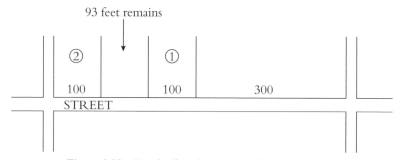

Figure 3.28   Result of random sequential conveyancing.

Boundaries of Source Parcel. When a boundary of an outlot is the same as, or is coincident with a boundary of its source parcel, it is not enough to search just the outlot back to its origin. The source parcel must also be searched, along with *its abutter*. For an example, see Figure 3.29.

While it is necessary to examine lot "y" as an abutter, it is also necessary to search both lots A and B, because it is the line between them that governs, or defines, the line between x and y. While this search could be taken to extreme, some practical judgment must be exercised. However, it would be wise to remember that an error in the description of either lot A or lot B will result in an error in lot x or lot y, or both.

Particular Boundaries Having an Effect. If one of the boundaries of the tract being searched is also a boundary for some other purpose, it must also be researched in its appropriate context. For instance, in Figure 3.30, one of the boundaries of Lot 43 is also the boundary line between two municipalities.

In addition to searching Lot 43 back to its origin, it will be necessary to research the town line to verify its correctness, to ensure that it is correctly located and marked, and to establish what monuments are the proper ones to define its location. Sometimes it is necessary to survey and locate long stretches of political boundary to verify or locate a short length of private boundary.[16]

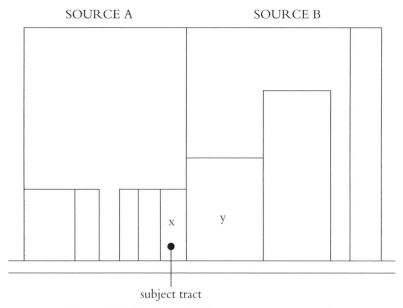

Figure 3.29   Outlots and their respective source parcels.

---

[16] Where lands of individuals are bounded on town lines, ancient maps of towns made by authority of the state are competent evidence in suits between such individuals. *Adams v. Stanyan*, 24 N.H. 405, 4 Fost. 405 (1852).

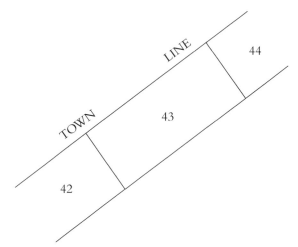

Figure 3.30   Property boundary is also town boundary.

A recent problem arose on a section of a state line where the original notes were transcribed several years ago and made available to the public. Rather than refer to the original notes which were in safekeeping, deteriorating, and difficult to read, people had relied on the transcribed set. When it became important to look at the originals in order to verify some discrepancy, it was found that *seventeen errors* had been made in the transcription. Many locations and surveys done using the transcribed notes are now incorrect due to those errors.

Changes in Related Titles.   Abutting roads, railroads, public utilities, and the like frequently change from time to time. Whether easement or fee, all related titles should be treated like abutting ownerships and thoroughly researched. Just because a lot abuts a road does not mean the street line was laid out correctly or that the road is even within the right of way limits. There is no way to know without researching it.

At times, fences are coincident with boundaries, especially along a railroad. That is because most of the fences were erected by the railroad corporations, or private fences were erected after the railroad was established. In a case illustrated in Figure 3.31, the stone wall existed before the railroad and, since it did not interfere with the grade, it was left in place. Several persons, including one surveyor, relied on the stone wall as being the property line between the railroad and the abutting lot.

The same can be said about roads and highways. Many times fences came after the road layout or construction. That was the case with this road, however, the town later chose to widen the road. The description read as follows:

Beginning again at a point further north on the west side of the highway in front of John Gordon's barn at a post in the fence 12 links from the road side fence thence north 18 rods to a stake 12 links from the road side fence, thence north 6 rods to a stake 8 links from the road side fence.

Figure 3.31   Railroad right of way.

Easements. Since easements are thought of and are treated the same as real property, they should be searched as if they were real property. Chains of title should be developed for them and their descriptions analyzed. Easements may limit the use of a piece of land to a significant extent. For example, if an entire parcel was subject to a flowage easement, and flooding was a distinct possibility, it would not be wise to invest in substantial improvements on it.

Buried in the Records. It would seem that easements would be discovered when a title search is done, but nothing could be further from the truth. While recent outsales will be detected, along with those which are carried forth in the instruments from when they were created, the majority of easements will not fall into either of those categories, *even if on record.*

Often an easement has been granted, but not carried forward in the subsequent conveyances. If such an easement was created at a time that falls outside the scope of the title search, it may never be found in the records, however, and it likely still encumbers the property, even if not mentioned in the deed.

Remember that the title search or its resulting certification is *subject to* items outside the period of search. However, an in-depth record search for boundary investigation or for survey purposes may go back far enough to uncover the easement.

Many easements that were created long ago are still valid and outstanding. They are a superior right conveyed out of the chain of title. Some are created and never mentioned thereafter; others are created, burden the subject tract, but are mentioned in the documents of the dominant estate but not the servient estate, or that parcel being searched. The most troublesome ones are those which are created in an early partition, affect parcels in the partition, and are not mentioned anyplace else. Only by having the partition can anyone know of their record, yet they are valid, and many of them in current and regular use.

Easements Not of Record. A lot of easements exist that are not part of any record. Prescriptive easements may be created independently of the record, while others, such as implied easements and those of necessity, may exist not part of, but because of, a record. On-site inspection and surveys sometimes discover these, particularly prescriptive easements. Evidence of use should be reported when known or observed. Again, the surveyor is the "eyes of the title examiner" as to conditions and evidence at the site itself.

PITFALLS IN RESEARCH

There are a number of common pitfalls, regardless of whether a title examination or an in-depth boundary or title investigation is being performed. The older the document or the further back in time the research extends, the more likely these pitfalls are to be encountered. However, they may appear any place in time.

The Calendar Change of 1752. Prior to 1752, England and its colonies used the Julian calendar, developed in 46–45 B.C. and that remained in effect until Pope Gregory issued a decree in 1582 to change it. The reason for this was that the Julian calendar was off by 11 minutes, or one day in 128 years. The Julian calendar began on March 25, the spring equinox, while the Gregorian calendar begins on January 1.

Some countries adopted the new calendar immediately, while others did not. England felt it should recognize both, and therefore, dates prior to 1752 falling between January 1 and March 25 (the beginning of the civil year) are designated twice, for instance, March 10, 1711/12. This practice, known as double dating, continued after the official adoption of the Gregorian calendar by England in 1752. Not every country made the change at the same time, and the orthodox countries of Russia and Greece converted as late as 1918 and 1923, respectively.

By 1752 the calendar was 11 days off, and it was decreed that the day after September 2, 1752 be called September 14, 1752. Many people living in England and the colonies at this time rectified their birthdates. George Washington was actually born on February 11, 1731 according to the Julian or Old Style calendar, but when the change occurred, he celebrated his birthday thereafter on February 22.

Number of the Month. Because the old calendar began in March, it was then designated as the first month, instead of January as it is now. Dates prior to 1752, or under the old calendar, are sometimes given as:

12th day, 4th month, in the year 1693

or

12 - IV - 1693

which is actually June 12, 1693 and not April 12, 1693. After 1752, it would be called April 12, 1693, but caution must be exercised, since that date may have been copied from an earlier document. Numerous mistakes exist in the records because dates were converted by people not conscious of the change in the calendar.

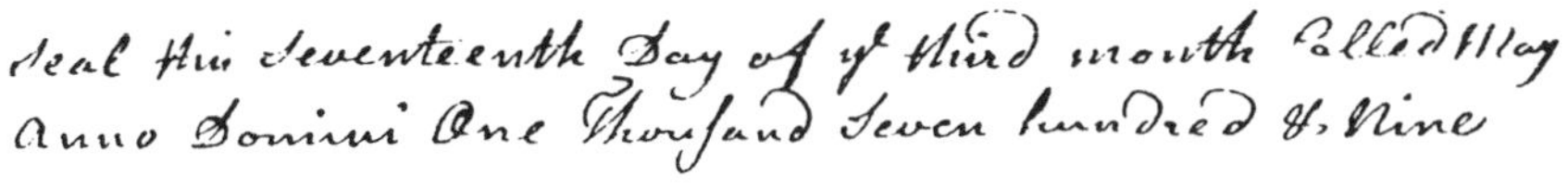

Figure 3.32   Example of a date from a 1709 deed:
"seventeenth day of the third month called May . . ."

Most Quaker records indicate the day first, followed by the month, such as

10, 4th, 1700

which is actually *July 10, 1700*. Be extremely careful when seeing the notation such as

7/10/1700

for, depending on who made that notation, it could signify either *July 10, 1700 or October 7, 1700.*

Is He Dead or Not? Sometimes the word "late" is found in the records, such as John Ford, late of Durham. It is often assumed that John Ford is deceased and that he was living in Durham at the time of his death. This is probably incorrect. The truth is that John Ford did live in Durham but has moved elsewhere and no longer lives there. It cannot be assumed that he died and that no probate was filed just because none was found, or worse yet, that a deed from John Ford was later found, but "it must be another, perhaps a son, because John Ford is dead." Proof of death comes in the form of a probate, a death certificate, a church record notation, or the like.

Don't Assume Who's Related to Whom. Sometimes the records, including deeds, carry designations of Senior, Junior, third, IV, and the like. Caution is advised because even if they were contemporaries, perhaps no two are related to one another. It was common in early records to separate two or more persons of the same name by attaching the designation according to when the person arrived in that town. For example, John III may have moved to town after John Jr. (or II), but may not be related, and he can be either younger or older than John Jr. If one were to move away, later designations would "move up one" to a lesser designation. For example, suppose that there are four Robert Jones in town, Robert Jones III moves away, so

Robert Jones IV now becomes Robert Jones III (it makes perfect sense, since there are now only three of them). If the former Robert Jones III were to move back, he would now become Robert Jones IV, and Robert III (formerly IV) would remain as Robert III. That is, of course, assuming that another Robert Jones did not appear on the scene in the meantime, in which case he (the newcomer) would be Robert the IV (whether a newborn or a move-in), and the former III who moved away and came back would become Robert Jones V.

To make matters worse, Robert Jones, son of Robert Jones may refer to his father as Robert Senior and himself as Robert Junior. For some purposes this may be permissible, but in official records they may be designated as Robert II and Robert IV, or even Robert IV (Senior) and Robert II (Junior) if they came to the town that way.

The term "junior" does not necessarily mean a father/son relationship. Not only is the term "junior" applied to a woman as well as a man, but may include two cousins, one older than the other, or even uncle and nephew.

Sometimes notations will be found, particularly in probate records, that are potentially misleading. A person may refer to others with different names as if they were family members. For example:

I, Robert Brown, leave to my son John Collins . . .

It should not be assumed that John is Robert's adopted son, or even that he is a son. John is probably a grandson or a nephew. Occasionally, there is no blood relationship whatever. Once in a great while, a woman will be referred to as a son, but be very careful because in very early records many names that seem to be female names actually belonged to a man, and some names were used for both.

Was She Married or Not? The term "Mrs." (originally "Mistress") was applied to both married and unmarried women and girls of a social position. If one finds a person designated as Mrs. so-and-so, it is important not to jump to the conclusion that she was a married woman.

Just Because They Are Listed Does Not Mean They Were Born There.
Sometimes an entire family would be listed in town records when (and after) the family moved to town, even though some of the children were born elsewhere. A person is ordinarily only baptized once and in one place, but that person may be registered in more than one town, leading to the erroneous belief that he or she was born there, if the search goes no further.

What Was That Word and What Did It Really Mean? Early Colonial documents contain a variety of terms that have a different meaning today than they did in the seventeenth or eighteenth centuries.

The word "gentleman" was reserved for a man entitled to bear a coat of arms. He did not perform manual labor or engage in trade, and he often owned an estate from which he derived his income. The usual abbreviation in records was "Gent." and the wife of a gentleman was a "gentlewoman." In the nineteenth century the word "gentleman" was often used to denote a man who had retired from business.

Sometimes the term "Goodman" or "Goodwife" appears. These were people who ranked below the gentry. It is important to understand that "Goodman" designated the position the person held in the community and may have had no connection with the Goodman family.

Terms of relationship found in records can also be confusing. The words "brother," "cousin," "uncle," "nephew," and "son," may not have had the same meaning in Colonial times as they have today. A brother may have been a full brother, but also may have been a half-brother, a stepbrother, a brother by adoption, a brother-in-law, or a brother in the church.

If a person referred to his "cousin," he may have been mentioning his uncle or nephew, for they once were interchangeable terms.

The word "nephew" does not always refer to the son of one's brother or sister. It is derived from the Latin word *nepos*, meaning grandson, and there are cases when it was used in its original sense.

When a man mentioned "my new wife" in his will, he was not implying that he had previously been married. He was referring to the woman who was his wife at the time he made the will, as opposed to someone he might marry at a later date. The purpose was often to safeguard a bequest for the issue of his present ("now") wife so that if she should die within his lifetime, he might not want that bequest to go to some future wife.

The term "spinster" may mean a woman who spins fibers into thread, or it may mean an unmarried woman, especially one who remains single beyond the ordinary marrying age. However, the term "spinster" in the eighteenth and early nineteenth centuries has also been found to designate wives and widows. In addition, many "spinsters" were engaged in transactions on their own behalf, such as selling or giving land they had received in their own right, or acquiring land in their own right. Therefore, "spinster" can better be defined as "a woman legally capable of transacting business or otherwise acting in her own behalf."

Sometimes there occurs a surname separated by the word "alias." It is common in England for an individual or a family to have two surnames separated by the "alias." Parish registers often used it to indicate illegitimacy, the father's name and the mother's surname with "alias" in between. Sometimes it meant an inheritance—a man marrying an heiress and adding her name to his, separated by "alias."

Chapter 8 is devoted to words and phrases and their meanings. An in-depth discussion appears there on how important it is to understand the law and the meaning of a word, term, or phrase at the time when it is used.

OVERCOMING PROBLEMS IN RESEARCH

Occasionally a seemingly impossible situation is encountered, whereby tracing a title chain back comes to a dead end. Things are going along smoothly, deeds have references to previous deeds, or where lacking, persons are found in the grantee index, until all of a sudden . . . the parcel cannot be found. There may be an unrecorded deed, it may be indexed erroneously (such as by misspelling of name), or the last grantor found may have inherited the parcel.

Other Spellings. Some names have various spellings, especially foreign names, some of which have been Anglicized. Also, names occasionally are misspelled in the documents and therefore become indexed that way. One should always consider other spellings and pronunciations. For example, the name we know as WHITTIER might be WITTIER, WITTER, WITCHER, and any other of a number of possibilities. A common name like SMITH, may also appear as SMYTH, SMYTHE, and so on. Early spellings of names are often changed to their more modern spelling; for example, CUMMINGS may have been CUMMINS, CUMMING, COMMINS, or COMINS, and ENGLISH may have once been ENGLIS, INGLISH, INGLIS, INGS, INGE, or ENGES. The name HAYWARD may have been HAWARD, HEYWARD, HAYWOOD, HEYWOOD, or HOWARD. It is important to consider what other possibilities may have existed.

Check Grantor Indexes. After the grantee index has been thoroughly exhausted without results, the grantor indexes should be checked. Sometimes the grantor sold other parcels and recites their source(s) or title. Sometimes these recitation(s) are, or include, your parcel or a clue as to where your parcel may have come from.

Check Probates. A check of all the probate records of the surname ended with sometimes will yield an estate which lists the last grantor as receiving property or title. Family records will also give relationships so that you can sometimes get an idea from whom the last grantor may have inherited or was left property. People do not always inherit from the father. Inheritance or title by will may come from anyone, a mother who may have remarried, an uncle or an aunt with different surnames, another relative, or even a friend. Sometimes further investigation is necessary beyond just checking probates with the same name.

Identify Former Owner. If earlier records can be located which identify ownerships, often the name of a predecessor in title can be used as a link in the chain of title. Then that ownership can be traced *forward* to connect with the dead end from before. As an example, many areas have at least one atlas from the nineteenth century, such as shown in Figure 3.33, which depicts residences and conditions at the time. Care must be taken however, because sometimes the names are not *owners*, but merely *residents*.

Additional information may be found in abundance. Town line descriptions, road layouts, town histories, genealogies, local (family) burial grounds, railroads, utility lines, and the like, frequently provide a researcher with possible names. These sometimes have to be verified for accuracy, but ordinarily in a thorough research project, proof becomes self-evident. Information can come from the most obscure places. For example the map in Figure 3.34 reveals the location of a horizontal control station.

The horizontal control station, set by the U.S. Coast and Geodetic Survey, now the National Geodetic Survey, named this station for the landowner. Descriptions of government control stations often have landowners' names on them, some from the late nineteenth century.

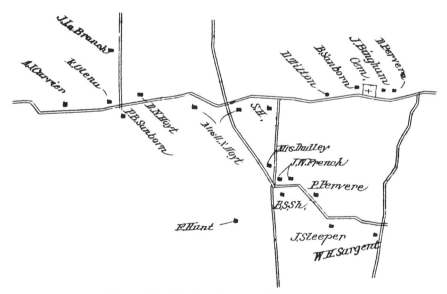

Figure 3.33 Section of nineteenth century atlas.

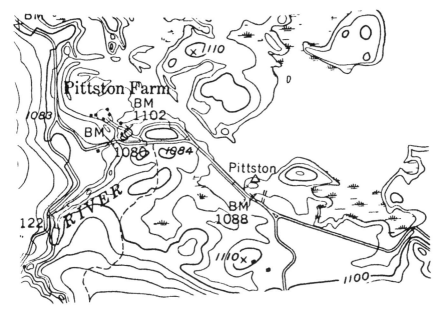

Figure 3.34   Horizontal control station "Pittston" located at Pittston Farm.

There are hundreds of examples just like this. Use of one's imagination will often help to overcome an otherwise insurmountable obstacle such as a gap in the chain of title.[17]

Use of Abutting Tracts. When it becomes impossible to trace backward, it is usually possible to search the abutting tract back to a point where it calls for an abutter (your tract), which is a previous owner in your chain of title. In doing this, always start with what appears to be the easiest parcel to search, or the most unusual one if there is a choice. Also, work with the most unusual name, or the name of a person who did not buy numerous parcels of land; otherwise, your search may become very lengthy. If a dead end is reached in that chain, work with its abutter, and if necessary, the next abutter and so on, playing one against the other, until you are able to get back to a point where you have a name for your parcel that you can trace forward.

In doing this, two things may have been accomplished. You may still have to go further back in the subject chain if you have not reached its beginning description,

---

[17] See 67 A.L.R. 1333, *Presumption or circumstantial evidence to establish missing link in chain of title.*

and if you are doing an in-depth study for either title or survey purposes, abutting tracts would have to be done anyway, so part of the work is now finished. And you have a check in verifying that abutting parcels do indeed abut one another. This is diagramed in Figure 3.35.

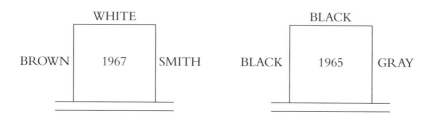

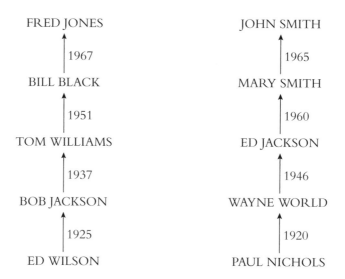

Figure 3.35 Chains of title for abutting parcels calling for each other.

In a recent case, it was necessary to examine a deed, but the parcel could not be found in either the grantee indexes, or in any probate of that name. Research of the abutting parcel yielded a common grantor and a description, which could be traced forward. Note the changes in name, even though they are all related.

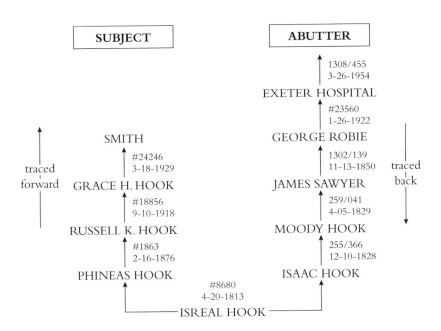

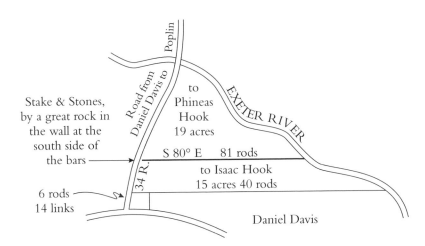

From: Seminar notes, *Case Studies in Boundary Research*, 38th Annual Conference of the Michigan Society of Land Surveyors, 1979, Kalamazoo. Short Course Proceedings: "Researching the Record," 23rd Annual Alaska Surveying and Mapping Conference, 1988, Anchorage.

When Grace H. Hook died, she left the property to her great grandson, Smith. There is no way to trace backward in probates, because the names are different. There is no way to find a deed to this property, because there has never been a deed. The parcel was created in a probate partition and has been willed, or inherited, every time since. This exercise is necessary to (1) determine the chain of title to the property and (2) to obtain a description.

Missing Link Not Found. Some courts and jurisdictions have accepted one or more documents as missing from a chain of title if the existence of such may be shown by parol, or if it can be shown that it is more probable that the document was executed than that it was not.[18]

In *Pratt v. Townsend*,[19] the court stated:

> The doctrine of the presumption of execution of a deed, or, more properly, the proof of its execution, by circumstances when no better evidence is obtainable, is well established, and especially so by the decisions in this state, and the disposition has been to extend rather than to limit it. . . . The destruction of the records in so many of the counties of this state, coupled with the carelessness, well-nigh universal at an earlier period of our history, on the part of the people in keeping the original deeds after they had been recorded, and the disposition now so prevalent to uncover, by reason of carefully prepared abstracts of title, the absence of such written muniments of title as are necessary to make a complete chain, require, in our opinion, a liberal application of this rule for the protection of titles long relied upon in good faith, the evidence of which has been lost through carelessness or accident, the destruction of records, and the death of all persons originally connected therewith, or likely to know anything about the facts. And especially is this true when such claim of title is accompanied by an entire absence of any assertion or claim of right or title inconsistent with the claim under the deed so sought to be established.
>
> It was clearly established in this case that diligent but unsuccessful search had been made for the originals of the deeds sought to be established by circumstances.

ADMISSIBILITY OF EVIDENCE

The question often arises as to how much of this information is admissible in a court of law. It would not be sufficient to base conclusions or decisions on information or evidence that would not later stand up in court. Much of the information of the

---

[18] *Hitchens v. Ellingsworth*, 94 A. 903, 5 Boyce, 497 (1915); *Melvin v. Props. of Locks & Canals*, 17 Pick. 255 (Mass., 1835); *Brooks v. Roberts*, 220 S.W. 11, 281 Mo. 551 (1920); *Earls v. Bennett*, 192 S.W. 916, 137 Tenn. 174 (1917).

[19] 125 S.W. 111(Tex.Civ.App., 1910).

nature discussed previously may fall under the category of hearsay.[20] However, there are a number of exceptions to the hearsay rule, several of which relate to, or include, records relating to boundary or title issues. A number of these exceptions follow.[21]

803(6):   *Records of Regularly Conducted Activity*

A memorandum, report, record or data compilation, in any form, of acts, events, conditions, opinions, or diagnoses, made at or near the time by, or from information transmitted by, a person with knowledge, if kept in the course of a regularly conducted business activity, and if it was the regular practice of that business activity to make the memorandum, report, record or data compilation, all as shown by the testimony of the custodian or other qualified witness, unless the source of information or the method of circumstances of preparation indicate lack of trustworthiness. The term "business" as used in this paragraph, includes business, institution, association, profession, occupation, and calling of every kind, whether or not conducted for profit.

Field notes, time sheets, payroll data, computer printouts, and anything that is done in the regular course of business would be included. Other things that might be included in this category could include town line perambulation returns (every so many years by statute), returns of road layouts (required each time a road is laid out), town meeting minutes, and similar documents.

803(7):   *Absence of Entry in Records Kept in Accordance With the Provisions of Rule 803(6)*

Evidence that a matter is not included in the memoranda, reports, records, or data compilations, in any form, kept in accordance with the provisions of Rule 803(6), to prove the nonoccurrence or non-existence of the matter, if the matter was of the kind of which a memorandum, report, record, or data compilation was regularly made and preserved, unless the sources of information of other circumstances indicate lack of trustworthiness.

---

[20] Federal Rules of Evidence, Rule 801(c) defines "hearsay" as "a statement, other than one made by the declarant while testifying at the trial or hearing, offered in evidence to prove the truth of the matter asserted."

[21] Federal Rules of Evidence. These rules have been adopted by many states. Some states may have their own rules, or variations of the Federal Rules.

Taken in part from: Wilson, Donald A. "The Rules Pertaining to Hearsay Evidence As They Apply to Land Surveying," *The Benchmark*, Volume 10, Number 3, Summer/Fall 1987, pp. 7-11.

By the same token, if the record does not appear where it customarily should, then it would seem logical that the activity did not take place.

803(8) Public Records and Reports

Records, reports, statements, or data compilations, in any form, of public offices or agencies, setting forth (A) the activities of the office or agency, or (B) matters observed pursuant to duty imposed by law as to which matters there was a duty to report, excluding, however, in criminal cases, matter observed by police officers and other law enforcement personnel or (C) in civil actions and proceedings and against the government in criminal cases, factual findings resulting from an investigation made pursuant to authority granted by law, unless the sources of information or other circumstances indicate lack of trustworthiness.

This exception will allow almost any document that is a matter of public record into evidence. In addition to those items found at the court house, are town and county records, as well as records kept by many public agencies such as the highway department, water resources, and the like.

803(9): Records of Vital Statistics

Records of data compilations in any form, of births, fetal deaths, deaths or marriages, if the report thereof was made to a public office pursuant to requirements of law.

In title matters, or in developing a chain of title for boundary research purposes, vital statistics may be very important where probate information or genealogical data are lacking.

803(10): Absence of Public Record or Entry

To prove the absence of a record, report, statement, or data compilation, in any form, or the non-occurrence or nonexistence of a matter of which a record, report, statement or data compilation, in any form, was regularly made and preserved by a public office or agency, evidence in the form of a certification in accordance with Rule 902, or testimony, that diligent search failed to disclose the record, report, statement, or data compilation, or entry.

Where those items provided for in 803(8) and 803(9) are lacking, that in itself is an indication that it may not have taken place, or may never have been in existence.

803(11): Records of Religious Organizations

Statements of births, marriages, divorces, deaths, legitimacy, and ancestry, relationship by blood or marriage, or other similar facts of personal or family history, contained in a regularly kept record of a religious organization.

Church records and records of other religious organizations fall into this category. The enormous collection made by the Church of Latter Day Saints should contain admissible items.

803(12): Marriage, Baptismal, and Similar Certificates

Statements of fact, contained in a certificate that the maker performed a marriage or ceremony or administered a sacrament, made by a member of the clergy, public official, or other person authorized by the rules or practices of a religious organization or by law to perform the act certified, and purporting to have been issued at the time of the act or within a reasonable time thereafter.

Included here might be certificates not included in other categories.

803(13): Family Records

Statements of fact containing personal or family history contained in family bibles, genealogies, charts, engravings on rings, inscriptions on family portraits, engravings on urns, crypts, or tombstones, or the like.

Many of these items are good sources of information for resolving title matters or completing chains of title. Occasionally, boundary and title information will be noted in books, such as a family bible. Families regularly keep maps and sketches as part of their personal and family papers.

803(14): Records of Documents Affecting an Interest in Property

The record of a document purporting to establish or affect an interest in property, as proof of the content of the original recorded document and its execution and delivery by each person

by whom it purports to have been executed, if the record of a
public office and an applicable statute authorizes the recording of
documents of that kind in that office.

Deed records such as deeds, mortgages, tax sales, and so on, and probate records, such as inventories, partitions, divisions, dower, and the like would all fall in this category.

803(15): Statements in Documents Affecting an Interest in Property

A statement contained in a document purporting to establish
or affect an interest in property if the matter stated was relevant
to the purpose of the document, unless the dealings with the property since the document was made have been inconsistent with
the truth of the statement, or the purport of the document.

This would include the same documents as the foregoing category, but refers to statements rather than the record.

803(16): Statements in Ancient Documents

Statements in a document in existence twenty (20) years or
more, the authenticity of which is established.

Three criteria must be satisfied for this rule to take effect, (1) the document must be twenty or more years of age, (2) free from suspicion on its face, and (3) in proper custody, or in a place where, if authentic, it would ordinarily be.

Many documents used by surveyors fall into this category, such as highway records, railroad plans, aerial photographs, other types of photos, ancient maps, and other documents.

803(19): Reputation Concerning Personal or Family History

Reputation among members of a person's family by blood,
adoption, or a marriage, or among his or her associates, or in the
community concerning a person's birth, adoption, marriage,
divorce, death, legitimacy, relationship by blood, adoption,
marriage, ancestry, or other similar fact of his or her personal or
family history.

Reputation received as parol evidence may be the only available, information concerning personal or family history and relevant to a chain of ownership.

803(20): Reputation Concerning Boundaries or General History

Reputation in a community, arising before the controversy, as
to boundaries of or customs affecting lands in the community,
and reputation as to events of general history important to the
community or state, or nation in which located.

Statements received from previous owners, owners of adjacent lands, or persons
familiar with an area may be an important source of information.

803(23): Judgment as to Personal, Family or General History, or Boundaries

Judgments as proof of matters of personal, family or general
history, or boundaries, essential to the judgment, if the same
would be provable by evidence of reputation.

Judgments may be admissible if provable by evidence of reputation. Persons familiar with an area or a particular situation, may be able to offer judgment relative to it.

FOR FURTHER REFERENCE

Cuomo, Paul, and Roy Minnick. *Advanced Land Descriptions.* Rancho Cordova:
Landmark Enterprises. 1993.

Kirkham, E. Kay. *The Handwriting of American Records for a Period of 300 Years.*
Logan: The Everton Press. 1973.

Patton, Rufford G., and Carroll G. Patton. *Patton on Land Titles*, three volumes.
St. Paul: West Publishing Company. 1957.

Stevenson, Noel C. *Genealogical Evidence.* Laguna Hills, CA: Aegean Park Press.
1989.

Stratton, Eugene Aubrey. *Applied Genealogy.* Salt Lake City: Ancestry, Inc. 1988.

REFERENCES

Rubincam, Milton. *Pitfalls in Genealogical Research.* Salt Lake City: Ancestry
Publishing, 1987.

Welch, William Lewis. *Francis Lyford of Boston, and Exeter, and some of his
descendants.* Salem: Essex Institute. 1902.

West Publishing Company. *Federal Rules of Evidence for United States Courts and Magistrates.* St. Paul.

Wilson, Donald A. "The Land Surveyor and His Relationship to Minimum Standards for Record Research," *Legal Topics in Boundary Surveying. A Compendium.* Bethesda: American Congress on Surveying and Mapping. 1990, pp. 61–70.

CHAPTER 4

RULES OF CONSTRUCTION

> Descriptions are not to identify land but to
> furnish the means of identification.
>
> —*City of North Mankato v. Carlstrom,*
> 2 N.W.2d 130, 212 Minn. 32 (1942).

Construction in General. The term "construction" means to build, piece, or interpret the intent or meaning of the document. The purpose of construction is to determine the intent of the parties from the real (legal) meaning of the words or phrases employed in the document.[1] The construction of a deed is governed by the law in force in the state were the land is situated,[2] and deeds are to be construed according to the law in force at the time they are executed.[3]

Rules of Construction. Every deed, otherwise valid, will be considered to have intended to convey an estate of some nature.[4] Therefore, every attempt should be made to uphold the deed whenever possible.

---

[1] Hermanson, Knud Everett. *Boundary Retracement Principles and Procedures for Pennsylvania.* General rule that purpose in construing deeds is to ascertain intention of parties applies to description of land as well as to other parts of deed. *Grimes v. Field*, Civ. App., 222 S.W.2d 697 (Tex., 1949).

[2] *Chidester v. City of Newark*, C.C.A.N.J., 162 F.2d 598; *McCraw v. Simpson*, C.C.A. Okla., 141 F.2d 789 (1944); *Alcorn v. Epler*, 206 Ill.App. 140 (1917); *Liles v. Pitts*, 82 So. 735, 145 La. 650 (1919); *Peter v. Peter*, 110 A. 211, 136 Md. 157 (1920); *In re Jenney's Estate*, 82 N.Y.S.2d 657, 193 Misc. 162 (1948); *Kline v. Mueller*, 276 P. 200, 135 Okla. 123 (1928).

[3] Ibid.; *New Haven Trust Co. v. Camp*, 71 A. 788, 81 Conn. 539 (1909).

[4] *Penienskice v. Short*, 194 A. 409, 38 Del. 526 (1937).

Where descriptions set forth in deeds are not ambiguous, they must be followed.[5] When, and only when, the meaning of a deed is not clear, or is ambiguous or uncertain, will a court of law or equity resort to established rules of construction to aid in the ascertainment of the grantor's intention by artificial means where such intention cannot otherwise be determined. Unlike a settled rule of property, which has become a rule of law,[6] rules of construction are subordinate and always yield to the intention of the parties, particularly the intention of the grantor, where such intention can be ascertained. Since all rules of construction are essentially only methods of reasoning that experience has taught are best calculated to lead to the intention of the parties, generally no rule will be adopted that tends to defeat that intention.[7]

The modern tendency is to disregard technicalities and to treat all uncertainties in a conveyance as ambiguities to be clarified by resort to the intention of the parties as gathered from the instrument itself, the circumstances attending and leading up to

---

[5] *Gawrylak v. Cowie*, 86 N.W.2d 809, 350 Mich. 679 (1957).

Rules of construction should not be used in construing a deed if intention of parties to the deed, especially the grantor, can be ascertained from the document itself. *Alabama Medicaid Agency v. Wade*, 494 So.2d 654 (Ala. Civ. App., 1986).

Rules for construing deeds are adopted for the sole purpose of removing doubts and obscurities so as to get at the meaning intended by the parties, and where there is no doubt or obscurity, there is no room for construction. *Kennedy v. Rutter*, 6 A.2d 17, 110 Vt. 332 (1939).

[6] Unlike a rule of construction, a settled rule of law or rule of property is one which attaches a specific significance and meaning on particular language used in a deed and states the legal effect that such language will have, attaching thereto a specific and irreproachable intention, even though the parties employing the language may have had and may have evinced quite a different intention. Such rules, therefore, ingraft certain meaning upon language used in a deed and determine what effect is given to such language in law. In other words, a rule of property is to be applied automatically as a result of the language used, and the court will not refuse to apply such rule merely on the assumption that the grantor did not intend that his phraseology operate in the way that the rule makes it operate. Van *Olinda v. Carpenter*, 19 N.E. 868, 127 Ill. 42 (1889), *Earnhart v. Earnhart*, 26 N.E. 895, 127 Ind. 397 (1891); *Long v. Holden*, 112 So. 444, 216 Ala. 81 (1927), *Doyle v. Andis*, 102 N.W. 177, 127 Iowa 36 (1905), *Weinbeck v. Dahms*, 107 A. 12, 134 Md. 464 (1919). See 23 *Am Jur 2d, Evidence, § 224,* Settled rules of property.

[7] *Re Estate of Fleck*, 154 N.W.2d 865, 261 Iowa 434 (1967); *Lassiter v. Goldblatt Bros., Inc.*, 41 N.E.2d 803, 229 Ind. 215 (1942).

There is an abstract distinction between "construction" and "interpretation," in that "construction" is the drawing of conclusions from elements known from, given in, and indicated by, the language used, while "interpretation" is the art of finding the true sense of the language itself, or of any form of words or symbols. In other words, "interpretation" is used with respect to language or symbols themselves, while "construction" is used to determine, not the sense of the words or symbols, but the legal meaning of the entire contract. However, in many contract cases it would be difficult, as well as unrewarding, to determine where "interpretation" ends and "construction" takes over. It is doubtful if the abstract distinction between the terms is of any vital significance, and certainly, in common and legal usage, "construction" and "interpretation" are used interchangeably. See 17 *Am Jur 2d, Contracts, § 240.* See 30 *Am Jur 2d Evidence, § 1065.*

A deed must be construed from its four corners. *Palos Verdes Corp. v. Housing Authority of Los Angeles County*, 21 Cal. Rptr. 225, 202 C.A.2d 827 (Cal.App., 1962).

A deed must be read as a whole and every part thereof given effect if possible in order to arrive at the true meaning of the parties, and until such rule has been exhausted resort should not be had to artificial and arbitrary rules of construction. *Burchfield v. Hodges*, 197 S.W.2d 815, 29 Tenn. App. 488 (1946).

its execution, and the subject matter and the situation of the parties as of that time. *Substance rather than form controls.* Hence, in the construction of deeds, surrounding circumstances are accorded due weight. In the consideration of these various factors, the court will place itself as nearly as possible in the position of the parties when the instrument was executed, and where the language of a deed is ambiguous, the intention of the parties may be ascertained by a consideration of the surrounding circumstances existing at the time of the execution of the deed. In this connection, it has been said that in interpreting a deed, the choice of words used to grant the land may show the aptitude of the scrivener and may be a pertinent factor.[8]

Rules of construction are particularly relevant where all clauses of a deed cannot be harmonized, and one of several possible meanings must by selected as expressing the true intent.[9]

The construction of deeds presents a question of law for a court to decide. It is the court's duty to interpret a deed in the light of the law in existence at the time of its execution and delivery, which must be read into and become an enforceable part thereof.[10]

The rules for the construction of deeds are essentially those applicable to other written instruments and to contracts generally. But while, as a general consideration, the same broad rules govern the construction of both deeds and wills in some jurisdictions deeds are more strictly construed than are wills.[11]

When the intention of the parties is uncertain, resort must be had to well settled but subordinate rules of construction to be treated as such and not as rules of positive law.[12] All rules of construction are but aids in arriving at the grantor's intention and rules of construction may be applied only where the application of the rule with respect to the intent of the parties does not expel all doubt concerning the conclusions to be drawn from the language of the conveyance and its surrounding circumstances. In construing a deed, it is not permissible to interpret that which has no need of interpretation.

[8] Ibid.

[9] Ibid.

[10] Ibid.

[11] *Rush v. Champlin Refining Co.*, 321 P.2d 697 (Okla., 1958); *Crow v. Thompson*, 131 S.W.2d 1064 (Tex.Civ.App., 1939); *Joseph Mann Library Ass'n v. Two Rivers*, 76 N.W.2d 388, 272 Wis. 441 (1952); *Swearingen v. McGee*, 198 S.W.2d 805, 303 Ky. 825 (1946), *Federal Land Bank v. Nicholson*, 251 P.2d 490, 207 Okla. 512 (1952), *Hicks v. Sprankle*, 257 S.W. 1044, 149 Tenn. 310 (1924); *Horne v. Horne*, 22 S.E.2d 80, 181 Va. 685 (1943); *Beasley v. Beasley*, 88 N.E.2d 435, 404 Ill. 225 (1949), *Noe v. Moseley*, 36 N.E.2d 240, 377 Ill. 152 (1941), *Barnett v. Barnett*, 83 A. 160, 117 Md. 265 (1912).

According to the court in *Roberson v. Wampler*, 51 S.E. 835, 104 Va. 380 (1905), "better reason favors the rule that in deeds, as well as in wills, the intention of the maker of the instrument, as gathered from all its parts, must prevail."

[12] *People v. Call*, 223 N.Y.S. 257, 129 Misc. 862 (1927).

Rules for construing deeds must be treated as subordinate, and not as rules of positive law. *Johnson v. Barden*, 83 A. 721, 86 Vt. 19 (1912), Ann Cas 1915A, 1243; see also *deNeergaard v. Dillingham*, 187 A.2d 494, 123 Vt. 327 (1962).

Each instrument must be construed in the light of its own language and peculiar facts, and it has been said that, since the language employed in deeds varies so materially and so much, precedents are rarely controlling in a concrete case, except as they may furnish general guidance. Nevertheless, it has been said that the construction of a deed must be governed by the strict rules of the common law, and where a deed expresses two conflicting intentions, it must be construed according to the rules of construction.[13]

Care must be used in construing a deed, and a construction that will leave the way open for repeated and indecisive litigation should be avoided. A deed will also be interpreted and construed as of the date of its execution. Deeds are to be construed as other written instruments.[14]

Conditions at the Time. In cases of ambiguity in descriptions, weight may be given to circumstances in which a conveyance was made, and to the physical characteristics of the property transferred, in determining what the parties intended.[15]

Conveyance of House. It is a general rule of construction that the grant of a house carries with it title to all land under the house, including the land on which the building stands and the land under the projecting eaves.[16]

Generally deeds are liberally construed in order to uphold them and give effect to the manifest intention of the parties.[17] Between deeds and wills, greater strictness is required in the construction of deeds than of wills.[18] In addition, greater latitude of construction is indulged in case of deed drawn by unskilled draftsman than in case of product of skilled scrivener.[19]

When a deed is ambiguous the construction most favorable to its validity will be adopted.[20]

---

[13] Ibid.

[14] Ibid.

[15] *Weintrebe v. Coffman*, 263 N.E.2d 454, 358 Mass. 247 (1970).

[16] *Forbush v. Lombard*, 13 Met. 109 (Mass., 1847); *Webster v. Potter*, 105 Mass. 414 (1870); *Sherman v. Williams*, 113 Mass. 481(1873).

[17] *Comi v. M. & M. Corp.*, 4 So.2d 389, 148 Fla. 422 (1941).

A liberal construction should be placed upon the language of written instruments, when by so doing they may be upheld, and when otherwise the plain intention of the parties will be defeated. *Dinkins v. Latham*, 45 So. 60, 154 Ala. 90 (1907).

[18] *Porter v. Henderson*, 82 So. 668, 203 Ala. 312 (1919).

[19] *Gamble v. Gamble*, 75 So. 924, 200 Ala. 176 (1917).

[20] *Earle v. International Paper Co.*, 429 So.2d 989 (1983).

INTENT OF THE PARTIES

Above all, the primary rule of construction is that the real intention of the parties, particularly that of the grantor, is to be sought and carried out whenever possible, so long as it is not contrary to a settled rule of property that specifically gives a particular meaning to certain language, or when not contrary to, or in violation of, settled principles of law or statutory provisions.[21]

The court in *Weisenberger v. Kirkwood*[22] summarized the procedure regarding an analysis of the evidence:

> Ambiguity in descriptions in deeds from common grantor as to location of common boundary line between adjoining lake lots required construction of deeds as to intent to be arrived at upon consideration of (1) language of deed descriptions; (2) nature of descriptions; (3) maps, surveys and similar documents; (4) positions of lake lot owners and common grantor concerning certain matters; and (5) other relevant and pertinent evidence of surrounding circumstances relating to and aiding determination as intent of parties.

The main object in construing a deed is to ascertain the intention of the parties from the language used and to effectuate such intention where it is not inconsistent with any rule of law.[23] "Intention" as applied to the construction of a deed is a term of art and signifies a meaning of the writing.[24] The intent of the grantor as spelled out in the deed itself must be interpreted, not the grantor's intent in general, or even what he *may* have intended. (*Wilson v. DeGenaro*, 415 A.2d 1334, 36 Conn.Sup. 200, 1979). In ascertaining the intent of the grantor in a deed, the court will not attempt to ascertain and declare what the grantor meant to say, but only the meaning

---

[21] *Gibson v. Pickett*, 512 S.W.2d 532, 256 Ark. 1035 (1974), *Phillips v. State*, 449 A.2d 250 (Del.Sup., 1982), *Re Estate of Fleck*, 154 N.W.2d 865, 261 Iowa 434 (1967), *Stutts v. Humphries*, 408 So.2d 940 (La.App., 1981), *Chesapeake Corp. of Virginia v. McCreery*, 216 S.E.2d 22, 216 Va. 33 1975); *Bauer v. Bauer*, 141 N.W.2d 837, 180 Neb. 177 (1966), *Larchmont v. New Rochelle*, 100 Misc.2d 463, 418 N.Y.S.2d 966 (1979).

"Intention" as applied to the construction of a deed is a term of art, and signifies a meaning of the writing. *U.S. v. 15,883.55 Acres of Land in Spartanburg County*, S.C., D.C.S.C., 54 F.Supp. 849.

The California Court stated in *Basin Oil Co. of California v. City of Inglewood*, that the intent that governs the construction is not the grantor's intent but the joint intent of the grantor and the grantee. After all, the grantor makes and signs the deed, but it must be delivered to the grantee in order to be valid. 271 P.2d 73, 125 Cal. App.2d 661 (1954).

[22] 151 N.W.2d 889, 7 Mich. App. 283 (1967).

[23] *Iselin v. C.W. Hunter Co.*, C.A.La., 173 F.2d 388.

Grantor's intention controls unless in conflict with some positive rule of law. *Leeper v. Leeper*, 147 S.W.2d 660, 347 Mo. 442 (1941), see 133 ALR 586.

[24] *U.S. v. 15,883.5 Acres of Land in Spartanburg County*, S.C., D.C.S.C., 54 F.Supp. 849.

of what the grantor did say. *Holloway's Unknown Heirs v. Whatley*, 104 S.W.2d 646 (Texas, 1937). Stated similarly, the Texas court said a year later that "generally, secret intentions of the grantor not communicated to the grantee are not binding on the grantee." (*Cawthon v. Cochell*, 121 S.W.2d 414, 1938).

The Colorado court in *Tilbury v. Osmundson* (352 P.2d 102, 1960) stated it this way: "A deed conveys the land actually described, regardless of the mistake of the parties. Land cannot be transferred by the intent of the parties alone, especially when the specific words used state less than what was intended."

The cardinal rule for construction of deeds is that the intention of the parties is to be ascertained and given effect, as gathered from the entire instrument, together with the surrounding circumstances, unless such intention is in conflict with some unbending canon of construction or settled rule of property, or is repugnant to the terms of the grant.[25] The only legitimate or permissible object of interpreting written contracts and conveyances is to determine the meaning of what the parties have said in them.[26]

It is especially important to ascertain the intention of the grantor, for it is the grantor who made and signed the instrument.[27] Such intention is to be ascertained from the words that have been used in connection with the subject matter and the surrounding circumstances. The presumption is that the intention of the parties is expressed by the language of the deed and the grantor is presumed to intend what his words communicate. It is the intent existing at the time the deed was made.[28]

The intent of the parties, or more specifically the intent of the grantor, when ascertained, is controlling and is to be given effect as long as it is not repugnant to any rule of law or inconsistent with settled rules of law or some principle of law, or in violation thereof, in violation of some rule of property, or when there are no expressions in the deed that positively forbid it or render it impossible.[29]

The intention sought in the construction of a deed is that expressed in the deed, and not some secret, unexpressed intention, even though such secret intention be that actually in mind at the time of execution.[30]

---

[25] *Hedick v. Lone Star Steel Co.*, Civ. App., 227 S.W.2d 925.

[26] *Yukon Pocahontas Coal Co. v. Ratliff*, 24 S.E.2d 559, 181 Va. 195 (1943).

[27] *Henry v. White*, 60 So.2d 149, 257 Ala. 549 (1952).

[28] *Corn v. Branche*, 240 P.2d 537, 74 Ariz. 356 (1952); *Aller v. Berkeley Hall School Foundation*, 103 P.2d 1052, 40 Cal.App.2d 31 (1940); *Pierce v. Freitas*, 280 P.2d 67, 131 Cal.App.2d 65 (1955) ; *Mason v. Peabody Coal Co.*, 51 N.E.2d 285, 320 Ill.App. 350 (1943); *Bruen v. Thaxton*, 28 S.E.2d 59, 126 W.Va. 330 (1943).

Even a clearly stated intention is ineffectual, if the instrument does not meet the legal requirements as to the language used. *Long v. Holden*, 112 So. 444, 216 Ala. 81 (1927), see 52 ALR 536.

The courts are compelled to identify the intent of grantor with his plain language. *Gaston v. Mitchell*, 4 So.2d 892, 192 Miss. 452 (1941).

Where parties to a deed are deceased, the court must be guided by the language contained in the deed in determining the intention of the parties. *Hoppes v. American Nat. Red Cross*, Com. Pl., 128 N.E.2d 851 (Ohio, 1955).

The nature of the subject matter may be an important factor in construing a deed when the language is such as to require interpretation. *Brooke v. Dellinger*, 17 S.E.2d 178, 193 Ga. 66 (1941).

[29] Ibid.

[30] *Sullivan v. Rhode Island Hospital Trust Co.*, 185 A. 148, 56 R.I. 253 (1936).

It is not what the parties meant to say, but the meaning of what they did say that is controlling.[31]

The California court stated in *Basin Oil Co. of California v. City of Inglewood*,[32] that the intent which governs the construction is not the grantor's intent, but the joint intent of the grantor and the grantee. After all, the grantor makes and signs the deed, but it must be delivered to the grantee in order to be valid.[33] Where a deed is susceptible of two constructions, the construction to be adopted is the one which will most effectively carry out the intentions of the parties.[34]

Four Corners of the Instrument. The intention of the grantor, as gathered from the four corners of the instrument, is the root of construction, and the court will enforce that intention no matter in what part of the instrument it is found.[35] It is the grantor's intention, as expressed in the deed, not as shown by extrinsic evidence, that governs in determining the title conveyed to the grantee.[36]

Intent May Appear Anywhere in the Instrument. The intention of the grantor may be expressed anywhere in a deed, and in any words, and the court will enforce the grantor's intention, no matter in what part of the deed it is found.[37]

Contemporaneous Instruments. A general rule of construction in ascertaining the intention of the parties is that separate deeds or instruments executed at the same

---

[31] *Urban v. Urban*, 18 Conn. Sup. 83 (1952).

[32] 271 P.2d 73, 125 Cal. App.2d 661 (1954); see also *Martin v. Robinson*, 11 P.2d 70, 123 Cal. App. 373 (1932).

[33] Title by deed passes only by delivery of the deed. *Paliotta v. Celletii*, 30 A.2d 108, 68 R.I. 500 (1943).

A deed in order to pass title to land, must be delivered by the grantor and accepted by the grantee. *Chattanay v. City of New London,* 51 A.2d 917, 133 Conn. 377 (1947).

Before deed becomes effective, there must be delivery and an acceptance. *Winick v.Winick*, 272 N.Y.S.2d 869, 26 A.D.2d 663 (1966).

[34] *Kupau v. Waiahole Water Co.*, 37 Hawaii 234 (1945).

[35] *Davidson v. Davidson*, 167 S.W.2d 641, 350 Mo. 639 (1943).

In construing deeds, as well as wills, the intention, as gathered from the four corners of the instrument, controls. *Boxley v. Easter*, 319 S.W.2d 628 (1959).

Cardinal rule in interpretation of deed is that grantor's intention governs and, as an aid to ascertaining that intent, all the words of the deed within its four corners must be considered together and given effect. *Lloyd v. Garren*, 366 S.W.2d 341 (Mo., 1963).

[36] *Rummerfield v. Mason*, 179 S.W.2d 732, 352 Mo. 865 (1944).

Auxiliary rules of construction should not be employed where intention of grantor is clearly expressed in the deed. *Knox College v. Jones Store Co.*, 406 S.W.2d 675 (Mo., 1966).

Generally, in interpreting deeds, the primary object is to determine the intention of the parties from the instrument itself, and, if the instrument indicates an ambiguity, then resort may be had to extraneous matter. *Curran v. Maple Island Resort Ass'n*, 14 N.W.2d 655, 308 Mich. 672 (1942).

[37] *Ott v. Pickard*, 237 S.W.2d 109, 361 Mo. 823 (1951); see also *Leeper v. Leeper*, 147 S.W.2d 660, 347 Mo. 442 (1941), see 133 ALR 586.

time and in relation to the same subject matter, between the same parties, may be taken together and construed as one instrument.[38] The rule will not be applied, however, to allow an unambiguous conveyance to be modified by contemporaneous instruments that are not a part of the identical transaction in which the deed was given.[39]

Where a deed almost identical with the one before the court was executed at the same time by the same grantor but to a different grantee, the other deed may be used to aid in construing the ambiguous one.[40]

Where a deed and a written agreement were executed at the same time, the agreement referring to the deed, the two instruments are construed together to determine the intention of the parties.[41]

Representation of Premises. Discrepancies between deed description and ground features are to be reconciled by inquiring into the intention of the parties that existed at the time of the conveyance.[42]

The portion of the description that best identifies the land in accordance with the intent of the parties is the controlling part of the description.[43]

DESCRIPTION HONORED IN ITS ENTIRETY

A deed must be construed as a whole, and a meaning given to every part thereof.[44]

And in construing a deed, each and every word used must be given a meaning, if possible, and the instrument must be construed as a whole and within the limits of its four corners.[45] This rule applies not only to deeds but also to all other instruments.[46]

---

[38] This rule applies to conveyances of realty even though they do not expressly refer to the contemporaneous agreement. *Rudes v. Field,* 204 S.W.2d 5, 146 Tex. 133 (1947).

[39] *Miles v. Martin,* 321 S.W.2d 62, 159 Tex. 336 (1959).

[40] *Hacker v. Carlisle,* 388 So.2d 947 (Ala., 1980).

[41] *Brock v. Brock,* 16 So.2d 881, 245 Ala. 296 (1944).

[42] *Pauquette v. Ray,* 397 N.Y.S.2d 442, 58 A.D.2d 950 (1977).

[43] *Mazzucco v. Eastman,* 263 N.Y.S.2d 986, 36 Misc.2d 648 (1958), affirmed 239 N.Y.S.2d 535, 18 A.D.2d 1138 (1962).

[44] *Lyford v. City of Laconia,* 72 A. 1085, 75 N.H. 220, 22 L.R.A., N.S., 1062, 139 Am. St. R. 680 (1909).

Whenever possible, each part of the scrivener's phraseology should be given some import. *Peckheiser v. Tarone,* 438 A.2d 1192, 186 Conn. 53 (1982).

Words in a deed must be given effect if reasonably possible. *Daniels Gardens v. Hilyard,* 49 A.2d 721, 29 Del. Ch. 336 (1946).

[45] *Schreier v. Chicago & N.W. Ry. Co.,* 239 N.E.2d 281, 96 Ill. App.2d 425 (1968).

[46] Contracts, deeds and wills must be construed as a whole, and effect must be given to every part, if it is fairly possible to do so, to determine the true intention of the parties. *Ott v. Pickard,* 237 S.W.2d 109, 361 Mo. 823 (1951).

The legal effect of a deed is not determined from a single word, or part, or the relative positions of different parts, but from the entire instrument.[47] All parts must be considered, and, unless conflicting, given effect.[48] And, when conflicts arise, or appear, other rules are resorted to for their resolution.

Even when dealing with a single sentence, the entire sentence must be read as a whole, and one part will not be read as contradicting another part.[49]

The same principles apply to wills in that if provisions are apparently inconsistent or revolting, every effort must be made to construe the instrument as to harmonize the conflicting words, phrases, or clauses.[50] Effect should be given, if possible, to all words, clauses, and provisions of the instrument.[51]

Cannot Insert Words. While the effect should be given to every word of a written instrument whenever possible, it is never permissible to insert meaning to which parties did not agree.[52] Not even the court can add words to modify the clear meaning of a deed.[53] The problem is that adding words not used by the grantor may change the effect of the instrument.[54]

Latter Words. Words that are added in the latter part of a deed, for the sake of greater certainty, may be resorted to explain the preceding parts that are not entirely clear.[55]

---

[47] *Reynolds v. McMan Oil & Gas Co.*, 11 S.W.2d 778 (Com.App., 1928), reversing 279 S.W. 939 (Tex., 1926), rehearing denied 14 S.W.2d 819 (1929).

[48] *Spiller v. McGehee*, 68 S.W.2d 1093 (Civ.App., 1934).

Deeds are to be construed to give effect to all words used. *Earle v. International Paper Co.*, 429 So.2d 989 (Ala., 1983).

[49] *Burchett v. Burchett*, 35 S.W.2d 557, 237 Ky. 411 (1931).

[50] *Smith v. Bell*, 31 U.S. 68, 8 L.Ed. 322 (1832), *Springer v. Vickers*, 66 So.2d 740, 259 Ala. 465 (1953), *Bowman v. Morgan*, 33 S.W.2d 703, 236 Ky. 653 (1930), *Moran v. Moran*, 106 N.W. 206, 143 Mich. 322, *McClure v. Keeling*, 43 S.W.2d 383, 163 Tenn. 251 (1931), *Pabst v. Goodrich*, 113 N.W. 398, 133 Wis. 43 (1907).

[51] *Scott v. Nelson*, 3 Port. 452 (Ala., 1836), *Archer v. Palmer*, 167 S.W. 99, 112 Ark. 527 (1914), *Re Miller's Estate*, 54 N.W.2d 433, 243 Iowa 920 (1952), *Church v. Mize*, 205 S.W. 674, 181 Ky. 567 (1918), *Re Convey's Estate*, 225 N.W. 17, 177 Minn. 266 (1929), *Tevis v. Tevis*, 167 S.W. 1003, 259 Mo. 19 (1914), *Fowler v. Whelan*, 83 N.H. 453 (1928).

[52] *Tubb v. Rolling Ridge, Inc.*, 214 N.Y.S.2d 607, 28 Misc.2d 532 (1961).

[53] *Seested v. Applegate*, 26 S.W.2d 796 (Mo.App., 1930).

The court can neither put words into a deed that are not there nor put a construction on words directly contrary to their plain sense. *Urban v. Urban*, 18 Conn. Sup. 83 (1952).

[54] Effect of deed cannot be changed by addition of words not used by grantor. *Triplett v. Triplett*, 60 S.W.2d 13, 332 Mo. 870 (1933).

[55] *Wallace v. Crow*, 1 S.W. 372 (Tex., 1886).

Written vs. Printed. The printed part of a deed is as much a part of the deed as the written part, and the parties are as much bound by one as they are the other.[56] When conflicts between them appear, however, the general rule is that when a deed is prepared on a printed form but in such manner that words inserted in blanks are inconsistent with printed words of the form, or produce ambiguity, the greater and commonly controlling weight is to be given to the inserted words.[57] While many cases refer to written matter, the same may be said about typewritten insertions,[58] because, as the Nebraska court said, when there is inconsistency "typewriting is writing."[59] The basic reason for such rule is that written or typed portions of an instrument drawn on a printed form, are more strongly indicative of the intent than seemingly inconsistent language of the printed form.[60]

Numbers. In a Missouri case, where the number of the lot conveyed was in both figures and writing and there was a variance between the figures and the written numbers, the written numbers controlled.[61]

Conflicting Clauses. The old rule that the earlier clause controls the later one is only applicable when reconcilement is impossible.[62]

Habendum Clause. In a number of instances printed granting clauses have been controlled by written or typewritten language inserted into habendum or other later portions of the deeds.[63]

---

[56] *Wallwork v. Derby*, 40 Ill. 527 (1866).

While written and printed parts of instruments, including deeds, are equally binding, if they are inconsistent, written part prevails over the printed form. *Hardee v. Hardee*, 93 So.2d 127, 265 Ala. 669 (1957).

Handwritten portions of deed take precedence over printed language when in conflict. *Hacker v. Carlisle*, 388 So.2d 947 (Ala., 1980).

[57] *Porter v. Henderson*, 82 So. 668, 203 Ala. 312 (1919); *Re Brookfield*, 68 N.E. 138, 176 N.Y. 138 (1903); *Graves v. Graves*, 3 Tenn.App. 439 (1926).

If the written and printed portions of a deed are repugnant to each other, the printed form must yield to the written expression. *Ham v. St. Paul Fire and Marine Insurance Co.*, 130 Pa. 113, 18 A. 621 (1886).

[58] *Kern v. Pawlega*, 146 N.W.2d 689, 5 Mich. App. 384 (1966).

[59] *New Masonic Temple Assoc. v. Globe Indem. Co.*, 279 N.W. 475, 134 Neb. 731 (1938).

[60] *Mathy v. Mathy*, 291 N.W. 761, 234 Wis. 557 (1940).

[61] *Bradshaw v. Bradbury*, 64 Mo. 334 (1877).

[62] *Waterman v. Andrews*, 14 R.I. 589 (1884).

There is no rule that, if clauses in a description of land are repugnant, the first necessarily prevails over the last. *Rathburn v. Geer*, 30 A. 60, 64 Conn. 421 (1894).

[63] *Thompson v. Thompson*, 46 N.W.2d 437, 330 Mich. 1 (1951), *Jacobs v. All Persons, etc.*, 106 P. 896, 12 Cal.App. 163 (1909); *Davidson v. Manson*, 48 S.W. 635, 146 Mo. 608 (1898); *In re Brookfield*, 68 N.E. 138, 176 N.Y. 138 (1903; *J.E. Hughes Oil Co. v. Mayflower Invest. Co.*, 193 S.W.2d 971 (Tex.Civ.App., 1946) error refused.

In *Miller v. Mowers*[64] where the deed, prepared by an unskilled person, contained in the printed granting clause the usual words, "do grant, bargain and sell, unto the said party of the second part," her "heirs and assigns" (the word "her" being here inserted by the draftsman), and the premises were described by the insertion of written matter, followed by words of the form, purporting to include in the conveyance all and singular the hereditaments and appurtenances and the reversions, remainder and remainders, and all the estate, right, title, interest, claim and demand whatsoever of the grantor, but the habendum read: "to have and to hold the premises above bargained and described," with the appurtenances "unto the said party of the second part, her heirs and assigns forever during her natural lifetime," the words "her" and "during her lifetime" having been inserted by the draftsman and the word "forever" crossed out by him as above indicated, the court, in holding that the grantee took only a life estate, declared broadly that "if there be any conflict in a deed between the printed part and the part written in, the latter will control in construing it."

In *re Brookfield*,[65] where the instrument included formal phrases in ordinary use and apt words for the conveyance of the land in fee, with all the hereditaments and appurtenances thereunto belonging, as well as the rents, issues, and profits, but then referred to the subject of conveyance as "being all the land on both sides of Byram River and Byram Pond that will be overflowed" by the waters of the river and pond in consequence of the erection of a dam across the river "southerly of lands hereby conveyed of sufficient height to raise the waters" of the pond eight feet above the existing level, "and the above-described land is conveyed by the party of the first part to the party of the second part only for the purpose of being flowed by said pond," the court noted that it was apparent that but for the provision last above noted the title to the grantor's whole farm would have passed by the conveyance, but the court observed that the provision limiting the grant to lands to be overflowed by water was inconsistent with the provision describing the whole farm as that conveyed, and likewise that the provision that the lands were conveyed "only for the purpose of being flowed by said pond" was inconsistent with the other provisions of the deed, which were of such a character as would ordinarily be construed as passing a fee, but that the inserted clauses were "the prominent and noticeable provisions of the deed," and were "its essential features, the real essence of the contract, and evidently . . . the result of the deliberate thought and agreement of the parties," and expressed their intention, and accordingly the court concluded that the inserted provisions "should be given force and effect in preference to the usual format provisions" and that the result was that the instrument could not be regarded as conveying the fee nor even the right of possession but at most a mere easement, leaving the title, possession, and use in the grantor, subject only to a right of flowage. The court stated that when inconsistent provisions are found in a deed "the rule is well settled that those provisions which are written or are unusual, or those which have received special

---

[64] 227 Ill. 392, 81 NE 420 (1907).
[65] 175 NY 138, 68 NE 138 (1903).

attention, will be deemed to express the intention of the parties rather than the printed or formal provisions of the instrument."

Later written or typed words may also impose a condition.[66]

Granting Clause. Even though a granting clause in a deed is regarded as of its very essence, there is no doubt that if written (or typewritten) words of such a clause are inconsistent with later printed words the former will control.[67]

In *Cummings v. Dearborn*,[68] where a blank form or warranty deed was used, but in the granting part, after the printed words "freely give, grant" etc. "unto," etc., the printed words, "forever, a certain piece of land lying and being," etc., were stricken out, and in lieu thereof the words "all my right, title, and interest in and unto" were written in, the court was of the opinion that the change significantly showed that the intention was to convey only the grantor's title and interest in the land, not the land itself, and accordingly held that the instrument was a quitclaim deed notwithstanding that the habendum made reference to the above granted and bargained "premises," and there were covenants that the grantor was the sole owner "of the premises" and that the same were free from every encumbrance and that the grantor engaged to warrant and defend the same against all lawful claims. The court concluded that under the circumstances the word "premises" should be construed to refer to the title and interest intended to be conveyed rather than to the land itself.

In *Keller v. Keller*[69] where the deed, prepared on a printed form, began with a recital that it was made by and between certain named persons, parties of the first part, "and Hobart Keller during his lifetime, and at his death to his heirs . . . party of the second part," and purported to grant, bargain, sell, convey, and confirm unto the said party of the second part "during his lifetime and at his death to his heirs and assigns" the property described, to have and to hold the premises with all and singular the rights, etc. "unto the said party of the second part, and unto his heirs and assigns forever," and the covenant to warrant and defend ran "unto the said party of the second part, and unto his heirs and assigns, forever," the court, in holding that the said Hobart took only a life estate, referred to *Davidson v. Manson* (1898) 146 Mo. 608, 48 SW 635, as a similar case in that there, also, the words indicate an intent to convey a life estate only were written in while those claimed to show an intent to convey a greater interest were simply parts of the printed form.

In *re Jones' Estate*[70] where the deed was prepared on the form of an ordinary warranty deed but the granting clause was made out to "Thomas T. Jones and heirs

---

[66] See *In Riverton Country Club v. Thomas*, 58 A.2d 89, 141 NJ Eq. 435 (1948), aff'd on op below, 64 A.2d 347, 1 NJ 508 (1948).

[67] *Porter v. Henderson*, 82 So. 668, 203 Ala. 312 (1919); *Lawless v. Caddo River Lumber Co.*, 223 S.W. 395, 145 Ark. 132 (1920); *McNear v. McComber*, 18 Iowa 12 (1864); *Mathy v. Mathy*, 291 N.W. 761, 234 Wis. 557 (1940).

[68] 56 Vt. 441 (1884).

[69] 343 Mo. 815, 123 S.W.2d 113 (1938).

[70] 64 N.E.2d 609, 44 Ohio L Abs 339 (App., 1945)

of his body," the quoted words being typed in, and in the warranty clause the typed reference was to "Thomas T. Jones and his heirs of his body," it was held that the effect was to grant a fee tail and constitute the said Thomas the first donee in tail notwithstanding the habendum clause ran to "Thomas T. Jones, his heirs an assigns forever," the words "heirs and assigns" in that clause having been printed. The court relied upon what was stated to be the "well-known rule of construction that type-written phrases are given effect over printed phrases if there is any inconsistency between them."

Additional Clauses. Where a description by metes and bounds is unambiguous, the added clause, "being the same," which is of doubtful meaning as embracing either "part of the same" or "all the same," cannot be taken to create an ambiguity in the antecedent unambiguous description.[71]

Conflicts within Clauses or Provisions. When a particular clause or portion of a deed contains both printed and written (or typewritten) words and the two are inconsistent, the written or typewritten words are to be given the greater weight.[72]

In *Haughn v. Haughn*,[73] where the deed, following the printed form, purported to grant "unto said party of the second part, her heirs and assigns," the lots described, but immediately following the description specified that the object and intent was to convey to the grantee a life estate only with remainder in fee simple to her legitimate children, etc., after which the granting clause concluded in the usual form ("together with all and singular the hereditaments and appurtenances" etc.), the court was of the opinion that the cases holding that the intention expressed in a granting clause takes precedence over later words attempting to cut down the estate did not apply, for the court said that the word "heirs" as used in the "earlier part of the clause" was a part of the printed form and that the words of the same clause limiting the estate granted were written in a blank space of the form, and that following the "well-established rule that written words must control where there is a conflict between the written and printed parts" it must be held that the estate granted was a life estate only.

In *Thompson v. Thompson*[74] limiting words inserted in the habendum controlled formal printed words of the habendum and also those of the formal granting clause.

---

[71] *Finlay v. Stevens*, 36 A.2d 767, 93 N.H. 124 (1944).

[72] *Loveless v. Thomas*, 38 N.E. 907, 152 Ill. 479 (1894); *Shephard v. Horton*, 125 S.E. 539, 188 N.C. 787 (1924); *Wilson v. Harrold*, 123 N.E. 563, 288 Ill. 388 (1919).

[73] 296 Ill. 305, 129 NE 807 (1921).

[74] 330 Mich. 1, 46 NW2d 437 (1951).

Construe Instruments Together. In ascertaining the intent of a deed, separate or different instruments may be construed together, where they are executed at the same time, and relate to the same subject matter, and are between the same parties, or are parts of the same transaction; where a deed is made to correct a mistake and supply an omission in a previous deed between the same parties; where a separate instrument contains a declaration of the grantor's objects and purposes; where a deed has been executed in compliance with a bond previously given by the parties for a deed; where one instrument is attached to, and made a part of, the other; and where there is an incorporation by reference to other instruments, records thereof, and to orders or judgments.[75]

However, two instruments that do not relate to the same subject matter will not be construed together, nor will they be so construed when not part of the same transaction. Even where they are executed at the same time and refer to the same subject matter, they cannot be construed together as one instrument or as constituting parts of one transaction where they are not between the same parties.[76]

Where instruments are construed together, the general purpose of the entire transaction should control, although in case of repugnancy, the deed should be given greater weight. Where two prior deeds are referred to in a conveyance, the first because it contained a full description of the land, it did not serve to destroy a reservation in the second deed.[77]

SPECIFIC DESCRIPTION CONTROLS GENERAL DESCRIPTION

The rule that *a particular description controls over a general one* is an old and universal rule. It is not, however, absolute and may be more correctly called a "guide post"; that is, useful phraseology to be resorted to when the intention of the parties to an instrument cannot be ascertained from the instrument itself.[78]

---

[75] *Metzger v. Miller*, D.C.Cal., 291 F. 780; *Arcari v. Strouch*, 158 A. 222, 114 Conn. 200 (1932); *Polk v. Carey*, Tex.Civ.App., 247 S.W. 568 (1922); *Huxford v. Trustees of Funds and Donations for Diocese of Iowa*, 185 N.W. 72, 193 Iowa 134 (1921); *Jackson v. Lady*, 216 S.W. 505, 140 Ark. 512 (1919); *Threadgill v. Bickerstaff*, 26 S.W. 739, 7 Tex.Civ.App. 406 (1894); *George v. Manhattan Land & Fruit Co.*, C.C.A.La., 51 F.2d 28; *Farrar v. Gessenden*, 39 N.H. 268 (1859); *Clough v. Bowman*, 15 N.H. 504 (1844); *Hacker v. Hoover*, 66 S.W. 382, 23 Ky.L. 1848 (1902).

Generally, when a description of land conveyed in a deed is ambiguous or uncertain, resort may be had in some circumstances to another instrument to explain the meaning of the description. *Goloskie v. Recorvitoz*, 219 A.2d 759, 101 R.I. 4 (1966).

[76] *Allen v. Parker*, 27 Me. 531 (1847).

[77] *Ford v. Belmont*, 30 N.Y.Super. 97 (1867), affirmed 30 N.Y.Super. 508 (1868); *Wilson v. Barney*, 9 P.2d 1058, 90 Colo. 461 (1932); *Ball v. Streeter*, 113 N.E. 1034, 225 Mass. 100 (1916).

[78] 72 ALR 410. *Rule that particular description prevails over general description.*

The Rule. Where a particular and general description in a deed conflict, and are repugnant to each other, the particular will prevail unless the intent of the parties is otherwise manifested on the face of the instrument.[79] It does not matter which one comes first in the deed.[80] In ascertaining the intent of the parties, general expressions must be read in the light of restrictive or more specific terms, for, presumably, the later words are added for greater certainty and to explain the premises.[81] The rule, however, that a later specific description controls a prior general one, is limited to the evident subject matter of the conveyance.[82]

In *Perry v. Buswell*,[83] the Maine court stated as follows:

> Of all rules of construction, none is more rigid than the one that, where the language describing the grant is specific and definite, as, for instance, by metes and bounds, the grant cannot be enlarged or diminished by a later general description, or by mere reference to deeds through which title was obtained. And this rule holds because the specific description is necessarily more indicative of intention than the general one.

Additionally, the Maine court stated in *Smith v. Sweat*:[84]

> It is too well settled to require the citation of authorities that a particular description of premises conveyed, when such particular description is definite and certain, will control a general reference to another deed as the source of title. So a clause in a deed, at the end of a particular description of the premises by metes and bounds, "meaning and intending to convey the same premises conveyed to me," etc., does not enlarge or limit the grant. . . . The exception to this rule is, where the particular description of land by metes and bounds is uncertain and impossible, then a general description in the same conveyance will govern.

A general description of land may be enlarged, and is controlled, by a subsequent particular description.[85] However, land cannot be added by this rule. As an example, the Maine court said in *Peasley v. Drisko*:[86]

---

[79] Ibid.

[80] *City of Rome v. Vescio*, 397 N.Y.S.2d 267, 58 A.D.2d 990 (1927), reversed 412 N.Y.S.2d 892, 45 N.Y.2d 980, 385 N.E.2d 629 (1928).

[81] *Parker v. Kane*, 63 U.S. 1, 16 L.Ed. 286 (1859), *Adams v. Law*, 58 U.S. 417, 15 L.Ed. 149 (1854), *Snadon v. Gayer*, 566 S.W.2d 483 (Mo.App., 1978), *Kilcullen v. Dery*, 334 A.2d 410, 133 Vt. 140 (1975).

[82] *Peasley v. Drisko*, 64 A. 24, 102 Me. 17 (1906).

[83] 113 Me. 399, 94 A. 483 (1915).

[84] 90 Me. 528, 38 A. 554 (1897).

[85] *Lord v. Wentworth*, 68 N.H. 610, 36 A. 17 (1895).

[86] 102 Me. 17, 65 A. 24 (1906).

"If A writes: 'I grant White acre, the same deeded to me by B,' and the deed of B included Black acre with White acre, it does not follow that A has granted Black acre also. So if A should write, 'I grant a certain parcel of flats, the same deeded to me by B,' and the deed of B included upland and flats in one description, it would not follow that A had granted the upland as well as the flats, especially if the upland was five times the extent of the flats." Thus, the rule that a later specific description controls a prior general description is limited to the evident subject matter of the conveyance. It does not require the inclusion of other matter. "The reference to another deed does not necessarily make the boundaries named in that deed the boundaries of the lot named in the first deed. The language was only to state the source of the title, or to identify the lot, and not for statement of boundaries."

In addition, where a deed contains a general description of the property, which is definite and certain in itself, and is followed by a particular description also, such particular description will not limit or restrict the grant that is clear and unambiguous by the general description.[87] Also, where a particular description will restrict the deed so that it will convey only a part of the land included in the general description, contrary to the declared intent of the grantor, the general description will prevail.[88]

The general rule, however, like most others in the law, is subject to exceptions, chief among which is that, if it appears from the conveying instrument, the surroundings of the parties, and their interpretation of it afterwards, it was the intention of the parties to give effect to the general description, and for it to prevail over the particular one, then that interpretation will be administered.[89]

Erroneous Particular Description. In *Guilmartin v. Wood*,[90] it was decided that a good general description would not pass the title to a lot of land, where an erroneous particular description followed the general description, because the general description must yield to the particular one.

In *Cornett v. Creech*,[91] the court said "It is true that, in the construction of deeds, particular or specific descriptions will control a general description, but this rule cannot be invoked when the particular description is evidently incomplete and defective, as in the deed before us. When it is apparent that a call from a deed has been omitted, or is set down erroneously, the court will read into the deed the omitted call, or correct the erroneous one, so that effect may be given to the intention of the parties, and the result intended by them be accomplished."

---

[87] *Barney v. Miller*, 18 Iowa 460 (1865).

[88] *Keith v. Reynolds*, 3 Me. 393 (1825).

[89] *Hatcher v. Virginia Min. Co.*, 282 S.W. 1102, 214 Ky 193 (1926).

[90] 76 Ala. 204 (1884).

[91] 100 S.W. 1188, 30 Ky L. Rep. 1265 (1907).

Intent of the Parties Controls. The general description will prevail over the particular description where there is a clear intent to have the general control.[92]

Words. Subsequent words in a deed of doubtful import cannot be construed so as to contradict preceding words that are certain.[93]

Numbers. In a deed containing two conflicting descriptions, one general by numbers, the other particular by reference to the plat of the town, the latter was held to control.[94]

References. A particular reference in a deed prevails over a more general reference.[95] Also, a particular description that is certain and definite will prevail over the identifying reference in a prior deed.[96]

REFERENCES PART OF THE DESCRIPTION

Any items referenced in a description are part of the description,[97] with as much effect as if copied into the description.[98] One of the reasons for employing references and not copying other information is so not to encumber the description, the reference being sufficient.[99] This is particularly true in the case of plans or plats.[100]

---

[92] *Sutton, S.S. Mfg. Mill. d Min. Co. v. McCullough*, 64 Colo. 415, 174 P. 302 (1918).

[93] *Johnson v. Harrison*, 130 So.2d 35, 272 Ala. 210 (1961).

[94] *Hannibal & St. J.R. Co. v. Green*, 68 Mo. 169 (1978).

[95] *Sorenson v. Wilson*, 476 A.2d 244 (1984).

[96] *Groth v. Johnson's Dairy Farm, Inc.*, 470 A.2d 399, 124 N.H. 286 (1983). Where the particular description in a deed is unambiguous, an identifying reference in a prior deed may be used to identify the land, but may not be used as evidence that the grantor intended to convey less than is precisely described, and may not be used to create an ambiguity.

[97] A reference in a deed to other deeds, when it appears that it is so intended, makes them a part of the description as if their language had been copied as part of the deed. *Perry v. Buswell*, 94 A. 483, 113 Me. 399 (1915).

[98] A deed may refer to another instrument, sufficiently identified with the same effect as if the document referred to was set forth in full in the deed. *Jacobs v. All Persons, etc.*, 106 P. 896, 12 C.A. 163 (1910).

Map or plan referred to in a deed as identifying land must be treated as if copied into the deed. *Goldsmith v. Means*, 158 A. 596, 104 Pa. Super. 571 (1932).

[99] One of the objects of filing maps, plats, plans, and field notes is to avoid the necessity of encumbering grants and deeds with lengthy descriptions of the land to be granted or conveyed. *Cragin v. Powell*, 128 US 691, 32 L Ed 566, 9 S Ct 203 (1888).

[100] *McElwee v. Mahlman*, 104 A. 705, 117 Me. 402 (1918).

In fact, not just *any* references, but *all* instruments in a chain of title when referred to in a deed will be read into it.[101]

The court, in arriving at the meaning of a deed, must consult the language of deeds linked together by express terms.[102] A description in a former deed repeated in terms in a later one retains the same meaning in the later deed that it had in the former.[103]

The doctrine of incorporation by reference permits the incorporation in a will of a prior will, an instrument operating *inter vivos*, such as a contract, a promissory note, a deed, or an informal writing.[104]

References to Maps and Plats. In order that a plat referred to in a deed may become a part of the deed for the purpose of description, the reference to such plat must be certain, so that, on production of the same, it will appear to be the plat referred to.[105]

When a description refers to a plat, the plat itself with all its notes, lines, descriptions, and landmarks becomes a part of and controls the grant.[106] Lines and figures are part of the description as if the courses and distances were set out in the deed.[107]

Where no other description is given of lands sold than by number of the lot in a survey of a tract of land, or the plan of a town, or an addition to the same, the authentic map of such survey is as much a part of the deed as though set out in it.[108]

Plat Not of Record, or Lost. A plan incorporated in a deed by reference need not be recorded, but it is enough to prove it and its contents.[109] In addition, a valid deed does not become void because, by reason of the loss of a plat referred to therein, it has become difficult to define the boundaries.[110]

---

[101] *Scheller v. Groesbeck*, 231 S.W. 1092 (1921), reversing 215 S.W. 353 (Tex., 1919); see also *Harris v. Windsor, Civ. App.*, 279 S.W.2d 648 (Tex., 1955).

Prior deed in plaintiff's chain of record title was relevant and could properly be used to assist in interpreting the meaning of an ambiguous deed to the plaintiff. *Hathaway v. Rancourt*, 409 A.2d 209 (1979).

[102] *Drake v. Russian River Land Co.*, 103 P. 167, 10 C.A. 654 (1909).

[103] *Wilson v. Underhill*, 108 Mass. 360 (1871).

[104] *Fifth Third Union Trust Co. v. Wilensky*, 70 N.E.2d 920, 79 Ohio App. 73 (1946), *Allday v. Cage*, 148 S.W. 838 (Tex.Civ.App., 1912); *Newton v. Seaman's Friend Soc.*, 130 Mass. 91 (1881); *Fickle v. Snepp*, 97 Ind. 289 (1884).

[105] *Kenyon v. Nichols*, 1 R.I. 411 (1851).

[106] *Gibson v. Johnson*, 244 So.2d 713 (1971), writ refused 246 So.2d 197, 258 La. 347.

[107] *Ford v. Ward*, 130 So.2d 380, 272 Ala. 235 (1961).

When a recorded map or plat of a subdivision is referred to in a deed conveying a portion of the tract, the map or plat becomes as much a portion of the deed as if it were fully incorporated therein. *Wolf v. Miraville*, 372 S.W.2d 28 (Mo., 1963).

[108] *Dolde v. Vodicka*, 49 Mo. 98 (1871).

[109] *Perkins v. Jacobs*, 129 A. 4, 124 Me. 347 (1925).

[110] *New Hampshire Land Co. v. Tilton*, 19 F. 598, (U.S.C.C., 1884).

Unrecorded Plat. Unrecorded plats have been considered part of the description. As the Wisconsin court said, an unrecorded plat or other map referred to in deed and identified by parol evidence will be treated at least between grantor and grantee as effectively incorporated into deed provided that the reference is definite and certain.[111] And the use of an unrecorded plat to make a property description sufficiently definite and certain may be recognized where such plat is specifically referred to and in effect incorporated in the instrument.[112]

Validity of Plat. Though a plat that fails to comply with statutory requirements is generally ineffective as notice and void as a conveyance or dedication of public easements or public grounds, a conveyance by reference to a recorded plat, or by merely drafting the description of the basis thereof, is not affected by the plat's being invalid. This is so because the plat, whether valid or invalid, if conclusively identified, is as much a part of the description as would be the case if it were copied into the instrument, or if the data furnished by it were set out in full.[113]

Reference to Atlas. Where a grantor who conveyed a house with the land belonging to the same was shown to have been familiar with an atlas in which a considerable part of the village was divided into house lots, on one of which the house conveyed stood, and he referred to the atlas several times in the deed, it may be presumed that the grantor meant to convey the lot on which the house stood as shown by that atlas.[114]

Government Surveys. When property is described in terminology of the United States government survey system, it is necessary to refer to survey to determine boundaries of land.[115]

---

[111] *Pavela v. Fliesz*, 133 N.W.2d 244, 26 Wis.2d 710 (1965).

[112] *Independent Gravel Co. v. Arne*, 589 S.W.2d 652 (Mo., 1979).

Fact that survey referred to in deed was not recorded with town clerk did not impair its utility in describing land conveyed. *Rambeau v. Barrows*, 225 A.2d 175 (Vt., 1969).

[113] *Patton on Titles*, 2nd Ed., § 120.

[114] *Webber v. Mixter*, 121 A. 677, 123 Me. 104 (1923).

[115] *Schulz v. McCracken*, 4 Ill. Dec. 73, 359 N.E.2d 906, 45 Ill. App.3d 152 (1977).

A conveyance of a lot according to a plat or by governmental subdivision carries all within the lines so run and platted. *Gary Land Co. v. Griesel*, 100 N.E. 835, 179 Ind. 220 (1913).

A reference in a description to the government patent makes the patent description and the government survey a part of the deed. *Miller v. Topeka Land Co.*, 24 P. 420 (Kan., 1890).

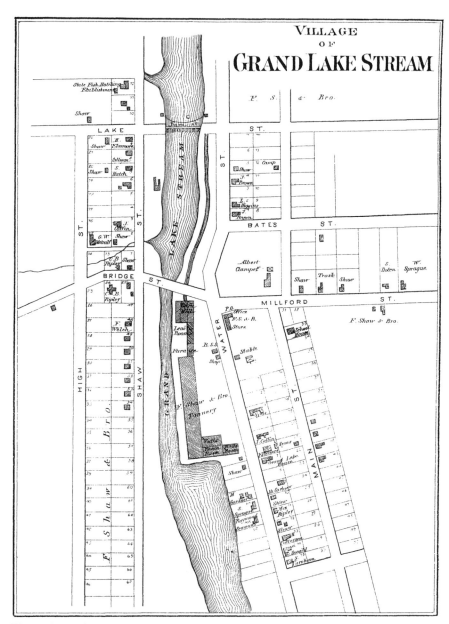

Figure 4.1 Atlas map referenced in *Webber v. Mixter*. From *Atlas of Washington County, Maine*, 1881. George N. Colby & Co. See also Figure 10.11.

Map Not Referenced. A map in existence but not referenced may also control. The Connecticut court said that a map in a deed should be treated as part of the description, when evidently intended to be so treated, even though it is not expressly referred to therein.[116]

Diagrams. In the case of discrepancy between the words of the description in a deed and a diagram in the deed, in general the diagram controls.[117]

Field Notes. Even field notes[118] of a grant referred to in a deed would be read into the deed.[119]

Sometimes a reference or a series of references can be an extremely valuable tool to the searcher in that if the present description is uncertain or inadequate, additional information may make it certain, or identify and locate the premises in question.[120]

Boundaries in References. Where one deed refers to another for description of the granted premises, reference may be had to the description in the latter.[121] In the description of land to be conveyed by a deed the expression "the same deeded to me by" a specified person may only indicate the source of the grantor's title, or locate and identify the parcel intended to be conveyed; it does not necessarily adopt all and singular the boundaries named in the deed referred to.[122]

---

[116] *Murray v. Klinzing*, 29 A. 244, 64 Conn. 78 (1894).

[117] *Maginnis Land & Improvement Co. v. Marcello*, 123 So. 653, 168 La. 997 (1929).

[118] The term "field notes" in its ordinary sense means "notes made by the surveyor in the field while making the survey, describing by course and distance, and by natural or artificial marks found or made by him, where he ran the lines and made the corners." *State v. Palacios*, Tex. Civ. App., 150 S.W. 229 (1912).

[119] *Temple Lumber Co. v. Mackechney*, 228 S.W. 177 (1921), affirming 197 S.W. 744 (1917).

[120] Uncertainty in description is immaterial, if premises can be identified by description, in connection with other conveyances, plats, lines, or records, well known or on file in public offices. *Pittsburgh, C.,C. & St. L. R. Co. v. Beck*, 53 N.E. 439, 152 Ind. 421 (1899).

[121] *Newmarket Mfg. Co. v. Prendergast*, 4 Fost. 54, 24 N.H. 54 (1851).

[122] *Peasley v. Drisko*, 65 A. 24, 102 Me. 17 (1906).

However, where a description is uncertain, but the land thereby conveyed forms a part of a larger tract previously owned by the grantors, reference may be made to prior deeds from the same grantors, conveying other portions of the larger tract, which make more certain the boundaries described in the later deed.[123]

Part of Lot Held by Adverse Possession. A deed described land by a specified lot number in a specified city block and added the recital "being the property purchased by me from" a specified person, "which deed is dated" as specified. It conveyed all of the lot so specified not withstanding the grantor, when the purchase by him referred to in the deed was made; it was held that a part of such lot was by adverse possession; the recital as to such purchase could not be so construed as to reduce the quantity conveyed and could serve no purpose other than that of further identification and of showing source of title.[124]

Extent of Conveyance. Where a reference is intended to show more than merely the source of title, it may require the grantee to resort to the deed referred to in order to determine the extent of his acquisition.[125]

Problems with References. References, however, are not without their own problems. A valid deed does not become void because, by reason of the loss of a plat referred to therein, it has become difficult to define the boundaries.[126] References do not have to be recorded to be valid.[127]

The only requirement is that the reference must be certain, so that, on production of the same, it will appear to be the item referred to.[128]

---

[123] *White v. Spahr*, 59 S.E.2d 916, 207 Ga. 10 (1950).

[124] *Trott v. Joslyn*, 192 N.W. 556, 222 Mich. 452 (1923).

[125] *Carter's Adm'r v. Quillen*, 39 S.W.2d 1012, 239 Ky 583 (1931).

[126] *New Hampshire Land Co. v. Tilton*, 19 F. 73 (C.C.N.H., 1884).

[127] Validity of a deed is not affected by fact that legal description of premises conveyed includes reference to unrecorded but identifiable plat or to a plat improperly accepted for record by register of deeds, and grantee in such a deed is estopped to deny legal existence of streets as shown on such plat. *Raines v. Village of Alden*, 90 N.W.2d 906, 252 Minn. 530 (1958).

[128] *Kenyon v. Nichols*, 1 R.I. 411 (1851).

Mere numbering of lots in deed is insufficient to make unrecorded private map part of the conveyance. *Nixon v. City of Anniston*, 121 So. 514, 219 Ala. 219 (1929).

Even if illegally or improperly done, a reference, if identifiable, is a material and essential part of the description.[129] A mistaken reference does nothing to invalidate the deed or weaken the description, so long as it is identifiable.[130]

Which One Controls. It is mostly fundamentally agreed that whether a description by reference to a map, plat, plan, sketch, or diagram will prevail over a conflicting or inconsistent description by words appears to depend, to some extent at least, upon the circumstances. Thus, while it may be said that the great majority of the cases hold that the map controls the worded description, it would be difficult to lay down a comprehensive general rule on the question, but general expressions of the control of the map or plat over the description by words are numerous.[131]

A very common decision is as the court stated in *Cook v. Hensler*:[132]

> If a party purchases a certain numbered block of land according to the official map of the city and his purchase is so described in the deed, a further description of the block by metes and bounds or courses and distances would be subordinate to the description of the block by its number, and would have to give way in case of conflict.

The Georgia court found, however, in *Thompson v. Hill*,[133] that the rule is not inflexible under some circumstances. For instance, a map or plat that is so inexact and extremely lacking in essential data that it would convey nothing if standing by itself will not control a verbal description in the deed, referring to such map, attached thereto, as showing the land conveyed.[134]

Where there is an inconsistency between the description contained in a will, deed, contract, lease, and so on and the description in another instrument referred to therein, the court will endeavor to ascertain and effectuate the intention of the testator or parties to the instrument.[135] In determining whether a description in a will

---

[129] *Noonan v. Lee*, 67 U.S. 499, 2 Black 499, 17 L.Ed 278 (Wis., 1862).

[130] *Prouty v. Rogers*, 164 P. 901, 33 C.A. 246 (1917).

[131] *A. Ken v. Wallace*, 68 S.E. 937, 134 Ga. 873 (1910).

[132] 57 Wash. 392, 107 P. 178 (1910).

[133] 73 S.E. 640, 137 Ga. 308 (1912).

[134] *Futrell v. Holloway*, 149 So. 167 (La.App., 1933).

See 130 ALR 643, *Description of land in deed or mortgage by reference to map, plat, plan, sketch, or diagram as prevailing over description by words.*

[135] *John L. Roper Lumber Co. v. Hinton* (D.C., 1919) 260 F. 996, affirmed in (1920.C.C.A., 4th) 269 F. 574; Doe ex dem. *Knight v. Roe*, 96 A. 32, 5 Boyce 570 (Del., 1915); *Hathorn v. Hinds*, 69 Me. 326 (1879).

or a conflicting description in an instrument referred to in the will must prevail as to the land to pass under the will, in case of uncertainty, the court will admit evidence of extrinsic circumstances and surroundings to discover the testator's meaning.[136] See later examples presented in Chapter 8.

A change in possession is very persuasive as showing the intention of the grantor, and the belief of his grantee and successors in interest; however, the court said in *Scheller v. Groesbeck*,[137] that neither the intention of the one nor the belief of the other can change the legal meaning of plain language or be perpetuated as notice by record. And accordingly, that although the grantor surrendered more land than was conveyed by his deed, being all that had been granted to him by another instrument referred to therein, only the amount described in his deed passed.

References are part of the description, but general references in a deed will not control the specific language of the deed. Nor will a reference serve to limit or enlarge a clear and complete description.[138] It has frequently been said that if the reference to another instrument is merely to show the chain of title, the inconsistent description contained in such other instrument is not controlling.[139] Nevertheless, where it is the evident intention of the grantor, a reference to another instrument will control a description by metes and bounds or courses and distances contained in the instrument from which the reference is made, although such instrument itself contains an adequate description of the same parcel of land.[140]

Cases do exist whereby a reference was found to be controlling. In *Keith v. Reynolds*,[141] the plaintiff's deed conveyed "a certain tract of land, or farm, lying in Winslow, it being included in that tract which was granted by the Plymouth company to Gamaliel Bradford, Esq. and five others, and which was granted by the proprietors of the above said tract to Ezekiel Pattee, Esq. and lately owned by John Bran; said farm is bounded as followeth—beginning on the easterly side of the mile brook so called, and thence on an easterly course parallel to . . . [a certain line described] one mile; thence northerly at right angles fifty rods; thence westerly, parallel with the first line until it comes to said mile brook; thence southerly, by said brook, fifty rods to the first mentioned bounds; containing one hundred acres." This description excluded a small triangular parcel covered by the Pattee deed, whereas that deed described the northerly course as parallel to the brook, which ran obliquely across the farm rather than at right angles. The court held that the entire

---

[136] *Knight v. Roe*, 96 A. 32, 5 Boyce 570 (Del., 1915).

[137] 231 S.W. 1092 (Tex., 1921).

[138] *Campbell v. United Fuel Gas Co.*, 130 S.E. 666, 100 W. Va. 508 (1925).

[139] Doe ex. dem. *Knight v. Roe*, 96 A. 32, 5 Boyce 570 (Del., 1915); *Brown v. Heard*, 27 A. 182, 85 Me. 294 (1893); *Smith v. Sweat*, 38 A. 554, 90 Me. 528 (1897); *Dow v. Whitney*, 16 N.E. 722, 147 Mass. 1 (1888); *Trott v. Joslyn,* 192 N.W. 536, 222 Mich. 452 (1923).

[140] See 134 ALR 1045.

[141] 3 Me. 393 (1825).

farm was conveyed, saying that the intention of the parties must be gathered from the whole of the deed, and that a reference to the source of title was not mere matter of recital but was part of the description.

In *Cochrane v. Harris*,[142] where it appeared that a deed to "the hereinafter described pieces or parcels of ground lying and being in the said city of Cumberland, county of Allegheny and state of Maryland," as conveyed by a certain deed further described the property by location, courses, and distances, and lot numbers, it was held that such deed did not convey land not otherwise described therein but that was described in the prior deed. The court pointed out that the conveyance was not of the land conveyed by the deed referred to but of the "hereinafter described pieces or parcel of ground," and not all of the parcels conveyed by the earlier deed.

Failure to Recite Reference. When a conveyancer makes a deed, he or she should always provide a reference to the source of the grantor's title.[143] However, failure to recite the source of title in a deed does not prevent the title to real estate from passing from the grantor to the grantee.[144]

CONSTRUE AGAINST THE GRANTOR

Generally, a grantee in a deed or other instrument, who accepts the instrument, is bound by the recitals therein, even though he or she does not sign it.[145] However, the general rule is that all grants, deeds, and leases are to be most strongly construed against the grantor if there is any doubt or uncertainty as to the meaning of the grant.[146] The cases are many and the rule is a prevalent one. But there must be doubt or ambiguity, for a grantee is not entitled to the benefit of an alternative interpretation favoring him where ambiguity is not established.[147]

However, while ambiguities in a grant are to be resolved in favor of grantees, such rule must yield to the paramount rule that the intention of the parties is to be given effect if it can be ascertained and if, as ascertained, it does not contravene the clear meaning of the words of the grant.[148]

---

[142] 84 A. 499, 118 Md. 295 (1912).

[143] Law Rev. 1961. 3 N.H.B.J. 225.

[144] *Blackburn v. Pond Creek Coal & Land Co.*, 287 S.W.2d 610 (Ky., 1956).

[145] *Webb v. British Am. Oil Producing Co*, Civ. App., 281 S.W.2d 726 (Tex., 1955).

[146] *Treharne v. Klint*, 58 N.E.2d 638, 324 Ill. App. 546 (1945).

[147] *Foslund v. Cookman*, 211 A.2d 190, 125 Vt. 112 (1965).

[148] *Maciey v. Woods*, 154 A.2d 901, 38 Del. Ch. 528 (1959).

One of the main reasons for this rule is that the grantor, or his representative, drafted the deed and its description and not only may say whatever he or she wishes to say but also has complete control over how it is said and what is to be included in the conveyance.[149] Whatever the grantor says should be read to have meaning if such reading is possible.[150] Since the language of the deed is that of the grantor, any doubt as to its construction should be resolved against him.[151] And, where uncertain or ambiguous, a deed is usually regarded as conveying the largest interest a grantor could convey.[152]

In the 1926 case of *Smart v. Huckins* (82 N.H. 342), the Supreme Court of New Hampshire stated, after applying many rules of construction to no avail, that construing against the grantor is a rule of last resort, but useful "when all other rules fail."

Reservations and Exceptions. Reservations or exceptions in a deed and language of doubtful meaning contained in deed also shall be construed against the grantor and in favor of the grantee.[153]

FALSE DESCRIPTION REJECTED

Where the land in a deed is so described that it can be ascertained, it will pass, even though some part of the description is false.[154] The principle is that *a mistake in the description of lands in a deed will not void the deed or defeat the legal title of the grantee or anyone claiming under his title.*[155] Where property intended to be conveyed can be ascertained from such parts of description in a deed as are found to be correct, the property will pass, and the incorrect parts of the description will be disregarded.[156]

---

[149] An ambiguous deed is construed most strongly against its author. *A.S. and W. Club v. Drobnick*, 187 N.E.2d 247, 26 Ill.2d 521 (1962).

[150] *Painter v. McDonald*, 427 S.W.2d 127 (Tex.Civ.App., 1968), reversed 441 S.W.2d 179 (Tex., 1969).

[151] *James v. Dalhart Consol. Independent School Dist.*, 254 S.W.2d 826 (Tex.Civ.App., 1953).

When language used in conveyance is susceptible of more than one interpretation, courts will look into surrounding circumstances and situation of parties, and ambiguity is to be construed most strongly against the grantor. *Rome v. Vescio* (4th Dept) 58 App. Div. 2d 990, 397 NYS 2d 267 (1977), reversed on other grounds, 45 NY 2d 980, 412 NYS 2d 892, 385 NE2d 629 (1978).

[152] Easement deeds must be construed most strongly against the grantor, and most favorable to the grantee, so as to confer the largest estate that a fair interpretation will permit. *Peterson v. Barron* (Tex. Civ. App. 5th Dist.), 401 S.W.2d 680 (1966).

[153] *Hidalgo County Water Control and Imp. Dist. No.16 v. Hippchen*, 233 F.2d 712 (C.A. Tex., 1956).

[154] *Scofield v. Lockwood*, 35 Conn. 425 (1868).

[155] *Bevans v. Menry*, 49 Ala. 123 (1873).

[156] *McClausland v. York*, 174 A. 383, 133 Me. 115 (1934).

A plainly erroneous boundary description will be rejected, and a reasonable meaning given to the deed which will conform to the intent of the parties.[157] Also, any particular of a description may be rejected if it is manifestly erroneous and enough remains to identify the land intended to be conveyed.[158]

Obvious Errors Ignored. A plain error in a deed will be rejected, and the deed construed reasonably to conform to the intent of the parties.[159] An impossible or senseless course given in a deed must be disregarded, and, if the other calls or parts of the description are sufficient to identify the land conveyed, the deed will be sustained.[160]

Where a deed contained a reservation, "always provided that there is a convenient landing place accepted on the north side of H. Brook," the word "accepted" should be construed as "excepted."[161]

Conflicting Documents. Where there is a conflict between boundaries described in deeds from the same grantor, the deed first executed has priority, and the grantee named therein has superior title.[162] However, where deeds were not from a common grantor, the deed earlier recorded would have no preference over the later deed in the sense that the description in the first would control the description in the record.[163]

A common grantor can only convey that property he has left to convey, regardless of what his deed may specify in metes and bounds.[164]

Conflicting Descriptions. Where there are two descriptions in a conveyance, one erroneous and the other correct, the latter will prevail.[165]

---

[157] *Wadleigh v. Cline*, 108 A.2d 35, 99 N.H. 202 (1954).

[158] *Lane v. Thompson*, 43 N.H. 320 (1861).

[159] *Reney v. Hebert*, 242 A.2d 72, 109 N.H. 74 (1968).

[160] *Brose v. Boise City Railway & Terminal Co.*, 51 P. 753, 5 Idaho 694 (1897).

[161] *Dougan v. Town of Greenwich*, 59 A. 505, 77 Conn. 444 (1904).

[162] *Wysinski v. Mazzotta*, Pa. Super. 472 A.2d 680 (1984).

Murrer v. American Oil Co., 359 A.2d 817 (Pa.Super., 1976).

A description in a junior conveyance cannot be used to locate lines called for in a prior conveyance. *Napoli v. Philbrick*, 173 S.E.2d 574, 8 N.C. App. 9 (1968).

[163] *Baker v. Miller*, 187 N.E. 699, 284 Mass. 217 (1933).

[164] *Mazzucco v. Eastman*, 236 N.Y.S.2d 986, 36 Misc.2d 648 (1958), affirmed 239 N.Y.S.2d 535, 18 A.D.2d 1138 (1962).

[165] *State v. Franco-American Securities*, 172 S.W.2d 731 (Civ.App., 1943).

Where a deed contained two descriptions of the land conveyed, one of which, when applied to the land owned by the grantor, is found to be not true, it must be rejected. *Hornet v. Dumbeck*, 78 N.E. 691, 39 Ind. App. 482 (1906).

Conflicting Clauses. Generally, when two clauses of a deed are so repugnant that there is no rational construction that will render them both effective, the earliest clause will stand and the latter one will be rejected.[166]

If recitals are ambiguous and the operative part is clear, the operative part must govern; if both recitals and operative part are clear but inconsistent with each other, the operative part is to be preferred.[167]

MEANING AND INTENDING CLAUSE

The "meaning and intending" clause in a deed, by itself, cannot be used to limit or enlarge the description contained within the deed.

A clause in a deed, at the end of a particular description of the premises by metes and bounds, "meaning and intending to convey the same premises conveyed to me," and so on, does not either enlarge or limit the grant.[168] The Maine court also said, in *Brown v. Heard*,[169] that such wording was "merely a help to trace the title." The court also said in *Sinford v. Watts*[170] that the clause "meaning to convey lot known as," etc., following a particular description by metes and bounds, did not enlarge the grant, unless the contrary appears, because such clause is ordinarily intended as a help to trace the title.

Sufficiency of Description. A deed will not be avoided because some particulars of the description of the premises are false or inconsistent, if it is sufficient to identify the premises.[171] However, if the description is so imperfect that it is impossible to know what land was intended to be conveyed, a deed may be void.[172]

> In this case, the only description was "the southwest part of the northeast fractional quarter of section thirty-six," without any specification of quantity. The deed was held to be void for uncertainty.

---

[166] *Martin v. Adams*, 62 So.2d 328, 216 Miss. 270 (1953).

[167] *Davidson v. Vaughn*, 44 A.2d 144, 114 Vt. 243 (1945).

Jefferson v. Davis, 95 A.2d 617, 25 N.J. Super. 135 (1953).

[168] *Smith v. Sweat*, 38 A. 554, 90 Me. 528 (1897).

[169] 27 A. 182, 85 Me. 294 (1893).

[170] 122 A. 573, 123 Me. 230 (1923).

[171] *Wing v. Burgis*, 13 Me. (1 Shep.) 111 (1836).

[172] *McChesney's Lessee v. Wainwright*, 5 Ohio 452 (1832).

Whenever an attempt is made to convey lands, and the description is so uncertain that the lands cannot be identified, or it calls for lands which have no existence, the deed will be void. *Carter v. Barnes*, 26 Ill. 454 (1861).

The basic rule is if the description is certain enough to enable a person to locate the land, it is sufficient.[173]

Courts have consistently ruled on the sufficiency of description when the lands can be located by a surveyor,[174] or by extrinsic evidence.[175] For example, in *Brenneman v. Dillon,*[176] the court said,

> Deed need not describe property by governmental survey lots and blocks, or metes and bounds, if it can be located from description given, and is not void for uncertainty, if, by extrinsic evidence, it can be made certain. Description is sufficiently definite if surveyor can locate property with or without extrinsic evidence. Description omitting town, county, or state may be sufficient if containing other means of identification, and general description as "tract" adjoining other lands is sustainable if land can be identified. Deed to "my house and lot" or to "one-half of my lot" may be made sufficient by extrinsic evidence.

The courts do not want to render an instrument void or defeat a description, if it is at all possible to sustain it. In the case of *Adams v. Adams,*[177] the court said that the fact that the grantor had no title to the land, and the deeds conveyed nothing, did not render the deeds void; the grantor being liable on the covenants and deeds conveying any title she had or may not have to the land.

Courts are extremely liberal in construing descriptions of land conveyed by deeds to determine whether such descriptions are sufficiently definite and certain to identify lands and render instruments operative as conveyances.[178] Generally, a deed will not be declared void for uncertainty in description of land conveyed, if it is possible, by any reasonable rules of construction, to ascertain from the description, aided by extrinsic evidence, what property is intended to be conveyed.[179] Under the maxim that that is certain which can be made certain,[180] courts properly lean against striking

---

[173] *Cunningham's Lessee v. Harper*, 1 Wright 366 (Ohio, 1833).

[174] Any description in a deed by which a parcel of property may be identified with reasonable certainty by a competent surveyor is sufficient. *Brunotte v. DeWitt*, 196 N.E. 489, 360 Ill. 518 (1935); see also *Smiley v. Fries*, 104 Ill. 416 (1882).

If a surveyor, by applying the rules of surveying, can locate the land, the description in the deed is sufficient. *McCullough v. Olds*, 41 P. 420, 108 C. 529 (1895).

[175] A description in a deed may be sustained, if, with the aid of the usual deed inquiries, there would be no difficulty in locating the land. *Persinger v. Jubb*, 17 N.W. 851, 52 Mich. 304 (1883).

A deed will only be held void for uncertainty where, after resort to oral proof, it still remains a matter of conjecture what was intended by the instrument. *Smith v. Crawford*, 81 Ill. 296 (1876).

[176] 129 N.E. 564, 296 Ill. 140 (1920).

[177] 133 A. 279, 80 N.H. 80 (1921).

[178] *City of North Mankato v. Carlstrom*, 2 N.W.2d 130, 212 Minn. 32 (1942).

[179] Ibid.

[180] A deed description is to be considered certain if it can be made certain. *Town of Brookhaven v. Dinos*, 431 N.Y.S.2d 567, 76 A.D.2d 555 (1980), affirmed 445 N.Y.S.2d 151, 54 N.Y.2d 911, 429 N.E.2d 830 (1981).

down a deed for uncertainty and generally will adopt liberal rule of construction to uphold a conveyance.[181] *Descriptions are not to identify land but to furnish the means of identification.*[182]

Only when it remains a matter of conjecture what real estate was intended to be conveyed by the deed, after resorting to such extrinsic evidence as is admissible, will the deed be held void for uncertainty of description of the realty.[183] A deed, a devise, or a reservation of real estate, the description of which is impossible of ascertainment, is void.[184]

DECISIONS RENDERING DESCRIPTIONS CERTAIN OR UNCERTAIN

The name by which a grantee is called in a deed is immaterial where his or her identity is known,[185] however, a conveyance to a fictitious person is inoperative and title remains in the grantor.[186]

A mistake as to the Christian name of the former owner of the land will not avoid the deed.[187]

The fact that the name of a former owner of the land was incorrect as used in the description of a deed was immaterial where, from the rest of the description, the premises could be sufficiently designated.[188]

A deed is not defeated by a manifestly erroneous statement of a monument, when the remaining description is sufficiently certain to locate the land.[189]

A description is not insufficient where one of the corners called for can be determined by a simple mathematical computation.[190]

---

[181] *City of North Mankato v. Carlstrom*, 2 N.W.2d 130, 212 Minn. 32 (1942).

[182] *Daly v. Duwane Constr. Co.*, 106 N.W.2d 631, 259 Minn. 155 (1960).

The office of the description is not to identify the land, but to furnish the means of identification.

Diffie v. White, 184 S.W. 1065 (Civ.App., 1916).

Sanderson v. Sanderson, 825 S.W.2d 1008, affirmed 109 S.W.2d 744, 130 Tex. 264 (1935).

[183] *State v. Rosenquist*, 51 N.W.2d 767, 78 N.D. 671 (1952).

A deed is void for uncertainty if, after taking all the evidence, it cannot be determined what land was intended to be conveyed. *Wooded Shores Property Owners Ass'n, Inc. v. Matthews*, 44 Ill. Dec. 840, 411 N.E.2d 1026, 89 Ill. App.3d 187 (1980).

[184] *Edens v. Miller*, 46 N.E. 526, 147 Ind. 208 (1897).

[185] *In re Buffs's Estate*, 248 N.Y.S. 332, 139 Misc. 298 (1931).

[186] *Johns v. Wear*, 230 S.W. 1008 (Tex. Civ. App., 1921).

Deed to fictitious person is void but one to person taking under assumed name is not. *Marky Inv., Inc. v. Arnezeder*, 112 N.W.2d 211, 15 Wisc.2d 74 (1962).

[187] *Greely v. Steele*, 2 N.H. 284 (1820).

[188] *Thompson v. Ela*, 58 N.H. 490 (1878).

[189] *Benton v. McIntire*, 15 A. 413, 64 N.H. 598 (1888).

[190] *Wall v. Club Land & Cattle Co.*, 88 S.W. 534 (Civ.App., 1905), reversed 92 S.W. 984, 99 Tex. 591, 122 Am. St. Rep. 666 (1906).

However, where the description specified in a deed involved a geometrical proposition that could be satisfied by an almost indefinite number of figures, the deed was void for uncertainty of description.[191]

In *Brittain v. Dickson*, a deed was held sufficient because the starting point could be reasonably ascertained.[192]

A description which did not entirely enclose the land did not invalidate the deed, where it otherwise sufficiently identified the land.[193]

A deed, to a place "known as the place built on by Thos. Davis and lastly occupied by G.N. Breckenridge" was not void for uncertainty, because the land could be identified by extrinsic evidence.[194]

The erroneous mention of an incident in the history of the title to a piece of land was held to have no force as against the mention of metes, bounds, courses, distances, and visible monuments, when the question is whether the deed is sufficient to convey the land intended.[195]

The fact that more realty is described in a deed than is owned by the grantor will not destroy the deed's operative effect as to the described land that is owned by the grantor.[196]

The mere fact that a description included land that the grantor had previously conveyed and no longer owned did not create a latent ambiguity.[197]

Where the metes and bounds definitely located a tract, the recital in the deed that placed the tract in a lake, was held to be a mistake not defeating the grant.[198]

A person, being seized of land lying partly in lot No. 10 and partly in lot No. 9, granted a tract of land which he described in the deed as lot No. 10, but as bounded on all sides by land of other persons. It was held that his whole land, lying in both lots, passed although, in the description of the premises, it appeared that there were mistakes as to the owners of the adjoining lots.[199]

In another situation, no number was given to designate a lot conveyed, but it was described as adjoining the land of four several individuals, which description, in full, would include three several lots, and a much larger quantity of land than that named in the deed, but, by rejecting the name of one of the individuals, one lot only would be clearly designated. It was held that such portion of the description might be rejected, as a mistake as to said persons being an adjoining owner.[200]

---

[191] *Wilson v. Carling*, 66 So. 188, 188 Ala. 543 (1914).

[192] 60 S.W.2d 1093 (Tex., 1933).

[193] *Randolph v. Lewis*, 163 S.W. 647 (Civ.App., 1916), affirmed 210 S.W. 795 (Tex., 1919).

[194] *Petty v. Wilkins*, 190 S.W. 531 (Tex.Civ.App, 1916).

[195] *Sherwood v. Whiting*, 8 A. 50, 54 Conn. 330 (1887), 1 Am. St. Rep. 116.

[196] *Miller v. Miller*, 38 N.E.2d 343, 110 Ind. App. 191 (1942).

[197] *Grosshans v. Rueping*, 153 N.W.2d 619, 36 Wis.2d 519 (1967).

[198] *Kidder v. Pueschner*, 247 N.W. 315, 211 Wis. 19 (1933).

[199] *Tenny v. Beard*, 5 N.H. 58 (1829).

[200] *White v. Gay*, 9 N.H. 126, 31 Am. Dec. 224 (1837).

A deed that failed to recite in what state the lands, described by section, town, and range, were situated, and that though the grantor's acknowledgment was in a foreign state, the certificate of the official character of the notary was appended to the acknowledgment, and the deed recorded in a county within the state where the lands were in fact located, the land was held to be sufficiently identified.[201] However, a deed that described a property by metes and bounds, without reference to any known government survey, and that failed to state the county and state, was held to be insufficient.[202]

And, a deed that described a certain acre out of a tract of land, without specifying the portion of the tract out of which it was to be taken, was void for uncertainty.[203]

A deed describing the land conveyed as the "southwest" quarter rather than the northwest quarter was sufficient to identify the land and was valid when the grantor owned no land in the southwest quarter and all other descriptions of the land were accurate.[204] This is a good example of where extensive research is necessary to determine that the grantor in fact did not own land in the southwest quarter. If he had, the decision might have been different. At the very least, it would be necessary to show which lands were intended to be conveyed and which were not, or could not be, intended to satisfy the description.

A description, reading "Also another piece or parcel of land, being all that part of" a certain quarter "lying outside of the town, containing 34 acres," is sufficient, given the previous descriptions in the same deed covering lots in a certain town.[205]

In the description of property in a deed according to the maps of the United States survey, an error in the number of the range is immaterial when the description otherwise identifies the property.[206]

The word "North" in description in deed reading "Township One (1), Range Two (2) North," was not in context with the word "range" and would be rejected as surplus age, since a range can be only east or west, and not north or south, of the applicable principal meridian.[207]

A deed of several parcels of land, describing the first as "the northwest 1/4 section 27, 11 S., 2 W.," but omitting the word "section" in the description of the others, was held to be sufficient to convey all.[208]

A mistake in a deed in locating the land conveyed in the "northwest quarter" instead of the "northeast quarter" of a given section, will not prevent the title from

---

[201] *Mee v. Benedict*, 57 N.W.175, 98 Mich. 260, 22 L.R.A. 641, 39 Am. St. Rep. 543 (1893).

[202] *Pfaff v. Cilsdorf*, 50 N.E. 670, 173 Ill. 86 (1898).

[203] *Hanna v. Palmer*, 61 N.E. 1051, 191 Ill. 41, 551 L.R.A. 93 (1901).

[204] *Podd v. Anderson*, 30 Cal. Rptr. 345, 215 C.A.2d 660 (1965), citing C.J.S. Deeds, § 90.

[205] *Smiley v. Fries*, 104 Ill. 416 (1882).

[206] *Willis v. Ruddock Cypress Co.*, 32 So. 386, 108 La. 255 (1902).

[207] *Tregloan v. Hayden*, 282 N.W. 698, 229 Wis. 500 (1939).

[208] *Bowen v. Prout*, 52 Ill. 354 (1869).

passing, where there is an additional designation by metes and bounds, which furnishes the means of identifying the land conveyed.[209]

Where the land intended to be conveyed by an executor's deed was located in range 23, but, by mistake, was described as in range 24, the legal title was not conveyed by the deed to the purchaser.[210]

A deed describing land as the W. 1/2 of the N.E. 1/4, the E. 1/2, of the N.W. 1/4, except 18 acres in the S.W. corner, and the N.W. 1/4 of the W. 1/2 of the S.E. 1/4, lying north of the property of a railroad company, which ran through several sections, no section, town, or range being given, is insufficient, nor will proof that the grantor owns in a designated township land otherwise corresponding with the description help it.[211]

A description, "The southeast part of the southeast fourth of the northeast quarter of section 36, township 4 south, and range 2 east, containing 32 acres," was found to be too indefinite to sustain a suit for possession of the land.[212]

The description of land as "the S. 1/2 of the N.E. 1/2 of the S.E. 1/2 of section 19" is void, for uncertainty.[213]

A description of land as "fractional township 20, of range 23," is sufficiently definite, as the courts judicially know that there is but one fractional township answering the description in the state.[214]

A description of land, "All that portion of fractional sections 31 and 32, in township 7 south, of range 2 west, in Perry County, Indiana," when corrected by the addition of, "of which Nathaniel Ewing was the owner at the time of his death, containing 331 acres, more or less," is sufficient.[215]

Where a description in a deed read "A part of fractional section number 19, being the half of the west half of the northwest quarter of section number 29, in township 7 south, of range 14 west, containing 40 acres," it was held that, when rejecting the words "A part of fractional section number 19" as contradicting a more particular description following, the conveyance was good to pass an undivided half of the W. 1/2 of the N.W. 1/4, and so on.[216]

A deed describing the property as "the south fractional one quarter of the northeast quarter," of a certain section, township and range, "north of G River, containing 18 acres," was found as sufficient to convey the S.E. fractional quarter of said N.E. 1/4, it being shown that both the S.E. and S.W. 1/4 of the N.E. 1/4 were fractional;

---

[209] *Frick v. Godare*, 42 N.E. 1015, 144 Ind. 170 (1895).

[210] *Gilmore v. Thomas*, 158 S.W. 577, 252 Mo.147 (1913).

[211] *Miller v. Beardslee*, 141 N.W. 566, 175 Mich. 414 (1913).

[212] *Shoemaker v. McMonigle*, 86 Ind. 421 (1882).

[213] *Pry v. Pry*, 109 Ill. 466 (1884).

[214] *Webb v. Mullins*, 78 Ala. 111 (1884).

[215] *Dunn v. Tousey*, 80 Ind. 288 (1881).

[216] *Gano v. Aldridge*, 27 Ind. 294 (1866).

the S.W. fractional quarter containing 34/40 acres, and the S.E. fractional quarter containing 18 acres.[217]

However, a conveyance describing land only as "27 acres, fractional section 15," in a town and range mentioned, such section containing a greater quantity of land, is void for uncertainty.[218]

In addition, a sale of real estate by the description, "the fractional east half of the southeast quarter of section 22, township 23, range 10 east, containing sixty-one acres more or less, in Blackford County, Indiana," is void, the description being too indefinite to furnish the means of identifying the land.[219]

It is settled that "part" descriptions are void for indefiniteness. *Miller v. Best*,[220] stated "Part of a quarter section, or other subdivision, without further description, is void." According to the courts, this type of description is insufficient, since "it cannot be ascertained just what part" of the subdivision is intended.

In *Montgomery and wife v. Johnson*, the court ruled that there is no such thing as the North quarter (N 1/4) of any part of a section.[221]

Correcting Mistakes. Earlier deeds may be used to correct mistakes in more recent deeds.[222]

Missing Course. Where a course is evidently omitted in a deed to land sold according to a plan of lots the missing course may be shown from the description in a deed to the adjoining lot.[223]

CONFLICT BETWEEN GRANTING AND HABENDUM CLAUSES

The granting portions, when clear and unambiguous, control over other words or phrases in introductory or other parts of the deed.[224] The habendum may modify, limit, and explain the grant, but cannot defeat it when expressed in clear and unambiguous language.[225] In some cases the habendum may diminish the estate granted. For example,

---

[217] *Prior v. Scott*, 87 Mo. 303 (1885).²

[218] *Craven v. Butterfield*, 80 Ind. 503 (1881).

[219] *Peck v. Sims*, 22 N.E. 313, 120 Ind. 345 (1889).

[220] 235 Ark. 737 (1962).

[221] 31 Ark. 74 (1876).

[222] *Cooper v. White*, 46 N.C.(1 Jo.) 389 (1854); *Campbell v. McArthur*, N.C. (2 Hawk.) 33 (1822).

[223] *Zervey v. Allan*, 64 A. 587, 215 Pa. 383 (1906).

[224] *Weber v. Nedin*, 246 N.W. 686, 210 Wis. 39 (1933).

Higginson v. Smith, App., 272 S.W.2d 348 (Tenn., 1954).

[225] *Blair v. Blair*, 10 A.2d 188, 111 Vt. 53 (1910).

The purpose of a habendum clause in a deed is to limit, enlarge, and define the estate indicated in the granting clause. *Hart v. Kanaye Nagasawa*, 24 P.2d 815, 218 C. 685 (1933).

In *Higgins v. Wasgatt,*[226] a conveyance "to A and his heirs" with a habendum "to A for life," was interpreted by the court to mean only a *life* estate, if, from other parts of the deed, such appears to have been the *intent* of the parties.

However, where the habendum purports to cut down the estate given by the granting clause, and there is nothing from which to determine the intent of the grantor, the granting clause will control.[227]

Where by the premises a fee simple is conveyed to a named grantee, and in the habendum the fee is conveyed to a different grantee, the conveyance in the premises takes effect and the habendum is void.[228]

But,

a grant to A., habendum to A and B, "their heirs and assigns forever" was held to vest a fee in A and B.[229]

Additionally,

a person who was named as one of the parties of the second part did not take any interest under the deed where his name was omitted both from the granting clause and the habendum.[230]

And, the Alabama court said,

If the granting clause does not "specifically define" the estate granted the habendum disclosing an intent to pass less than the fee will be given effect.[231]

---

[226] 34 Me. 305 (1852).

[227] 3 Am. L. Prop., § 12.47.

[228] *Hafner v. Irwin*, 20 N.C. 433, 34 Am. D. 390 (1839).

[229] *McLeod v. Tarrant*, 17 S.E. 773, 39 S.C. 271 (1892), 20 L.R.A. 846.

[230] *Hardie v. Andrews*, 13 N.Y. Civ. Proc. 413 (1888).

[231] *Prudential Ins. Co. v. Karr*, 3 So.2d 409, 241 Ala. 525 (1941).

A construction which requires that an entire clause of a deed should be rejected will be adopted only from unavoidable necessity.[232] The general rule is that, of two repugnant clauses in a deed, the first will prevail,[233] and, in cases of conflict between the granting and habendum clauses, the former will control,[234] especially if the latter is ambiguous.[235]

The habendum may be entirely rejected if repugnant to other clauses of the conveyance,[236] however, even though there is repugnancy, all parts of the deed will be permitted to stand where possible.[237]

The office of the habendum in the deed at common law was to limit, restrain, lessen, enlarge, explain, vary, or qualify, but not totally to contradict or be repugnant to the estate granted in the premises. Although the habendum may be resorted to for the purpose of restraining, lessening, explaining, varying, or qualifying the estate indicated in the grant, it will not be permitted to defeat a clear intention of the grantor expressed in the granting clause.[238]

The reasoning behind this is that a grantor cannot destroy his own grant and, once having granted an estate in his deed, no subsequent clause even in the deed itself can operate to nullify it.[239] General words in the habendum cannot control or govern special words of limitation used in the grant or premises of a deed.[240]

The other important consideration is that the habendum clause is not an essential part of a deed.[241] And, since it is immaterial in what order clauses appear in a deed,[242] conditions, though following the habendum clause, are good in any other place.[243]

---

[232] *City of Alton v. Illinois Transp. Co.*, 12 Ill. 38, 52 Am. Dec. 479 (1850).

[233] *Finney v. Brandon*, 135 N.E. 10, 78 Ind. App. 450 (1922).

[234] *Richards v. Richards*, 110 N.E. 103, 60 Ind. App. 34 (1915).

[235] *In re Schoemaker*, 295 P. 830, 211 C. 457 (1931).

[236] *Moore v. Proctor*, 234 S.W.2d 479 (1950).

[237] *Bailey v. Mullens*, 313 S.W.2d 99 (Tex.Civ.App., 1958).

[238] *Stambaugh v. Stambaugh*, 156 S.W.2d 827, 288 Ky. 491 (1941).

Though the habendum in a deed can never operate as a grant, it may sometimes enlarge or diminish the grant, when it is so worded as to show a clear intention to do so. *Sumner v. Williams*, 8 Mass. 162, 5 Am. Dec. 83 (1811).

[239] *Inhabitants of Town of Canton v. Livermore Falls Trust Co.*, 3 A.2d 429, 136 Me. 103 (1939).

[240] *Hunter v. Patterson*, 44 S.W. 250, 142 Mo. 310 (1898).

[241] *Flynn v. Palmer*, 70 N.W.2d 231, 270 Wis. 43 (1955), see 51 A.L.R. 1000.

[242] *Birk v. First Wichita Nat. Bank of Wichita Falls*, 352 S.W.2d 781 (Tex.Civ.App., 1961).

[243] *Lawe v. Hyde*, 39 Wis. 345 (1876).

FOR FURTHER REFERENCE

72 ALR 410. *Rule that particular description in deed prevails over general description.*

130 ALR 643. *Description of land in deed or mortgage by reference to map, plat, sketch, or diagram as prevailing over description by words.*

134 ALR 1041. *Inconsistency between description of land in instrument conveying same or affecting title thereto and description in another instrument referred to therein.*

37 ALR2d 820. *Written matter as controlling printed matter in construction of deed.*

12 ALR4th 795. *Which of conflicting descriptions in deeds or mortgages of fractional quantity of interest intended to be conveyed prevails.*

REFERENCES

American Jurisprudence 2d. Volume 12. *Boundaries*. Rochester: Lawyers Cooperative Publishing Company.

American Jurisprudence 2d. Volume 23. *Deeds*. Rochester: Lawyers Cooperative Publishing Company.

American Jurisprudence 2d. Volumes 79 & 80. *Wills*. Rochester: Lawyers Cooperative Publishing Company.

Corpus Juris Secundum. Volume 11. *Boundaries*. St. Paul: West Publishing Company.

Corpus Juris Secundum. Volume 26. *Deeds*. St. Paul: West Publishing Company.

Corpus Juris Secundum. Volumes 95, 96 & 97. *Wills*. St. Paul: West Publishing Company.

Hermanson, Knud Everett. *Boundary Retracement Principles and Procedures for Pennsylvania*. The Pennsylvania Society of Land Surveyors. 1986.

Patton, Rufford G., and Carroll G. Patton. *Patton on Land Titles*. Three volumes. St. Paul: West Publishing Company. 1957.

CHAPTER 5

RELATIVE IMPORTANCE OF CONFLICTING ELEMENTS

> By the aid of other facts the weakest may in a given case
> overcome the force of the call of highest dignity.
>
> —*Goodson v. Fitzgerald*
> 90 S.W. 898, 40 Tex. Civ. App. 619 (1905)

Because the area of resolution of conflicting elements is relied upon so heavily in both surveying and title fields, it is given separate treatment in this book. For years, it has been the fall-back rule, not only for resolving conflicts generated by a description but also between conflicting descriptions and between various forms of evidence in the records or on the ground. Because the courts have not viewed it in such manner, the entire area of conflicting elements deserves discussion.

Basically, the general, often-quoted rule is that the priority of conflicting elements is as follows:

- Monuments: natural, controlling, artificial
- Courses*
- Distances*
- Area

---

* In recent years there has been much controversy over which of these should control the other, and there are valid arguments for either one. However, it is the rare case where it comes to just one controlling the other, the usual instance being both of them controlling, or being controlled by, some other element.

In determining boundaries of a tract of land, it is not permissible to disregard any of the calls if they can be applied and harmonized in any reasonable manner.[1] In so doing, conflicts may arise within a description such that it is not possible to harmonize all of the calls. However, the courts have agreed upon a classification of and gradation of calls in a grant, survey, or entry of land, by which their relative importance and weight are to be determined. These rules are not artificial built on mere theory but are the results of human experience. They are not conclusive, imperative, or universal but, like those in Chapter 4, are *rules of construction*, adaptable to circumstances, or only rules of evidence, or merely helpful in determining which of the conflicting calls should be given controlling effect. A call that would defeat the intention of the parties will be rejected regardless of the comparative dignity of the conflicting calls, and, where calls of a higher order are made by mistake, the calls of a lower order may control, as indicating the intention of the grant.[2]

The rules for the order of conflicting description elements are founded upon the principle that those elements are to control in which error is least likely to occur.

Intent of the Parties.   The object of all rules for the establishment of boundaries of their original creation is to ascertain the actual location of the boundary as made at the time of the original creation. The controlling consideration, where there is a conflict as to a boundary, is the intention of the parties, whether express or shown by the surrounding circumstances;[3] in other words, rules for the ascertainment of boundaries are controlling when, under the facts of the particular case, they best enable the court to arrive at the intention of the parties.[4]

Boundaries.   Simply, a boundary is a dividing line between two contiguous estates, or two entities. More explicitly, a boundary is a line or object indicating the limit or farthest extent of a tract of land or territory; a separating or dividing line between countries, states, districts of territory, or tracts of land. The term [boundary] is used to denote the physical object that divides, as well as the line of division itself.[5] Figure 5.1 illustrates various types of boundaries.

---

[1] *Boardman v. Reed*, 6 Pet. (U.S.) 328, 8 L.Ed 415 (1832), *Bostick v. Pernot*, 265 S.W. 356, 165 Ark 581 (1924).

[2] *Miller v. Southland Life Ins. Co.* Civ. App. 68 S.W.2d 558 (Tex., 1934).

[3] The intention that the law provides shall be ascertained and given controlling effect in locating a boundary line is not an intention that the description of the land fails to express, but the intention that it does express. *Ellison v. Humble Oil & Refining Co.*, Tex. Civ. App., 106 S.W.2d 1083.

[4] *Gill v. Peterson*, Civ. App. 868 S.W.2d 629, reversing Peterson v. Gill, Civ.App. 51 S.W.2d 1057 (Tex., 1932).

[5] *Board of Com'rs of Crook County v. Board of Com'rs. Of Sheridan County*, 100 P. 659, 17 Wyo. 424 (1908).

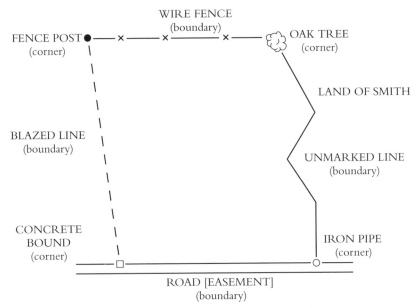

Figure 5.1   Boundaries of a parcel of land.

Corners and Marked Lines. Calls in a survey for established corners and marked lines, being based on monuments, take precedence over calls such as those for courses and distances, over calls for adjoiners, and over field notes, since established corners and marked lines represent the survey as it was actually made, while courses and distances and field notes are merely descriptions of the acts done by the surveyors.[6] The rule is therefore founded upon the principle that those particulars are to be regarded in which error is least likely to occur. This rule, that a marked line controls a call for course and distance will not apply, however, unless the line is so connected with the deed either through intrinsic or extrinsic evidence that a presumption is created that the grantor intends to adopt it.[7]

Marked lines also control distances indicated on survey plats. The designation that a line is [so many] feet in a given direction from a starting point will control only in the absence of direct evidence actually concerning the line marked by the surveyor, and a call must surrender to judicially determined corners in litigation between the same parties or their privies.[8]

---

[6] Lines of sections and parts of sections lie just where the government surveyor lined them out on the face of the earth and are to be determined by the monuments in the field. *Verdi Development Co. v. Dono-Han Mining Co.*, 296 P.2d 429, 141 Cal. App.2d 149 (1956).

[7] *Elliott v. Jefferson*, 45 S.E. 558, 133 N.C. 207 (1903), *McDowell v. Carothers*, 146 P. 800, 75 Or 126 (1915).

[8] Ibid.

A corner established by government surveyors with reference to which patents have been issued should be accepted as the true corner and given controlling effect without regard to whether or not it was originally located with mathematical certainty. It is a general rule that *the original corners as established by government surveyors, if the corners can be found established, if that can be determined or the places where they were originally are conclusive on all persons owning or claiming to hold with reference to such surveys and the monuments placed by the original surveyors, without regard to whether or not they were correctly located.* An established corner between sections of land must govern even as against other surveys or evidences tending to prove that such corner should be located elsewhere.

Accordingly, an established corner will control the field notes of the survey where a discrepancy exists between the field notes and such corner as is marked on the ground. Where, however, a corner cannot be located by clear and satisfactory evidence, the field notes may then be taken as the next best evidence as to the location of boundaries.[9]

Where the lines of a survey have been run, and can be found, they constitute the true boundaries, and must not be departed from, or made to yield to, any less certain and definite matter of description or identity.[10] The often-quoted statement is essential, that where the footsteps of the surveyor can be followed with certainty on the ground, all conflicts in calls or ambiguities in field notes are subordinated to the actual location; the surveyor's actual survey of the land, where it can be ascertained with certainty, is conclusive. So, where a conveyance is about to be made, and the parties go upon the land and have the line marked and surveyed, the line so fixed and intended will prevail over any inconsistent description in the conveyance.[11]

Control over Objects.   Ordinarily, lines actually surveyed and marked will, when found, control calls for natural or other fixed objects. However, it has been held that natural objects having a definite and distinct identification are to be preferred to calls for marked lines.[12]

Control over Maps, Plats, and Field Notes.   Where there is discrepancy between lines actually marked or surveyed and those called for on maps, plats, or in field notes, the lines marked or surveyed will control, provided, however, that the lines are in fact definitely located and fixed.[13]

---

[9] Ibid.

[10] The position of lines,corners, fences, roads, and well-marked outlines, which have been recognized by the people during the years, must control. *Texas Co. v. McMillan*, D.C. Tex., 13 F.Supp. 407.

[11] The rule was adopted for the sole purpose of executing the intention of the parties at the time the deed is delivered. *Watson v. Pierce*, 124 S.E. 838, 188 N.C. 430 (1924).

[12] *Lillis v. Urrutia*, 99 P. 992, 9 Cal.App. 557 (1908); *Gray v. Coleman*, 88 S.E. 489, 171 N.C. 244.

[13] Ibid.

In the case of *Susi v. Davis*,[14] the court stated,

> The plan is a picture, the survey the substance. The plan may be all wrong, but that does not matter if the actual survey can be shown.

Control over Calls for Adjoiners. Generally, lines marked on the ground for the survey, or adopted by the surveyor, control calls for adjoiners. However, when lines have never been surveyed, or, if surveyed, their location upon the ground cannot be ascertained, resort may be made to the lines of adjacent lots to determine their location. In the case of *Booker v. Hart*,[15] the court said,

> Known calls in other surveys can be appealed to, in order to locate a tract of land, only in the absence of other means of identification.

There are several cases that state that marked lines are superior when in conflict with an adjoiner call. Metes and bounds of a tract that are represented by visible marked lines cannot be extended beyond those lines although a boundary may be called for which lie beyond them. This would be especially true where the lines are merely designated as being near the land of a named person.[16]

Control over Courses and Distances. To have lines marked or surveyed control, they must have been established at or before the time of the grant. Further, they must either be made for, or be adopted by the grantor in his deed, or called for in the deed, and they must be identified on the ground. Where there is a claim that the lines do not agree with the courses and distances, the evidence of their actual location should be so clear and satisfactory as to establish that fact to the entire satisfaction of the court, and to place beyond question the actual location of the line.[17]

[14] 177 A. 610, 133 Me. 354 (1935), see 97 A.L.R. 1222.

[15] 12 S.W. 16, 77 Tex. 146 (1889); *McAninch v. Freeman*, 4 S.W. 369, 69 Tex. 445 (1887); *Taft v. Ward*, Tex. Civ. App. 124 S.W. 437 (1909).

[16] *Hall v. Tanner*, 4 Pa. 244 (1846), *Matheny v. Allen*, 60 S.E. 407, 63 W.Va. 443 (1907).

[17] *May's Lessee v. Sanders*, 6 J.J.Marsh. (Ky.) 350 ((1831), *Galloway v. Brown*, 16 Ohio 428 (1866); *Murphy v. Campbell*, 4 Pa. 480 (1846); *Missouri, etc., R. Co. v. Anderson*, 81 S.W. 781, 36 Tex.Civ.App. 121(1904); *Boynton Lumber Co. v. Houston Oil Co. of Texas*, Civ. App., 189 S.W. 749 (1916); *King v. Brigham*, 25 P. 150, 19 Or. 560 (1890).

The court stated in *Andrews v. Wheeler*:[18]

It has ever been held that the marks on the ground constitute the survey; that the courses and distances are only evidence of the survey.

Control a Call for Area. Lines marked or surveyed, when definitely located, control calls for quantity.[19] In the Louisiana case of *Hunter v. Forrest*,[20] the court noted,

One who acquires title to property in good faith and purchases according to certain definite and fixed boundaries, which can be located, "is entitled to all the land between such boundaries, although it gives him a greater quantity than that called for in his title."

Monuments. When relating to land, a monument is some tangible landmark established to indicate a boundary.[21] According to some decisions, in order for an object to be considered as a monument, it must be visible, permanent and stable, and have definite location, independent of measurements.

Monuments are of the highest value in a question of boundary.[22] Generally, it is considered that a monument must be mentioned in the conveyance, however, in the case of *Nelson v. Lineker*,[23] the court stated,

Where a stone was marked with plaintiff's initials and placed as a corner of the land at the time the survey was made, with the view and purpose of making the deed under which he claimed, and the deed was made, intending to convey the land so surveyed, the stone is the proper boundary, whether called for in the deed or not.

When monuments that designate the boundaries of land are obliterated or disturbed, they are to be relocated by the field notes and plats of the original survey. If a lost monument cannot be relocated properly, resort must be made to other calls.[24]

---

[18] 103 P. 144, 10 Cal. App. 614 (1909).

[19] *Sims v. Capper*, 112 S.E. 676, 133 Va. 278 (1922).

[20] 164 So. 163, 183 La. 434 (1935).

[21] A "monument" is any physical object on ground that helps to establish location of line called for; it may be either natural or artificial, and may be a tree, stone, stake, pipe, or the like. *Delphey v. Savage*, 177 A.2d 249, 227 Md. 373 (1961).

[22] *Post v. Wilkes-Barre Connecting R.R.*, 133 A. 377, 286 Pa. 273 (1926).

[23] 90 S.E. 251, 172 N.C. 279 (1916).

[24] *Hiltscher v. Wagner*, 273 P. 590, 96 Cal.App. 66 (1928); *Nelson v. Lineker*, 90 S.E. 251, 172 N.C. 279 (1916).

A very significant case regarding boundaries is *Diehl v. Zanger*,[25] wherein the court said the following:

> Where the monuments of the original survey have disappeared, the question where they were located is to be determined by the practical location of the lines, made at the time when the original monuments were presumably in existence, and probably well known.
>
> A resurvey, made after monuments of the original survey have disappeared, is for the purpose of determining where they were, and not where they ought to have been.

Mistakes in courses and distances are more probable and frequent than in calls for objects, which are capable of being clearly designated and accurately described. The accuracy of courses and distances depends upon the skill and experience of the surveyor to the extent where some courts have stated that the superiority of calls for monuments will be affirmed regardless of how great is the discrepancy shown by courses and distances.[26] However, in the Pennsylvania case of *Howarth v. Miller*,[27] the court said,

> Though monuments are very important in determining questions of boundary, where the monuments are doubtful, resort will be had to courses, distances, and quantity.

Control of Natural or Permanent Objects. A *natural monument* is an object, permanent in character, which is found on the land as it was placed by nature. Examples are lakes, ponds and rivers, trees, rocks and boulders, and the like.

Except where there are lines actually marked and surveyed, calls for natural or permanent monuments or objects prevail, and control other and conflicting calls or descriptions in the determination of boundaries, so long as the objects can be definitely located on the ground, unless a different intention of the parties is indicated, or the call for such monument is clearly erroneous or less certain.[28] The reason is that there is greater certainty and less likelihood of a mistake in calls of this nature.[29]

This rule, like any of the other rules, is not imperative, absolute, or inflexible, nor is it of universal application. It is a rule of construction, adaptable to circumstances, or one of evidence. Therefore, a call for a natural monument will yield to other and conflicting calls when application of the general rule would be contrary to the intention of

---

[25] 39 Mich. 601 (1878).

[26] *White v. Luning,* 93 U.S. 514, 23 L.Ed. 938 (1876); *Matheny v. Allen,* 60 S.E. 407, 63 W.Va. 443 (1908); *Higuera v. United States,* 5 Wall (U.S.) 827, 18 L.Ed. 469 (1864). See also *Riley v. Griffin,* 16 Ga. 141 (1854) and *Tompkins v. Vintroux,* 3 W.Va. 148 (1869).

[27] 115 A.2d 222, 382 Pa. 419 (1955).

[28] *Howarth v. Miller,* 115 A.2d 222, 382 Pa. 419 (1955).

[29] *Vermont Marble Co. v. Eastman,* 101 A. 151, 91 Vt. 425 (1915); *Fisher v. Pittsburgh Coal Co.,* 29 Pa. Dist. 885 (1920).

the parties, or when the call for the natural monument is clearly erroneous, or when it does not appear to be the more certain call.[30] Two such examples follow.

In the case of *Woodbury v. Venia*,[31] the Michigan court stated,

> The rule that monuments control courses and distances does not apply where the monuments were not placed by some authorized person, since then there is no sufficient evidence of the proper location of the monuments.

And in the case of *White v. Luning*,[32] the court said,

> The general rule that monuments, natural or artificial, referred to in a deed control on its construction, rather than courses and distances, is a flexible one. It yields whenever, taking all the particulars of the deed together, it would be absurd to apply it. For instance, if the rejection of a call for a monument would reconcile other parts of the description and leave enough to identify the land and render certain what was intended to be conveyed it would certainly be absurd to retain the false call, and thus defeat the conveyance.

Natural monuments take precedence over artificial monuments in case of conflict when the natural monuments have definite identification, especially if the artificial monuments are at remote distances. However, the rule is otherwise if it is clear that the artificial monument is the more reliable, if there was a mistake in the call, or if the artificial monuments were established with the express purpose of executing a deed in conformity therewith.[33]

Natural monuments will control maps, plats, and field notes as well as calls for adjoining lines or owners, except where there is a contrary intention, or a mistake.

---

[30] *U.S. v. Redondo Devbelopment Co.*, N.M., 254 F. 656, 166 C.C.A. 154, *Nattin v. Glassell*, 100 So. 609, 156 La. 423 (1924), *Ewart v. Squire*, W.Va., 239 F. 34, 152 C.C.A. 84, *Barrataria Land Co. v. Louisiana Meadows Co.*, 84 So. 334, 146 La. 999 (1920), *W.T. Carter & Bro. v. Collins*, Tex.Civ.App., 192 S.W. 316, error refused (1916), *Kelekolio v. Onomea Sugar Co.*, 29 Hawaii 130 (1926), *Darlington v. Lane*, 46 App.D.C. 465, reversed on other grounds, *Lane v. Darlington*, 39 S.Ct. 299, 249 U.S. 331, 63 L.Ed 629 (1918).

All of these ordinary rules [that natural monuments control], however, relating to conflicts between areas, courses and distances, and natural monuments, and between two or more natural monuments, are susceptible of exceptions and need not be followed or may be reversed when under the circumstances and the evidence the contrary rule better serves to carry out the intention of the parties as to the location or extent of the land described in the instrument under consideration. *Kelekolio v. Onomea Sugar Co.*, 29 Hawaii 130 (1926).

[31] 72 N.W. 189, 114 Mich. 251 (1897).

[32] 93 U.S. 514, 23 L.Ed 938 (1876).

[33] *Felder v. Bonnett*, 2 McMull (S.C.) 44, 37 Am.D. 545 (1841); *Schnackenberg v. State*, Civ.App., 229 S.W. 934 (Tex., 1921); *Gray v. Coleman*, 88 S.E. 489, 171 N.C. 344 (1916).

However, it has been decided that an adjoining line that is well known and well established will not yield to a conflicting call for a natural object.[34]

Generally, calls for natural objects will control calls for metes and bounds and calls for course and distance, especially if the distance is followed by the words "more or less." It does not matter that, by relying on the called-for monuments, a different area results or that the courses and distances differ from those called for in the description. The reason for the rule is that the monuments are more certain than computation of course and distance, and are less subject to errors in calls.[35]

The court stated in the case of *Lessee of McCoy v. Galloway*,[36]

> When a natural object is distinctly called for, and satisfactorily proved, it becomes a landmark not to be rejected, because the certainty which it affords, excludes the probability of mistake, while course and distance, depending for their correctness on a great variety of circumstances, are constantly liable to be incorrect. The difference in the instruments used, and in the care of surveyors and their assistants, must lead to different results. Hence it is, that this rule has been established.

This rule, like many others, is not inflexible or absolute, and is subject to exceptions. Defeat of the parties' intention, absurd results, or mistaken calls are all reasons for exceptions to the rule.[37]

As an example, the court said in *Day v. R.E. Wood Lumber Co.*,[38]

> When the court has nothing before it but the inconsistent calls, the call for the monument is allowed to prevail because it is deemed to be a more definite and certain expression of intention than the call for course and distance. When other circumstances appear, conclusively showing the call for the monument does not express the real intention and will not effectuate the plain purpose of the instrument, and that the calls generally regarded as inferior and less certain do express the real intent and purpose, the case falls within an exception to the rule, and the latter calls are allowed to prevail.

To control courses and distances, natural monuments must be referred to or designated in the title papers or the survey. Therefore, evidence of the location of

---

[34] *State v. Mounts*, 150 S.E. 513, 108 W.Va. 53 (1929); *Alden v. Pinney*, 12 Fla. 348, *Erskine v. Moulton*, 66 Me. 276 (1877), *Lincoln v. Wilder*, 29 Me. 169 (1848); *New York, etc, Land Co. v. Thomson*, 17 S.W. 920, 83 Tex. 169 (1892).

[35] *Johnson v. State*, 244 S.W. 609, 92 Tex.Cr. 418 (1922), *Wilson v. Connor*, 122 So. 404, 219 Ala. 344 (1929); *Hutchinson v. Little Four Oil & Gas Co.*, 119 A. 534, 275 Pa. 380 (1923), *Neill v. Ward*, 153 A. 219, 103 Vt. 117 (1930).

[36] 3 Ohio 282, 17 Am. Dec. 591 (1827).

[37] The different calls in a deed ought to be taken together in determining the boundaries of the property conveyed thereby; a call for a natural boundary may be controlled by the other calls if it appears that such call was inserted through inadvertence or mistake. *Barclay v. Howell*, 6 Pet (US) 498, 8 L.Ed. 477 (1832).

[38] 88 S.E. 452, 78 W.Va. 19 (1916).

monuments not called for in the title documents cannot be considered to control courses or distances designated in the title papers.[39]

A statement of the quantity of land supposed to be conveyed, as part of the description, will yield to natural objects called for in the conveyance, so long as they are specified in the deed, and are fixed and certain. It is immaterial that the objects called for define a different amount of land than that called for. A call for quantity, however, can be an important consideration if the evidence of the monuments or their location is in conflict, or if the boundaries cannot be located with reasonable certainty by the call for the monument. Quantity may also control if it better serves the intent of the parties or may be used to decide which of two conflicting monuments is the better choice.[40]

Control of Artificial Monuments of Marks. An *artificial monument* is a landmark erected by the hand of man.[41] Except for natural objects, calls for artificial monuments and objects ordinarily control other and conflicting calls; however, this rule is not inflexible or absolute. It does not apply to a false or mistaken call or to monuments not mentioned in the conveyance, but a monument not existing at the time a deed is made, and afterward set by the parties with intent to conform to the deed, will control.[42]

The public lands of the western territories of the United States were, under acts of Congress, surveyed and divided into townships, sections, and subdivisions thereof, and the boundaries of these sections and the quarter and half sections were marked by artificial monuments, which constituted the corners of such tracts. The original stakes and monuments placed by the government surveyors govern in determining boundaries, and mark the true corner when their location can be found or established, regardless of the fact that resurveys may show that such monuments should have been located elsewhere.[43]

---

[39] *Trimmer v. Martin*, 126 S.E. 217, 141 Va. 252 (1925), *Kenmont Coal Co. v. Combs*, 48 S.W.2d 9, 243 Ky. 328 (1932); *Trimmer v. Martin*, Ibid.

[40] *Snodgrass v. Snodgrass*, 101 So. 837, 212 Ala. 74 (1924); *Williams v. Bryan*, 73 So. 372, 197 Ala. 675 (1916); *Moranda v. Mapes*, 280 P. 713, 100 Cal. App. 629 (1929); *Alexander v. Hill*, 108 S.W. 225, 32 Ky.L. 1147 (1908); *Moranda v. Mapes*, Ibid.; *Benavides v. State*, Civ.App., 214 S.W. 568 (Tex., 1819); *Kelekolio v. Onomea Sugar Co.*, 29 Hawaii 130 (1926).

[41] *Timme v. Squires*, 225 N.W. 825, 199 Wis. 178 (1929); *Parran v. Wilson*, 154 A. 449, 160 Md. 604 (1931).

[42] *Pritchard v Rebori*, 186 S.W. 121, 135 Tenn. 328 (1916); *Cates v. Reynolds*, 228 S.W. 695, 143 Tenn. 667 (1920), *Talbot v. Smith*, 107 P. 480, 108 P. 125, 56 Or. 117 (1910); *Bemis v. Bradley*, 139 A. 593, 126 Me. 462 (1927), 69 ALR 1399, *Martin v. Nance*, 3 Head, Tenn. 649 (1859).

[43] *Myrick v. Peet*, 180 P. 574, 56 Mont. 13 (1919); *Oster v. Muhlhauser*, 249 N.W. 777, 63 N.D. 671 (1933); *Iverson v. Johnson*, 239 N.W. 757, 59 S.D. 313 (1931), *Lawson v. Viola Twp.*, 210 N.W. 979, 50 S.D. 555 (1926).

Government corners on township lines are primary, while subdivision corners are secondary. *Morse v. Breen*, 182 P. 887, 66 Colo. 398.

In *McClintock v. Rogers*, 11 Ill. 279 (1849), the court stated that " monuments found at the two extremes of a township line are entitled to no more controlling influence than any two intermediate monuments."

Artificial monuments generally control plats, although there have been decisions to the contrary. For a monument to control, it must be mentioned in the conveyance.[44] Such mention may be by reference to another instrument, wherein the actual mention is made. If the monument called for is found to be lost or obliterated, so long as the place where it was can be established, that place will control. If neither can be located, then the field notes, or a reproduction of the original survey according to the notes, will prevail.[45]

Calls for artificial objects will control calls for adjoiners, metes and bounds, courses and distances,[46] and quantity. The reason for the rule is the same as that of natural monuments, that there is less possibility of mistake in a call for an artificial monument than in a call for course and distance.[47]

In the case of government corners, some cases are noteworthy:

In *Mathews v. Parker*,[48] the court said that, "center of a section, although not a physical monument, 'is a point capable of being mathematically ascertained, thus constituting it, in a legal sense, a monument call of the description.'"

Also, in *Cordell v. Sanders*,[49] it was stated that, "quarter-section corners established by the government control, regardless of mistakes in length of lines between section corners."

However, like other rules of construction, this rule will not be followed where the parties' intention was otherwise, where a monument call was included by mistake,

---

[44] Markers not mentioned in the description have not the authentication qualifying them as monuments, in the true sense, and so have no force to dispute courses and distances set out in the deed. *Haklits v. Oldenburg*, 201 A.2d 690, 124 Vt. 199 (1964).

[45] *Cooper v. Quade*, 182 N.W. 798, 191 Iowa 461(1921), *Harris v. Harms*, 181 N.W. 158, 105 Neb. 375 (1920), *Iverson v. Johnson*, 239 N.W. 757, 59 S.D. 313 (1931).

[46] It has been said that "It is very generally recognized that a call for 'stakes' in the descriptive terms of a deed is not sufficient to control course and distance unless the deed itself affords data from which the term 'stake' can be given a definite placing . . . . They are usually of such a transitory nature, so liable to be destroyed or in some way removed, by chance or otherwise, that they are not regarded as monuments of boundary at all, but are considered only as imaginary points to be fixed and determined by a correct survey of course and distance, if such calls appear also in the deed." *Nelson v. Lineker*, 90 S.E. 251, 172 N.C. 279 (1916).

[47] Why will a call for an "artificial object" prevail, as a general rule, over a call for distance? Because an artificial object is visible, and when a surveyor says that he ran to such object he was probably not mistaken, and if such object can be found and identified, it indicates the point to which the surveyor arrived, though the distance called for may not reach such point. *Findlay v. State*, Civ. App. 238 S.W. 956 (Tex., 1921).

[48] 299 P. 354, 163 Wash. 10 (1931).

[49] 52 S.W.2d 834, 331 Mo. 84 (1932).

or where a call for distance appears to be more reliable. The court said in *State v. Stanolind*,[50]

> Where a surveyor, in accordance with instructions from the land commissioner to take up excess in block which was system of connected surveys made by the same surveyor at the same time, established corners by appropriating to each survey certain excess and returned corrected field notes, distance calls prevailed over field note calls for stakes and mounds in locating the boundary.

In addition, the court stated in *Pritchard v. Rebori*,[51]

> It is not true . . . that there is such magic in a monument called for that it will be made to control in construction invariably. If it controls, it is only because it is to be regarded as more certain than course or distance. If it should in a given case be less certain, the rule would fall with the reason for it and the monument would yield to the course and distance. . . .

Also, the rule has been stated to be applied with greater force when large boundaries in remote areas are involved, since mistakes in the use of a surveyor's chain may easily occur, but there is less chance of this happening in towns or developed areas. Such a consideration applies more so with earlier descriptions, or measurements, than with more recent ones, and need be considered in boundary retracement work and in construing earlier descriptions, or recent descriptions based on, or related to, earlier ones.[52]

In order for an artificial monument to control, it must have been actually placed as a boundary, and must be found or located, identified, or established, or its existence definitely determined; or the original location of a monument that cannot itself be found may be properly established or accounted for, so as to control course and distance.[53] In *Myrick v. Peet*,[54] the Montana court stated,

> All means of ascertaining the location of lost monuments must first be exhausted before courses and distances can be used to determine a boundary.

---

[50] 96 S.W.2d 297 (Tex.Civ.App. , 1936) .

[51] 186 S.W. 121, 135 Tenn. 328 (1916).

[52] Where the lines are so short as evidently to be susceptible of entire accuracy in their measurement, and are defined in such a manner as to indicate an exercise of care in describing the premises, such a description is regarded with great confidence as a means of ascertaining what is intended to be conveyed. *Pritchard v. Rebori*, 186 S.W. 121, 135 Tenn. 328 (1916).

[53] *Nelson v. Lineker*, 90 S.E. 251, 172 N.C. 279 (1916), *Polk v. Reinhard*, 193 S.W. 687 (Tex. Civ.App., 1917); *Holmes v. Barrett*, 169 N.E. 509, 269 Mass. 497 (1929); *Cordell v. Sanders*, 52 S.W.2d 834, 331 Mo. 84 (1932); *Lawson v. Viola Twp.*, 210 N.W. 979, 50 S.D. 555 (1926); *Oliver v. Muncy*, 89 S.W.2d 617, 262 Ky. 164 (1936); *Finley v. Spinks*, 8 La.App. 769 (1928).

[54] 180 P. 574, 56 Mont. 13 (1919).

In controlling quantity, or area, of land stated when a conflict exists with artificial monuments, the latter usually control, but in order to do so the monument must be fixed and certain. If the monument has been actually placed as a boundary, it may even control although appearing to have been located in error. Referring once again to *Myrick v. Peet*,[55] the court also said,

> The question is not whether the monuments were correctly placed, but whether they were placed by authority.

A call for quantity may become an important fact to be considered, however, where there is substantial conflict in the evidence as to the location of the monuments, and in some circumstances may even become the controlling factor in determining the boundary. Additional examples of this will be found in the discussion of quantity, following.

Where there are two conflicting artificial monuments, the one corresponding to the calls for course and distance will control. In addition, in several cases the courts decided that a boundary will be determined by a course from a known artificial monument rather than by a reestablishment of a lost monument by courses and distances, which would conflict therewith.[56]

Control of Maps, Plats, and Field Notes. One of the objects of filing maps, plats, plans, and field notes is to avoid the necessity of encumbering grants and deeds with lengthy descriptions of the land to be granted or conveyed. Accordingly, when a patent or deed refers to such a map or plat for a more particular description of the premises or describes the land as a certain lot or parcel as shown thereon, the map or plat becomes a part of the instrument and will assist the description therein. In locating land upon the ground from the calls and descriptions in the map, plat, or field notes referred to, the same primary rules apply as exist in the locating of calls and descriptions in a deed containing no such reference; that is, the various calls are given the same order of preference. Thus, where the lines have in fact been located and designated by monuments and there is a discrepancy between the calls for these monuments and courses and distances shown by a plan referred to in the conveyance, the normal rule as to the controlling effect of calls for monuments will be followed. Where natural monuments are designated on a plat referred to in a conveyance but no survey is made upon the ground, the same rule is followed and the tract must be considered to be run by the monuments without regard for other elements of description, even though neither the certificate of survey nor the conveyance refers

---

[55] Ibid.

[56] *Zeibold v. Foster*, 24 S.W. 155, 118 Mo. 349 (1893); *U.S. v. State Inv. Col.*, C.C.A.N.M., 285 F. 128, affirmed 44 S.Ct. 289, 264 U.S. 206, 68 L.Ed. 639 (1924).

specifically to the monuments. If no survey has been made or lines drawn with reference to monuments, the length of line given on a plan, measured accurately according to the scale in which the plan is drawn, may be resorted to in determining the true location of the boundaries of the lots. In case of a conflict, the declaration of distance shown by a plan or plat is to be used in preference to distances ascertained by measuring lines on the plan according to scale.[57]

Plats, maps, and field notes[58] fall into two categories when considering their dignity.[59] Those referenced in a grant or conveyance are generally regarded as incorporated, as though they were written into the instrument and as furnishing the true description of the boundaries. In some cases, they have been held as having the same standing as monuments. However, in some circumstances they have been treated as controlled by lines marked and surveyed, and by natural and artificial monuments. Whichever category will rule will depend on the circumstances, or particular facts of the case, because the description that best identifies the land according to the intent of the parties is the one that must control. The following two examples illustrate court rulings on this subject.

Two examples illustrate this. In the case of *Bower v. Earl*[60] the Michigan court stated:

> Since a description referring to a plat is governed, as to its boundaries, by the plat, calls for the points of the compass must be determined in accordance with the lines and angles as shown on the plat rather than in accordance with the true meridian.

---

[57] *Cragin v. Powell*, 128 U.S. 691, 32 L.Ed. 566, 9 S.Ct. 203, *Hoffman v. Van Duzee*, 65 P.2d 1330, 19 Cal.App.2d 577 (1937); *United States v. Sutter*, 21 How (U.S.) 170, 16 L.Ed. 119 (1858), *M'Iver v. Walker*, 9 Cranch. (U.S.) 173, 3 L.Ed. 694 (1815), *Rowell v. Weinemann*, 93 N.W. 279, 119 Iowa 256 (1903); *Heaton v. Hodges*, 14 Me. 66, *Ripley v. Berry*, 5 Me. 24 (1827), *Anderson-Prichard Oil Corp. v. Keyokla Oil Co.*, 299 P. 850, 149 Okla. 262 (1931); *Pereles v. Gross*, 105 N.W. 217, 126 Wis. 122 (1905).

[58] Plat: A map, or representation on paper, of a piece of land subdivided into lots, with streets, alleys, and so forth, usually drawn to a scale. *McDaniel v. Mace*, 47 Iowa 509 (1877); *Burke v. McCown*, 47 P. 67, 115 Cal. 481 (1896).

A representation of the earth's surface, or of some portion of it, showing the relative position of the parts represented, usually on a flat surface. *Banker v. Caldwell*, 3 Minn. 94 (Gil. 55) (1859).

Plan: A map, chart, or design; being a delineation or projection on a plane surface of the ground lines of a house, farm, street, city, and so forth, reduced in absolute length, but preserving their relative positions and proportion. *Jenney v. Des Moines*, 72 N.W. 550, 103 Iowa 347 (1897); *Wetherill v. Pennsylvania R. Co.*, 45 A. 658, 195 Pa. 156 (1900).

See Chapter 10 for further definitions and comparisons of these and related terms.

[59] A "map" is a picture of a survey, "field notes" constitute a description thereof, and the "survey" is the substance and consists of the actual acts of the surveyor, and, if existing established monuments are on the ground evidencing such acts, such monuments control because they are the best evidence of what the surveyor actually did in making the survey and are part at least of what the surveyor did. *Outlaw v. Gulf Oil Corp.*, 137 S.W.2d 787 (Tex., 1940), 150 S.W.2d 777, 136 Tex. 281 (1941).

[60] 18 Mich. 367 (1869).

And, in the case of *Mayo v. Mazeaux*,[61] the California court said:

Degrees of latitude and other imaginary lines referred to in a description will be dis-carded as less certain and reliable than the map to which the description also refers.

Control over adjoiner calls. A description referencing a map or plan will usually control a call for an adjoiner, unless the latter appears more reliable, however if the map is not referenced, it may not control.

In the case of *Atkinson v. Anderson*,[62] the court said,

A call for an adjacent tract as a boundary must follow that line however much it may meander, even though the surveyor's plat showed it to be a straight line, it appearing that this line was not actually surveyed by him; it is the boundary as it really exists which must govern.

In controlling metes and bounds, the calls that are in best agreement with the intent of the parties will be the controlling ones. Ordinarily, a call for a lot and block number on a plan will control metes and bounds, except when it is incorrect.[63]

Map, plat, and field note references will also control courses and distances as well as quantity, unless shown to be erroneous or in disagreement with the intent of the parties. A New York court[64] stated,

They [map and notes of survey] are products of technical engineering skill, and practical knowledge gained by laying out the many separate parcels and defining their relations to each other and to the whole section. Therefore, departure from their terms is tolerable only in instances where boundary line and measurements are so inconsistent that one or the other must yield something to perfect or to approximate the identity of the grant.

Control of Calls for Adjoiners. Calls for adjoiners, in the absence of calls for superior, or controlling, monuments, generally control other calls when conflicts

---

[61] 38 Cal. 442 (1869).

[62] 3 McCord (S.C.), 137 (1825).

[63] Description by dimensions, referred to in the deed as a description by metes and bounds is controlled by a description by lots and reference to a map. Mechler v. Dehn, 191 N.Y.S. 650, 117 Misc. 591 (1922).

[64] *In re Tremont Housing Corporation*, 242 N.Y.S. 128, 229 App. Div. 739, 137 Misc. 141 (1930).

occur,[65] since when adjoiners are certain they are considered monuments of the highest dignity.[66]

In *Fagan v. Walters*, the Washington court stated,

> Authorities are numerous which hold that, where the call for a boundary in another deed or for the boundary of another tract is expressed as 'by such a line' or 'by the north line of such a tract,' the line or boundary referred to is locative and fixes the boundary definitely.

In *McCausland v. York*,[67] a Maine case, the court stated that whether the adjoiners' deed was recorded or not was immaterial.

> Where deeds to owner of lot and her predecessor in title bounded conveyance by land conveyed to owner of adjacent lot, that latter's deed was unrecorded so as not to give notice that it included a triangular piece of land between lots is immaterial on the issue of ownership of the triangle, since the land referred to was the controlling monument.

Calls for adjoiners, even though not marked, generally control calls for courses and distances.[68] The reason is that a call for an adjoining tract, when proven, is more certain than one for mere course or distance.

However, to control, an adjoining parcel must be called for, must have existed at the time of the conveyance, and must be clearly established and identified and accurately located. Adjoiners may yield to courses and distances where not definitely ascertainable, when they are the result of a mistake or where such interpretation appears to be the intent of the parties.[69]

---

[65] *Fagan v. Walters*, 197 P. 635, 115 Wash. 454 (1921).

[66] A line called for is quite as controlling as any natural or artificial boundary. *Parran v. Wilson*, 154 A. 449, 160 Md. 604 (1931); *Ramsay v. Butler, Purdum & Co.*, 129 A. 650, 148 Md. 438 (1925).

[67] 174 A. 383, 133 Me. 115 (1934).

[68] Where an established line of a senior survey is called for, or a line of a senior survey that may be definitely ascertained and located from other established corners of the same or adjacent established surveys, such call should be given the dignity of an artificial object and control calls for course and distance. *Phillips Petroleum Co. v. State*, Tex.Civ.App., 63 S.W.2d 737.

Where the senior line is established, the junior begins, and calls for distance of the junior must yield accordingly, since a call for adjoiner is like a call for a natural or artificial object. *Kirby Lumber Co. v. Gibbs Bros. & Co.*, 14 S.W.2d 1013 (Tex.Com.App., 1911).

[69] *Love v. Jones*, 138 S.W. 1128 (Tex.Civ.App.), *Waters v. Dennis Simmons Lumber Co.*, 72 S.E. 284, 154 N.C. 232 (1911); *Land v. Dunn*, Tex. Civ.App. 241 S.W. 580, *Pritchard v. Rebori*, 186 S.W. 121, 135 Tenn. 328 (1916).

Control of Metes and Bounds. Metes and bounds[70] will control courses and distances, and, unless there is a covenant,[71] will control quantity as well.[72] In the New Hampshire case of *Cotton v. Cotton,*[73] the court stated,

> Construing deed most strongly against grantor, the clause 'said piece of land is to contain one acre by measure' may be understood as warranting at least one acre to grantee, but grantee gets all that is conveyed to him, the definite description fixing the disputed boundary not being limited by the quantity clause, in the absence of competent evidence establishing intention to grant no more.

Concerning government surveys, the Alabama court stated,[74]

> Where the land conveyed is described by the government surveys, and as containing so many acres more or less, it is a sale by metes and bounds; and, in the absence of fraud, the actual quantity, whether more or less than the estimate at the purchase, will not avail either party.

So, when a conveyance describes land by metes and bounds, all the land contained within the described boundaries will pass, regardless of any discrepancy in quantity.

The *presumption* is that the quantity was computed from the description, not that the parties intended to convey the quantity mentioned at all events.[75]

Control of Directions and Distances. Courses and distances have generally been held to be unreliable, but they will govern when no superior calls are available.[76]

> It is only in the absence of all monuments and marks upon the ground and in the total failure of evidence to supply them that recourse can be had to calls for courses and distances as authoritative.[77]

---

[70] Metes and bounds mean the boundary lines or limits of a tract. *People v. Guthrie*, 46 Ill.App. 124 (1891), reversed on other grounds 38 N.E. 549, 149 Ill. 360 (1894), *Moore v. Walsh*, 93 A. 355, 37 R.I. 436 (1915).

[71] An additional statement as to the quantity conveyed will be considered as descriptive and not as a covenant of quantity. *Ellenberg v. Barksdale*, 97 So. 54, 210 Ala. 11 (1923); *Smith v. Hartung*, 166 A. 168, 110 N.J.Law 543 (1933).

[72] It is a familiar law that a designation of acreage must yield to definite boundaries. *Rose v. Agee*, 104 S.E. 827, 128 Va. 502 (1920).

[73] 112 A. 245, 79 N.H. 507 (1920).

[74] *Dozier v. Duffe*, 1 Ala. 320 (1840).

[75] *Thompson v. Bracken*, Tex.Civ.App., 93 S.W.2d 614 (1936).

[76] In determining correct boundaries of land, it is only when there are neither monuments nor calls for adjoiners that courses and distances govern. *Will v. Piper*, 134 A.2d 41, 184 Pa. Super. 313 (1957).

[77] *M'Iver v. Walker*, 4 Wheat. (U.S.) 444, 4 L.Ed. 611(1819), *M'Iver v. Walker*, 9 Cranch. (U.S.) 173, 3 L.Ed. 694 (1815), *Bryan v. Beckley*, Litt.Sel.Cas. (Ky) 91 (1809), *Budd v. Brooke*, 3 Gill (Md) 198 (1845), *Collins v. Clough*, 71 A. 1077, 222 Pa. 472 (1909).

Courses and distances have also been said to be the most unsatisfactory of all calls in a survey because historically, chain carriers were liable to err, and instruments were not always reliable. Even though modern instruments are generally more precise, other forms of evidence, as a rule, tend to be more reliable. Courses and distances yielding to other calls is not a universal rule and is subject to exception, as are all other rules of construction.

Quantity will yield to courses and distances where there is a conflict, especially when quantity is qualified by the words "more or less." An important exception to this however, is where there is a clearly expressed intention to the contrary.[78]

Control of Distances by Courses. There is some authority to the contrary, but generally courses prevail over distances, since distances as a rule are more uncertain. Several reasons have been cited for this. One such rule is:

> . . . there is a greater probability that a chain carrier may lose count or report distances incorrectly by mistake or design than that there was an error in the compass which gave the course.[79]

Another rule states:

> The greater the distance, the greater the probability of error.[80]

The rule, however, is subject to many limitations and exceptions, so it must be understood and applied reasonably. Which one is the better choice is often dependent on the circumstances of the case and the intent of the parties, instruments, and procedures used when making the measurements.[81]

---

[78] *Ewell v. Weagley*, C.C.A. Md., 13 F.2d 712, *Strunk v. Geary*, 288 S.W. 1053, 217 Ky. 113 (1926), *Weniger v. Ripley*, 293 P. 425, 134 Or. 265 (1930), *Bosler v. Sun Oil Co.*, 190 A. 718, 325 Pa. 411(1937).

[79] *Matador Land & Cattle Co. v. Cassidy-Southwestern Commission Co.*, Tex.Civ.App., 207 S.W. 430 (1918).

[80] Ibid.

Strong v. Delhi-Taylor Oil Corp., 405 S.W.2d 351 (Tex., 1966).

[81] The law cannot determine whether the courses or distances shall govern, in cases where the ascertaining of the boundaries of land are in question, where such courses or distances do not correspond, but that must be settled by concurring testimony and the circumstances of each particular case. *McClintock v. Rogers*, 11 Ill. (1 Peck) 279 (1849).

If it be evident, from the calls of a deed, that distance is the material and controlling object, course must yield to distance. *Blight v. Atwell*, 27 Ky. (4 J.J. Marsh.) 278 (1830).

Angles. Angles lead to directions and result either from direct observation or from a computation involving two observations (courses). They are generally given the same dignity as courses and are to be honored except in cases of mistake or when not in keeping with the parties' intention. General directions or angles may be deflected so as to harmonize with the language of the instrument, depending on the situation of the land and the circumstances of the case.

Quantity. Quantity, or area, is ordinarily the least reliable element in a description and, therefore, the last to be considered.[82] However, there are special circumstances in which quantity may become more valuable, or may even control, such as where superior calls are lacking or leave the boundaries doubtful, or where there is a clear intention to convey a specific quantity only.

Statements such as "meaning and intending to convey exactly one acre," "conveying one acre and no more," "[a certain distance and direction] to complete ten acres" and the like could make quantity a controlling element.

When a specified tract of land is sold, the entire tract ordinarily passes, even though it exceeds the quantity mentioned in the deed. This is especially true when the land is represented to contain a certain number of acres "more or less." These words when applied to distances and quantities indicate precaution and are intended to cover some slight or unimportant inaccuracy; while aiding in an adjustment to the demands of fixed monuments, they do not weaken or destroy the value of indications of distance and quantity where no other guides are furnished.[83] The words "more or less" may also indicate that the area is dependent on superior calls in the description.

It has frequently been stated that all other elements of description must lose their superior value through ambiguities and uncertainties before resort may be had to quantity.

The various courts have long held that distances and area are among the most unreliable calls. The opposite is true in other sectors however, since landowners and town officials place great weight on road frontage and acreage.

---

[82] A call in a deed for acreage is the least reliable of all calls, and, while the mention of quantity may aid in defining the premises conveyed, it cannot control the rest of the description in the absence of an express covenant that the acreage mentioned is the only land conveyed; in the absence of such a covenant, a clause as to quantity is considered simply a part of the description and will be rejected if it is inconsistent with the actual land conveyed. *Texas Pacific Coal & Oil Co. v. Masterson*, 160 Tex. 548, 334 S.W.2d 436 (1960).

[83] *Innis v. McCrummin*, 12 Mart. (La.) 425; *Jackson v. McConnell*, 19 Wend. (N.Y.) 175 (1838), *Peay v. Briggs*, 9 S.C.L. (2 Mill. Const.) 98 (1818); *Ingelson v. Olson*, 272 N.W. 270, 199 Minn. 422 (1937), 110 ALR 167, *Oakes v. DeLancey*, 30 N.E. 974, 133 N.Y. 227 (1892), *Parrow v. Proulx*, 15 A.2d 835, 111 Vt. 274 (1940).

Conflicting Grants. When there are conflicting grants from the same grantor, the title of the grantee in the conveyance first executed is controlling, or superior, at least to the extent of the conflict and notwithstanding other factors, or circumstances. This is true whether or not the conveyances were made with reference to a map or plat.[84]

When two tracts of land are simultaneously conveyed by the same grantor to different persons, the grantee whose survey was first in time is the controlling one when there is an overlap. This problem is treated in greater detail as part of the subject matter of Chapter 4.

Conflicting Surveys. The general rule is that a senior survey controls a junior survey when the two do not agree, particularly when the junior makes reference to the senior. However, the senior survey, in order to control, must be valid and definite.[85]

The same applies to field notes, in that senior notes, control. Ordinarily, the notes of a junior survey cannot be relied upon to establish the boundary of the senior survey. However, there have been cases where the junior may be used such as where the controlling corner of the senior was a common corner to both surveys,[86] or when both surveys were part of the same system.[87]

Expanded List of Conflicting Elements. In summary, the initial list may be expanded to contain all of the elements. Keep in mind that it is not a fixed set of rules of priority, but is subject to parties' intention and the surrounding circumstances. Because of intent, any one element can control any of the others, depending on the circumstances.

---

[84] *Snodgrass v. Snodgrass*, 101 So. 837, 212 Ala. 74 (1924), *Sorenson v. Mosbacher*, 230 N.W. 656, 210 Iowa 156 (1930), *Dunn v. Stratton*, 133 So. 140, 160 Miss. 1 (1931); *Mechler v. Dehn*, 196 N.Y.S. 460, 203 App.Div. 128, reversing 191 N.Y.S. 650, 117 Misc. 591 (1922), affirmed 142 N.E. 288, 236 N.Y. 572 (1923).

Where there is a clash of boundaries in two conveyances from the same grantor, title of grantee first executed is, to extent of conflict, superior, and this is so even though conveyances were made with reference to a map or plat. *Merlino v. Eannotti*, 110 A.2d 783, 177 Pa.Super. 307 (1955).

[85] To extent there is conflict between earlier and later surveys, the earlier survey controls. *State ex rel. Buckson v. Pennsylvania R. Co.*, 228 A.2d 587 (Del.Super., 1967).

[86] *Plowman v. Miller*, Civ. App., 27 S.W.2d 612 (Tex., 1930).

Where the field notes of the junior survey establish its northwest corner at the southwest corner of the senior survey, a call for a natural object in the field notes of the junior survey, which establishes its northwest corner, may be controlling. *Jackson v. Graham*, 205 S.W. 755 (Tex.Civ.App., 1918).

[87] That a surveyor who had already surveyed a plot returned later to survey subsequent locations which he tied on to his previous work does not make his later survey a part of the same system. *Brooks v. Slaughter,* Tex.Civ.App., 232 S.W. 856 (1921); *Brooks v. Slaughter*, Tex.Civ.App., 218 S.W. 632 (1920).

Corners & Marked Lines

- Natural Monuments
- Artificial Monuments

    Maps, Plats and Field Notes
    Adjoiners
    Metes and Bounds

- Courses

    Angles

- Distances
- Quantity, or Area

Rejecting Calls. In the process of determining boundaries of a parcel of land, it is not permissible to disregard any of the calls if they can be applied and harmonized in any reasonable manner.[88] If there is, however, an actual contradiction between calls in the description, such that they are in fact irreconcilable, a court may reject or disregard the one which is false or mistaken. Calls that cannot be complied with because they are vague or contradictory may be rejected or controlled by other calls that are consistent and certain. An inconsistent call should be disregarded if all the rest of the calls are reconciled and the description perfected. In such case the call or description that is most certain and least likely to be incorrect, and is most consistent with the intent of the grant, should be retained, and as few calls as possible should be disregarded.[89]

REFERENCES

American Jurisprudence 2d. Volume 12. *Boundaries*. Rochester: Lawyers Cooperative Publishing Company.

---

[88] Calls, if they can be applied and harmonized in any reasonable manner in determining the boundaries of a tract of land, cannot be disregarded, and as few calls or descriptions as possible should be disregarded. *Gauley Coal Land Co. v. O'Dell*, 110 S.E.2d 833, 144 W.Va. 730 (1959).

[89] *Boardman v. Reed*, 6 Pet. (U.S.) 328, 8 L.Ed. 415, *Bostick v. Pernot*, 265 S.W. 356, 165 Ark 581 (1924), *Matheny v. Allen*, 60 S.E. 407, 63 W.Va. 443 (1908); *Shipp v. Miller*, 2 Wheat. (U.S.) 316, 4 L.Ed. 248 (1817); *White v. Luning*, 93 U.S. 514, 23 L.Ed. 938, *Vose v. Handy*, 2 Me. 322 (1823); *More v. Massini*, 37 Cal. 432 (1869), *Hill v. Smith*, 25 S.W. 1079, 6 Tex.Civ.App. 312 (1894).

American Jurisprudence 2d. Volume 23. *Deeds.* Rochester: Lawyers Cooperative Publishing Company.

Brown, Curtis M., Walter G. Robillard, and Donald A. Wilson. *Brown's Boundary Control and Legal Principles,* fourth edition. New York: John Wiley & Sons. 1995.

Corpus Juris Secundum. Volume 11. *Boundaries.* St. Paul: West Publishing Company.

Corpus Juris Secundum. Volume 26. *Deeds.* St. Paul: West Publishing Company.

Robillard, Walter G., and Lane J. Bouman. *Clark on Surveying and Boundaries,* fifth edition. Charlottesville: The Michie Company. 1987.

Skelton, Ray Hamilton. *The Legal Elements of Boundaries and Adjacent Properties.* Indianapolis: The Bobbs-Merrill Company. 1930.

CHAPTER 6

EXCEPTIONS AND RESERVATIONS

> The commonly recognized difference between a reservation
> and an exception is that the former creates or reserves
> something out of the thing granted, while an exception
> removes from the grant something already in existence.

> —*Adkins v. Arsht,*
> D.C. Ill. 50 F.Supp 761.

> The terms "reservation" and "exception" are often used
> interchangeably, and their distinction is technical, slight
> and shadowy.

> —*Victory Oil Co. v. Hancock Oil Co.,*
> 270 P.2d 604, 125 Cal. App.2d 222 (1954).

While the two terms are generally used together as one term, for example ". . .
excepting and reserving from the foregoing . . .," and they are less often used separately, they each have different meanings, separate and distinct from one another. An
exception in a description pertains to something included in the description but to be
taken from it, often the result of something previously conveyed. A reservation, in

contrast, is something being kept from the conveyance or the description by the grantor(s), and generally creates a new right in them.[1]

Reservation. A *reservation* has also been defined as a new condition arising from the taking back of something out of that which is granted,[2] or, more specifically, a new right in the thing granted, which was not in existence at the time of the making of the grant and was originated by it.[3] In the technical sense, there can be no reservation where the thing reserved is not a new right, a right not before existing.[4] Strictly speaking, a reservation does not affect the description of the property conveyed at all, but retains to the grantor some right upon the property, such as an easement, for example.[5]

Reservation Distinguished from Exception.

Provisions in deeds for right of way for siding was held to be a "reservation" for the benefit of the lots conveyed as well as others, and not an "exception."[6]

In view of the circumstances, and the language of deeds in plaintiff's chain of title, reserving to the successive grantors fenced cemetery on the farm conveyed, which had been in their family for years, reservation of such cemetery, rather than the exception from the conveyances, held intended and accomplished.[7]

A clause that the conveyance is subject to a right of the grantors to remove sand from the land is not an exception with respect to part of the thing granted, but a reservation to the grantors, such right not formerly existing separately, but having been called into being by the deed.[8]

Right to enter upon premises to remove building was technically "reservation."[9]

---

[1] *Thornhill v. Ford*, 56 So.2d 23, 213 Miss. 49 (1952):

A reservation reserves to the grantor some new thing issuing out of the thing granted and not existing before, and an exception excludes from the operation of the grant some existing portion of the estate or parcel granted, which would otherwise pass under the general description of the deed. However, since each is something to be deducted from the thing granted, it is often difficult to distinguish between them, and the use of the term "reservation" or "exception" is not conclusive as to the nature of the provision, which will be determined according to its actual character.

[2] *United Gas Public Service Co. v. Roy*, 147 So. 705 (La. App., 1933).

[3] *Rall v. Purcell*, 281 P. 832, 131 Or. 19 (1929).

[4] *Lewis v. Standard Oil Co. of California*, C.C.A.Cal. 88 F.2d 512.

[5] *Moore v. Davis*, 117 S.W.2d 1033, 273 Ky. 838 (1938).

[6] *Lauderbach-Zerby Co. v. Lewis*, 129 A. 83, 283 Pa. 250 (1925).

[7] *Picotte v. Smith*, 170 N.Y.S. 726, 110 Misc. 144 (1920).

[8] *Stanton v. T. L. Herbert & Sons*, 211 S.W. 353, 141 Tenn. 440 (1918).

[9] *Kennedy v. City of Hood River*, 259 P. 911, 122 Or. 531 (1927).

"Reserving" or "excepting and reserving" land for a public street creates a reservation proper and not an exception, for it does not require the exclusion of the fee, but only of an easement, or grant.[10]

Deed conveying forty-five acres off the southwest corner of certain lot, with three acres reserved for church and school purposes, passed title to entire forty-five acres, including three acres with reservation for such purposes.[11]

Reconveyance.

Reconveyance. A reservation has sometimes been called a reconveyance to the grantor.[12]

Implied Conveyance.

Implied Conveyance. Technically, a reservation in a deed is some newly created right, which the grantee impliedly conveys to the grantor.[13]

Purpose.

Purpose. The purpose of a reservation is to save to the grantor something that otherwise would pass by the grant.[14]

Exception.

Exception. The purpose of an *exception* is to take something out of the thing granted that would otherwise pass.[15] An exception of an interest in land constitutes an estate in the land.[16] It is a part of the realty itself, and is a visible tangible thing, that is, an incorporeal hereditament.[17]

Exception Distinguished from Reservation

Deed containing language "reserving however (20) twenty feet from top bank of lake which is not conveyed but reserved by . . . out of this conveyance" was held sufficient to except out of grant strip twenty feet wide measured from top bank of lake.[18]

---

[10] *Winston v. Johnson*, 45 N.W. 958, 42 Minn. 398 (1890); *Elliot v. Small*, 29 N.W. 158, 35 Minn. 396 (1886), 59 Am. R. 329.

[11] *Powell v. Harris*, 147 S.E. 189, 39 Ga. App. 295 (1928).

[12] *Koval v. Carnahan*, D.C. Ill., 45 F.Supp. 357.

[13] *Goss v. Congdon*, 40 A.2d 429, 114 Vt. 155 (1945); *Nelson v. Bacon*, 32 A.2d 140, 113 Vt. 161 (1943).

[14] *Hansen v. Bacher*, 299 S.W. 255 (Tex. Com. App., 1927).

[15] *Weyse v. Biedebach*, 261 P. 1092, 86 Cal.App. 736 (1927), *City of Jacksonville v. Shaffer*, 144 So 888, 107 Fla. 367 (1932), *Thornhill v. Ford*, 56 So.2d 23, 213 Miss. 49 (1952), *Nelson v. Bacon*, 32 A.2d 140, 113 Vt. 161(1943).

[16] *Brown v. Mathis*, 41 S.E.2d 137, 201 Ga. 740 (1947); *Cox v. Colossal Cavern Co.*, 276 S.W.2d 540, 210 Ky. 612 (1925).

[17] Ibid.

[18] *Pitts v. Zavala-Dimmit Counties Water Improvement Dist. No. 1*, 81 S.W.2d 801 (Tex. Civ. App., 1935).

Provision of deed reserving to grantor the right to bring water from spring on land conveyed to grantor's dwelling house on adjoining land, held to be an "exception" and not a reservation.[19]

Recital that warranty deed was subject to rights of third party to spring and land for reservoir constituted exception from grant of fee.[20]

Warranty deed conveying land but "reserving, however, all the lumber now standing on the premises, that has been previously sold, to be removed in twenty years from the date of sale."[21]

A deed created an "exception" rather than a "reservation," where grantor's right to use switch track in adjoining street antedated deed, and grantor sought only to retain such right rather than to create new right.[22]

Clause in deed, reserving and retaining from sale all caves and rights of way therefor, etc., for benefit of grantor, was held an exception, not a reservation.[23]

The word "retain" has been held to create an exception.[24]

Nature of an Exception. That which is excepted is something not granted at all.[25] Sometimes it is difficult to distinguish between an exception and a reservation in a deed. The terms "reservation" and "exception" are often used interchangeably or indiscriminately and are not conclusive as to the nature of the provision, but the technical meaning will give way to the manifest intent, even though the technical term to the contrary is used. Further, where the words "reservation" and "exception" are used together, without showing any definite knowledge of their technical meaning, the intention of the parties must be ascertained from the instrument interpreted in the light of the surrounding circumstances. While there cannot be with technical correctness both a reservation and an exception of the same thing, yet both terms may properly be used in the same deed where the nature of the things reserved and excepted shows that there is a separate subject matter on which each can operate.[26]

---

[19] *Haldiman v. Overton*, 115 A. 699, 95 Vt. 478 (1922).

[20] *Akron Cold Spring Co. v. Unknown Heirs of Ely*, 18 Ohio App. 74 (1923).

[21] *Gates v. Oliver*, 139 A. 230, 126 Me. 427 (1927).

[22] *Stout v. Frick*, 69 S.W.2d 677 (Mo. App., 1934).

[23] *Cox v. Colossal Cavern Co.*, 276 S.W. 540, 210 Ky. 612 (1925).

[24] *Stephan v. Kentucky Valley Distilling Co.*, 122 S.W.2d 493, 275 Ky 705 (1938).

[25] *Las-Daub Realty Corporation v. Fain*, 210 N.Y.S. 623, 214 App. Div. 8 (1925).

[26] *Peck v. McClelland*, 225 N.W. 514, 247 Mich. 369 (1929), *Studebaker v. Beck*, 145 P. 225, 83 Wash. 260 (1915).

The distinction between the words "excepting" and "reserving" is so rarely observed by those who draw deeds, they being generally used as synonyms, that perhaps but little weight ought to be attached to their use. *Keeler v. Wood*, 30 Vt. 242 (1858).

Whether particular provision in deed is intended to operate as exception or reservation is to be determined by its character, rather than by particular words used. *Foster v. Lee*, 171 N.E. 229, 271 Mass. 200 (1930).

Creation.   In order to create a reservation or exception, the intent to do so must be clearly disclosed by appropriate words, although no specific form or technical wording is required. Either a reservation or an exception may appear in any part of the deed.[27]

Implied Reservations or Exceptions.   As a general rule there can be no reservation or exception by implication except in a case of necessity. In order for an exception to be created, it must appear in express terms or be necessarily implied.[28]

Reservation to Grantor or Stranger.   A reservation must be in favor of the grantor and cannot be in favor of a stranger[29] to the deed, except in some jurisdictions in the case of a grantor reserving a life estate to himself and his spouse.[30]

Conversely, an exception may be in favor of the grantor or of a stranger if it confirms and recognizes an existing outstanding right or interest.[31]

Reservations Held to Be Void

Ordinarily a reservation in a deed conveying real property must describe the property reserved to the grantors with as much definiteness and certainty as the land conveyed.[32]

Where a deed conveys in full the fee in land, any reservation tending to destroy the fee granted is void because of repugnancy.[33]

Reservation not delineating limits of right of way and giving it no situs was void because description was too general and terms were too vague.[34]

---

[27] *Deckenbach v. Deckenbach*, 130 P. 729, 65 Or. 160 (1913), *Elkhorn Coal Corporation v. Slone*, 276 S.W. 826, 210 Ky. 761(1925), *White Flame Coal Co. v. Burgess*, 102 S.E. 690, 86 W.Va. 16 (1920).

A reservation may be contained in the warranty clause, habendum, or redendum clause, or, within the four corners of the instrument. *Rose v. Cook*, 250 P.2d 848, 207 Okla. 582 (1952). Under the ancient doctrine the proper place for an exception was in the granting clause, and for a reservation in the redendum clause. *Marias River Syndicate v. Big West Oil Co.*, 38 P.2d 599, 98 Mont. 254 (1934).

[28] Ibid.

[29] Persons not named in the premises or granting clause of a deed of conveyance are strangers thereto. *Collins v. Stalnaker*, 48 S.E.2d 430, 131 W.Va. 543 (1948).

[30] A reservation in favor of a stranger to the instrument is invalid as a reservation. *Martin v. Cook*, 60 N.W. 679, 102 Mich. 267 (1894); *Burns v. Bastien*, 50 P.2d 377, 174 Okl. 40 (1935); *Berkley Nat. Exchange Bank v. Lilly*, 182 S.E. 767, 116 W.Va. 608 (1935), see 102 A.L.R. 462.

[31] *Arnett v. Elkhorn Coal Corporation*, 231 S.W. 219, 191 Ky. 706 (1921), *Martin v. Cook*, 60 N.W. 679, 102 Mich. 267 (1894), *Dalton v. Eller*, 284 S.W. 68, 153 Tenn. 418 (1925).

An exception in a deed can only recognize and confirm rights already existing in strangers. *Deaver v. Aaron*, 126 S.E. 382, 159 Ga. 597 (1924), see 39 A.L.R. 126.

[32] *Russell v. Brown*, 260 S.W.2d 257, 195 Tenn. 482 (1953); *Haynes v. Morton*, 222 S.W.2d 389, 32 Tenn. App. 251 (1949).

[33] *Associated Oil Co. v. Hart*, 277 S.W. 1043 (Tex. Com. App., 1925).

[34] *Justice v. Justice*, 39 S.W.2d 250, 239 Ky. 155 (1931).

Exception Requirements. It is competent for the grantor to except from the conveyance of the fee in land some part of the land conveyed, but in order to be valid, any exception must be susceptible of establishment and must be of a thing that is severable from that thing granted, a part only of the thing granted, and a particular thing out of a general one. It must describe the property excepted with certainty and should not be in conflict with, or inconsistent with, the grant. An exception attempting to appropriate to the grantor property that he does not own will be ineffectual.[35]

Exceptions Held to Be Void

An exception in a deed, which incorrectly described the amount of land excepted from that conveyed by deed, was void.[36]

Where deed described part of the premises as the southwest and west half of the southeast quarter of Section 28, two hundred forty acres, less forty-six acres east of the railroad, and over one hundred acres were east of the railroad, the exception was void.[37]

An exception of about twelve acres in a section lying south of a river is void for uncertainty, where there were more than one hundred acres lying south of the river, and, the exception not being described to a certainty, the title to the whole tract passed, the exception alone being void.[38]

Where a deed conveys in full the fee in the land, any exception tending to destroy the fee granted is void because of repugnancy.[39]

May Have Been Created by Another. The exception may be a description of a portion of the thing granted, which prior to the grant had been conveyed to another, and not necessarily so conveyed by the grantor, but by a prior grantor.[40] That person may be a stranger to the description.[41]

---

[35] *Bartlett v. Bartlett*, 48 P.2d 560, 183 Wash. 278 (1935), *Hornsby v. Bartz*, Tex.Civ.App. 230 S.W.2d 360, *Johnson v. Mason*, 31 N.W.2d 910, 226 Minn 23 (1948), *Bennett v. Smith*, 69 S.E. 42, 136 W.Va. 903 (1952), *Mason v. Jackson*, 106 S.W.2d 610, 194 Ark. 236 (1937), see 111 ALR 1071.

The same certainty of description is required in an exception out of a grant, as in the grant itself. *Parker v. Cherry*, 193 S.W.2d 127, 209 Ark. 907 (1946).

[36] *Ballentine v. Bradley*, 191 So. 618, 238 Ala. 446 (1939).

[37] *Glasscock v. Mallory*, 213 S.W. 9, 139 Ark. 83 (1919).

[38] *Seavey v. Williams*, 191 P. 779, 97 Or. 310 (1920).

[39] *Associated Oil Co. v. Hart*, 277 S.W. 1043 (Tex. Com. App., 1925).

[40] *Arnett v. Elkhorn Coal Corporation*, 231 S.W. 219, 191 Ky. 706 (1921).

[41] Property cannot be conveyed by reservation, which is effective only in favor of the grantor, but an exception to recognize and confirm outstanding rights and interests in a third person will be effectuated and carried out. *Ogle v. Barker*, 68 N.E.2d 550, 224 Ind. 489 (1946), *Mott v. Nardo*, 166 P.2d 37, 73 Cal.App.2d 159 (1946), *Kronoff v. City of Worcester*, 125 N.E. 394, 234 Mass. 254 (1919).

Intention of Parties Controls. In construing a reservation or exception, the intention of the parties is to be pursued, if possible, and emphasis is given to the determination of what the grantor meant by the language of the reservation or exception. A reasonable construction should be given, and the entire instrument and the surrounding circumstances should be considered, including the purpose for which the property or right excepted or reserved is intended to be used.[42]

Plain Meaning Interpretation. A reservation or exception will be given effect according to the plain meaning and intent of the language used, but it cannot be extended beyond its terms, and although the practical construction given to it by the parties has weight where the reservation is ambiguous, such construction is immaterial where the language is clear and unambiguous. Property that is excepted is not granted and does not pass to the grantee.[43]

Construe against Grantor. Although there is some authority holding otherwise, reservations and exceptions expressed in doubtful or ambiguous language are, as a general rule, construed most strongly against the grantor and in favor of the grantee.[44]

Certainty of Description. Excepted property is described with sufficient certainty if the exact location thereof is left to the election of the grantor or is otherwise capable of subsequent ascertainment. The fact that an exception is void for uncertainty does not destroy the grant, but makes the conveyance operative as to the whole tract.[45] This rule, however, does not apply to an easement of such magnitude that it all but destroys the servient estate.[46] The cases are practically uniform in holding that uncertainty of or insufficiency in the description of an area of land that

---

[42] *Aiken v. McMillan*, 106 So. 150, 213 Ala. 494 (1925), *Dierks Lumber & Coal Co. v. Meyer*, D.C.Ark., 85 F.Supp. 157, *Brown v. Mathis*, 41 S.E.2d 137, 201 Ga. 740 (1947).

[43] *Lasyone v. Emerson*, 57 So.2d 906, 220 La. 951(1952), *Settegast v. Foley Bros. Dry Goods Co.*, 270 S.W. 1014, 114 Tex. 452 (1925), *City of Elkhart v. Christiana Hydraulics*, 59 N.E.2d 353, 223 Ind. 242 (1945), *Foster v. Lee*, 171 N.E. 229, 291 Mass. 200 (1930), *Wilson v. Gerard*, 56 So.2d 471, 213 Miss. 177 (1952), *McCue v. Berge*, 52 N.E.2d 789, 385 Ill. 292 (1944).

[44] *Green Bay & Mississippi Canal Co. v. Hewitt*, 29 N.W. 237, 66 Wis. 461 (1886), *Barker v. Lashbrook*, 279 P. 12, 128 Kan. 595 (1929).

In *Boring v. Filby (3d Dist.)*, 311 P.2d 869, 151 Cal App 2d 602 (1957), the court held that the following statutory language applied to an exception also: "A grant is to be interpreted in favor of the grantee, except that a reservation in any grant, and every grant by a public officer or body, as such, to a private party, is to be interpreted in favor of the grantor."

[45] *Sherman v. Arnold's Neck Boat Club*, 13 A.2d 272, 64 R.I. 485 (1940), *Ballentine v. Bradley*, 191 So. 618, 238 Ala. 446 (1939).

[46] *Justice v. Justice*, 39 S.W.2d 250, 239 Ky. 155 (1931).

the grantor or mortgagor attempts to except from the operation of a deed or mortgage affects only the validity of the exception and not the validity of the instrument as a whole. Accordingly, the exception, if so uncertain as to be void, will be ignored and the deed or mortgage will have the effect of conveying the entire tract described, including the part sought to be excepted.[47]

Descriptions Held to Be Sufficient

An exception in a deed of a tract of "eighty acres sold to W.S. Oates before" is not void, because it might be made certain by parol.[48]

An exception clause in deeds involving 220-acre tract of land reading, "save and except about 1/2 acre of land in the south end of this tract known at the McCutcheon Graveyard," was held not void for uncertainty.[49]

An exception in a deed that cited "excepting the school house lot" from the property conveyed was not void for uncertainty when an area could be shown by parol evidence to have actually been used at the time of deed for school purposes.[50]

Streets and Alleys. By merely reserving a strip for a public street, a deed has the effect of conveying the underlying title to the grantee subject to the easement so dedicated, but the title of the fee of a street does not pass under a deed where it is expressly reserved. A provision reserving the right to establish an alley for the use of the public across the lot conveyed relates to a public, rather than merely a private, alley.[51]

Right of Way. In general, an exception of a right of way retains or excludes from the operation of the grant merely an easement and not the fee to the real estate used for the right of way, but an exception of the land for a right of way, or on which a right of way is situated, covers the fee.[52]

---

[47] *Shell Petroleum Corp. v. Ward*, 100 F.2d 778 (C.C.A. 5th, 1939) (writ of cert. denied in 307 U.S. 632, 83 L.Ed. 1514, 595 S.Ct. 834, 1939); *Frank v. Myers*, 11 So. 832, 97 Ala. 437 (1892); *Mooney v. Cooledge*, 30 Ark. 640 (1875); *Stephens v. Terry*, 198 S.W. 768, 178 Ky. 129 (1917); *Harrill v. Pitts*, 193 So. 562, 194 La. 123 (1940); *Darling v. Crowell*, 6 N.H. 421 (1833); *Thayer v. Torrey*, 37 N.J.L. 339 (1875); *Butcher v. Creel*, 9 Gratt. 201 (Va., 1852). see, also, 162 A.L.R. 288. *Deed or mortgage as affected by uncertainty of description of excepted area.*

[48] *Loyd v. Oates*, 38 So. 1022, 143 Ala. 231 (1904), 111 Am. S.R. 39.

[49] *Houston Oil Co. of Texas v. Williams*, Civ.App., 57 S.W.2d 380, error refused (Tex., 1933).

[50] *Powell v. Trustees of Schools of Tp. 16*, 112 N.E.2d 478, 415 Ill. 236 (1953).

[51] *McCue v. Berge*, 52 N.E.2d 789, 385 Ill. 292 (1944), *St. John v. Quitzow*, 72 Ill. 334 (1874), *White v. Meadow Park Land Co.*, 213 S.W.2d 123, 240 Mo.App. 683 (1948).

[52] *Barker v. Lashbrook*, 279 P. 12, 128 Kan. 595 (1929), *Hall v. Wabash R. Co.*, 110 N.W. 1039, 133 Iowa 714 (1907).

Easement or Fee. A deed purporting to reserve or except a portion of the land from the conveyance for a specified use raises the question whether the parties intended to entirely except that portion from passing under the deed or merely to reserve an easement therein to the grantor. A reference in the deed to the future use intended to be made of the portion reserved or excepted has been held to indicate that the parties intended to reserve a mere easement rather than to except the fee. Thus, reservations for "street purposes," "highway purposes," or the like have been held to indicate the retention of an easement. In other instances, however, where similar language was used, the courts have concluded that there was an exception from the fee. In any event, however, where the deed is ambiguous in such respect, the intent may be shown by extrinsic evidence.

Whether used in a grant or in a reservation or exception, the words "right of way" are generally held to denote an easement or servitude rather than an interest in fee simple.[53]

Rights or Privileges. A reservation of a right or privilege should be construed in the same way as a grant of the same right or privilege. It may be inoperative until the exercise of the right reserved.[54]

Life Estate. A reservation in a deed permitting the grantors to occupy the premises until their death creates an interest and estate in the real estate. A deed reserves a life estate, which conveys a present interest but provides that the conveyance shall not take effect until the grantor's death, or which reserves to the grantor during his life the use, control, and profits of the property.[55]

Parol or Extrinsic Evidence. Parol evidence is held by most courts to be admissible for the purpose of aiding in the construction of the provisions of deeds as to reservations or exceptions by the grantor were the meaning or application of the provision in question is doubtful or ambiguous. While physical conditions or objects existing at the time of the grant are most frequently referred to for the

---

[53] *Chournos v. D'Agnillo*, 642 P.2d 710 (Utah, 1982).

[54] *French v. Carhart*, 1 N.Y. 96 (1847), *Provost v. Calder*, 2 Wend. (N.Y.) 517 (1829).

What will pass by words in a grant will be excepted by the same words in an exception. *French v. Carhart*, 1 N.Y. 96 (1847).

[55] *Zaveski v. Kish*, 46 A.2d 665, 138 N.J.Eq. 61(1946), *Carter Oil Co. v. McQuigg*, C.C.A.Ill., 112 F.2d 275 (1940), *Wise v. Wise*, 184 So. 91, 134 Fla. 553 (1938).

purpose of identifying land that is excepted from a grant, references to acts or agreements of parties relating to the excepted part or to identifying occurrences are recognized if they serve as means of determining the location and boundaries of the parcel that is excepted. Assuming competency, it is not material whether the extrinsic evidence consists of official records, written contracts, or parol evidence of agreements.[56]

Statement As to Use of Exception or Reservation. Statements as to the use being made or intended to be made of lands excepted or reserved may or may not limit the effect of the exception or reservation. Frequently, a question will arise as to whether the language used qualifies or limits the effect.

Roads and Rights of Way. One of the most frequent questions is whether a road or right of way reserved or excepted from a conveyance is intended to be fee or easement. Generally speaking, unless particular language qualifies it, an exception or reservation is merely an easement.

In the case of *Moakley v. Blog,* "except right of way of the [designated] Railway Company over strip thirty-five feet wide as granted by deed recorded in [designated book]" was deemed to reserve merely the right of way rather than the fee.[57]

In *Bolio v. Marvin*, "saving and preserving, however, from the operation hereof, the road" was deemed to reserve an easement, rather than except the fee.[58]

In *Stotzenberger v. Perkins*, "except there from a roadway twenty feet in width over and across said land described [by metes and bounds]" was deemed merely to reserve an easement for a roadway.[59]

The case of *Jones v. Sun Oil Co.,* where it stated that the grantor "is retaining free and unobstructed use of a twelve-foot lane, the entire length of the west side of this tract" was decided to be a reservation of an easement for road purposes, rather than the exception of the fee.[60]

---

[56] *McDonald v. Antelope Land & Cattle Co.*, 294 N.W.2d 391 (N.D., 1980), *Blythe v. Hines,* 577 P.2d 1268 (Okla, 1978), *Chesapeake Corp. of Virginia v. McCreery*, 216 S.E.2d 22, 216 Va. 33 (1975); *Texas Co. v. Wall (C.A.7 Ill.)*, 107 F.2d 45.

Where, to give effect to exception in a deed, it is necessary to resort to extrinsic evidence to determine written contracts, or parol evidence of agreements. *Texas Co. v. Wall*, C.C.A.III., 107 F.2d 45.

Parol evidence may be resorted to to designate the property contained in an exception, where the terms of the exception so identify it that parol proof can be applied to the language of the exception, and thus make it certain. *Prewitt v. Wilborn*, 212 S.W. 442, 184 Ky. 638 (1919).

[57] 90 Cal. App. 96, 265 P 548 (1928).

[58] 130 Mich. 82, 89 N.W. 563 (1902).

[59] 332 Mo. 391, 58 S.W.2d 983 (1933).

[60] Tex. Civ. App., 110 S.W.2d 80 (1937).

In those decisions in which it was determined that the fee was retained, other information was at hand. For instance,

> In construing the clause, "saving and excepting there from 15 feet by 15 feet on the north easterly corner of said lot, reserved as a way to Reade's cellar" was decided to be an exception of the fee rather than a reservation of an easement. This case relied on an earlier, similar case in which the defendant and his predecessors claiming under the grantor in such deed had utilized the tract for purposes other than those of a mere way of access and had placed first a one- and later a two-story structure upon it and had used it precisely as if they owned it and had fee to the land.[61]

Further, in the case of *Youngerman v. Polk County (Iowa)*, the clause: "the public square, as represented on said plat, is reserved for the purpose of building a court house thereon, and such other public buildings as the county commissioners deem proper for the use of the county of Polk," appearing in the county commissioners' acknowledgment of their dedication of "all the streets, alleys, and public ground in the town of Fort Des Moines, as represented on the above plat, to the general public," was construed as excepting the fee of the public square from such dedication. Noting that on the plat a certain area was designated as "public ground," while the land in controversy was designated thereon as "public square," that the acknowledgment expressed the lands dedicated to the public as "streets, alleys, and public ground" and did not mention "public square," so that under the maxim, *Expressio unius est exclusio alterius*," such public square would seem not to be included and that the acknowledgment also contained a specific description of the "public ground," such description not including the public square, the court proceeded to answer the argument that the clause amounted merely to a reservation of a use or easement in the county to occupy the ground of the public square with its courthouse and other necessary buildings, and said: "Now, we inquire, what was reserved from the dedication? The only possible answer is, the land, not merely its use, designated 'public square.' From what reserved? From the operation of the plat as a conveyance to the City of Des Moines for the public use. By whom reserved? The County of Polk. Why was this done? Because of the county having need of it as a place on which to erect its public buildings. But the mere fact that these commissioners stated their reason for reserving the square did not place a limitation on the absolute title the county had retained. The uses mentioned were practically the only ones Polk County had for the land. Clearly, then, the clause quoted operated as an exception rather than a reservation."[62]

School Site. In the case of *Thomas v. Jewell*,[63] where the owner of a quarter section leased to a school district a piece of land "to have and to hold the same for a school purposes," and thereafter by warranty deed conveyed the quarter section

---

[61] *Mount v. Hamble* , 22 Misc. 454, 50 NYS 813 (1898), affirmed 1898, 33 App. Div. 103, 53 NYS 1110 (1898), which is affirmed, 59 NE 1127, 164 NY 601 (1900).

[62] 81 NW 166, 110 Iowa 731 (1899); 2 N.W.2d 501, 300 Mich. 556 (1942).

[63] See 139 ALR 1335. *Statement as to use being made or intended to be made of lands excepted or reserved from conveyance as limiting effect of exception or reservation.*

"except 10 rods square lying in the northeast corner, which is leased to the school district as long as occupied for school purposes, with the right to occupy only," and many years thereafter the school site was abandoned by the school district, it was held that one deriving title from the grantee in such deed had no interest in the schoolhouse site, inasmuch as the provision in the deed expressly excepted such land from the conveyance, and title thereto remained in the grantor. So holding, the statement in *New York v. New York C. & H. R. R. Co.* (1893) 69 Hun 324, 53 NYSR 383, 28 NYS 562 (affirmed on opinion below in (1895) 147 NY 710, 42 NE 724):

> "A grantor who states in his deed that he excepts a certain portion of the land because he wants it for a certain purpose cannot be held to have conveyed that which he has expressly excluded because he afterwards devotes it to a different purpose."

Graveyards. Concerning a restrictive covenant, or "condition of use," the following cases have to do with graveyards, or burial grounds.

In *Brown v. Anderson*,[64] it was held that under the clause in a deed of a farm "excepting and reserving one-half acre of land of said tract, being the old family graveyard of the grantor, together with the right of way to said graveyard," descendants of the grantor did not have the right to license others not descendants of the grantor to use it for burial purposes. Stating that the main inquiry was as to whether the word "excepting" was to be understood and applied with or without qualification, the court said that it did not agree with counsel that the words "old family graveyard" were used simply to describe and identify the parcel of land excepted from the tract but that on the contrary they were evidently intended, and according to their natural import did indicate with reasonable certainty the purpose for which the half-acre was excepted, and the person for whose benefit it was excepted. This was an action wherein descendants of the grantor sought to enjoin a subsequent grantee of the farm from interfering with or preventing burials in the graveyard by person licensed by the grantor's descendants to make such use thereof. In affirming a dismissal of the bill the court observed that "though the fee simple title may be in appellants [descendants of the grantor], about which it is not necessary to decide, the parties to the deed had the right to annex a condition of the use and enjoyment by the grantor and his descendants of the estate in the half acre."

In *Damron v. Justice* (1915) 162 Ky 101, 172 S.W. 120, the clause in the deed was "excepting one-half acre of land at the grave, including the graveyard, with privileges of passing to and from said graveyard; unto the second party, her heirs and assigns forever, together with all its appurtenances thereunto belonging, except the said half an acre at the graveyard, to her heirs and assigns forever." Subsequently, the grantor in this deed brought an action against one deriving title under the grantee to enforce his right of passway to and from the graveyard and to have laid off to him one-half acre of ground at the graveyard. The grantor cross-appealed from the chancellor's ruling that he was entitled to have laid off for graveyard purposes alone one-half acre in a square

---

[64] 88 Ky 577, 11 S.W. 607 (1889).

including the graveyard, contending that he should have been adjudged the title to the half-acre of ground without any restriction as to its use. The court held the grantor's contention was not well taken inasmuch as considering the language of the deed and the circumstances of the parties, it was clear that the intention of the parties was to indicate with reasonable certainty the purpose for which the half-acre was excepted. "While it may be true," court said, "that the title to the half acre remains in the plaintiff [the grantor], yet the parties had a right to annex a condition to its use and enjoyment."

In *Johnson v. Elkhorn Gas Coal Min. Co.*,[65] a real estate contract provided "there is a reserve of one fourth of an acre square around the graveyard on the hillside," and the deed issued pursuant thereto contained the provision "there is excluded from the above-described tract the following described boundary around the grave [describing the boundary containing the one fourth of an acre]." Thereafter, the grantor leased the one-fourth acre to others and buildings were erected thereon and occupied by the lessees as a store. The court, (affirming a judgment whereby, upon action of the grantee in the deed, the grantor and his lessees were enjoined from using the lot for any purpose other than as a graveyard and from maintaining any buildings thereon), construed the provision in the deed as a reservation of the lot for graveyard purposes only, it being plain, from a consideration of the writings, that Johnson [the grantor] had in mind at the time he sold the tract of land, the purpose to save from intrusion, by the grantees or persons claiming under them, this piece of ground as a graveyard, and the wording of the reservation clearly expressed this intention, the court deeming that the word "graveyard" was not used merely for the purpose of identifying the location of the lot. The court said: "When there is a reservation of a lot for a specific purpose, and that purpose is expressed in the writing, it would be manifestly unfair to permit the grantor to use and occupy the lot for purposes entirely disconnected with and outside of the object of the reservation. The grantee might be perfectly willing to permit the grantor to reserve a lot for a specified purpose when he would not be willing that it should be reserved for another purpose, and when the use to which the lot is to be put is plainly expressed in the contract, the grantor should not be permitted, over the objection of the grantee, to use the lot for an entirely different purpose and one that was not in the contemplation of the parties at the time the transaction was entered into."

In *Salsburgh v. Hynds*,[66] a provision in a warranty deed of a farm "always excepting and reserving out of the above described premises the burying grounds containing about one-fourth of an acre of land, for the purpose only of burying the dead," was construed as an absolute exception out of the grant under which the grantee took no title to the burying ground. However, it was pointed out that the deed imposed a quasi-easement on the excepted tract, in favor of the land conveyed by the deed, and that any acts in violation of the terms of the reservation and exception, to wit, any use of it not necessary and proper as a burying ground, could undoubtedly be restrained by injunction.[67] In addition, in *Houston Oil Co. v. Williams*,[68] construing the clause in a deed of a farm: "save and except about 1/2 acre of land, known as the

---

[65] 176 Ky 676, 197 S.W. 409 (1917).

[66] 13 NY Week Dig 359 (1881), also reported, without opinion, in 25 Hun 566 (N.Y., 1881).

[67] See also *In re 122d Street*, 159 Misc 617, 288 NYS 697 (1936), affirmed in 257 App Div 940 (1939).

[68] 57 S.W. 2d 380 (Tex. Civ. App., 1933).

McCutcheon Graveyard," as an exception of a fee simple title with an annexed right of use of the graveyard by the descendants of those buried therein, under which such descendants had an equitable right to restrain the use of the graveyard for an inconsistent purpose.

Church. In the case of an exception for the duration of a designated use resulting in a determinable fee,

The disputed clause in a deed of a farm "reserving, however, the building situated on the last-described premises known as the chapel, together with the right to the land on which said building stands, said building to remain as long as the association owning the same may want it," and it appearing that the grantor had herself constructed and was the owner of the chapel building, but for a long period of time had permitted it to be used by an informal religious group, of which the grantor was the promoter and leader, the court in *Weed v. Woods*,[69] holding primarily that under such provision the property retained by the grantor was the chapel lot as it existed at the time of the execution of the deed, including certain horse sheds situated thereon as well as the chapel building proper, had occasion also to say:

> "It matters not whether this clause is technically a reservation or an exception; such a classification lends no aid to its interpretation. The estate retained by the plaintiff in the lot is a fee, not because as a matter of law it 'is an exception and not a reservation,' but because the clause, 'understood in the ordinary and popular sense of its terms,' reserves an estate which may be of perpetual continuance. It is not an estate for the life of the plaintiff, for the particular limitation agreed upon by the parties might happen either before or after her decease, or it might never happen. For the same reasons, it is not an estate for years or for any shorter period. It is an estate in fee, determinable upon the association ceasing to want it for chapel purposes. It is not 'an absolute fee, nor an estate on condition, but an estate which is to continue till the happening of a certain event, and then to cease. That event may happen at any time, or it may never happen. Because the estate may last forever, it is a fee. Because it may end on the happening of the event, it is what is usually called a determinable or qualified fee.' By such a construction, the intention of the parties will be carried out and effect given to this clause of the deed. So long as this estate continues, and the plaintiff and her successors in title retain the possession, they will have all the rights in respect to it which they would have if they were tenants in fee simple."

Public Purpose. Where there is a clause indicating that the land be used for school, church, or other public or quasi-public purpose,

In *Hartwig v. Central-Gaither Union School District*,[70] the provision in a deed of a certain quarter section "containing 160 acres, less one acre and one hundred and six square rods deed by said [grantors] in the year 1873, to Gaither School District, said lot being the northwest corner of the above parcel of land," was held a valid exception of

[69] 71 NH 581, 53 A. 1024 (1902).
[70] 252 P. 733, 200 Cal. 425 (1927).

the tract which grantors had previously conveyed to the school district by a deed providing for its reversion to grantors if it should cease to be used for educational purposes, the court holding that under the quoted provision the grantee acquired no interest in the school site, and that after it ceased to be used for school purposes neither those claiming under the grantee nor the school district had any interest in such site, it having reverted to the heirs of the grantors. Here, the plaintiff, claiming under the grantee, contended that the deed granted the entire quarter section and that the excepting clause was, in effect, meaningless, because at variance with the grant, his theory being that the exception was an attempt to describe the property by acreage and that such description was secondary to a description by boundaries. However, the court thought the rule of construction contended for was inapplicable to the facts and said: "The grant was not of the entire quarter section, but of the quarter section, less one acre, more or less, which was more particularly described by reference to a previous conveyance. There is no ambiguity about the deed."[71]

Private Passageway

In *Pritchard v. Lewis*,[72] the provision "excepting and reserving from the above-described premises a strip of land two (2) rods in width off the north side thereof, to be used as a right of way. The party of the second part, however, to have the privilege of fencing said right of way into his inclosure, and being required only to maintain a gate at each end thereof for the use of said party of the first part, his heirs and assigns," was construed as excepting from the grant the fee of such strip of land, in view of the fact that under a deed bearing the same date the grantor conveyed such strip of land to a third person and in such deed described it by metes and bounds and recited that it was "the same premises described as a right of way two (2) rods wide reserved by the said parties of the first part in a deed this day executed by them to one [naming grantee in first deed]," and therein excepted and reserved from such strip all of the timber situated thereon; and the further fact that one of the grantors refused to sign the deed conveying the entire tract until the other grantee received his deed to the one-acre strip; the court deeming that sufficient ambiguity existed to warrant the admission of such testimony indicative of the parties' intention.

While the foregoing case dealt with the reservation of the fee, the following case deals with an easement:

In *Terry v. Tinsley*,[73] construing a deed providing, "there is reserved a strip of land twenty (20) feet in width" describing the course of such strip as leading from a point on a turnpike, along an old road to a gate, thence along a fence to a certain division line, and providing that the grantors in locating said "roadway" should have the right to make convenient location, and providing that the grantee should have the right to use the "roadway" in common with the grantors, as merely reserving an easement for a roadway rather than as excepting the fee of such strip from the grant, the court remarked: "The use of apt words in the restrictive clause is necessary to except a part

---

[71] This case was expressly followed in *Hartwig v. Proper*, 253 P. 734, 200 Cal 796 (1927).

[72] 104 NW 989, 125 Wis. 604 (1905), 1 LRA (NS) 565, 110 Am. St. Rep. 873.

[73] 124 SE 290, 140 Va 240 (1924).

of the land from the conveyance. The use of the words 'roadway' and 'roadway reservation'[this latter term being used in a deed contemporaneously executed by the grantee, conveying the land subject to the 'roadway reservation twenty (20) feet wide'] are inapt for that purpose, and signify an intent merely to create an easement. In resolving this question, the repeated use of the words 'roadway' and 'roadway reservation' in the clauses under review is, we think, significant of the intention of the parties not merely to limit the use, but also to designate its character, and in all such controversies the intention, if it can be ascertained, is controlling."

Private Burial Rights

In view of the circumstances and the practical construction adopted by the parties, it was held in *Blackman v. Striker*,[74] that a clause in a deed "saving, excepting and reserving unto the heirs of the said John Hopper of the Out Ward, deceased, and to their and each of their heirs out of this present demise, all that certain burying ground now in fence consisting of forty-eight feet square parcel of the said lot of ground and commonly called the family burying ground, with free ingress, egress and re-gress into, out of and from the same to bury the dead, &c., forever," must be construed as reserving and easement for burial purposes rather than as excepting from the conveyance the fee of the cemetery lots.

The foregoing discussion and case examples serve to illustrate how each set of circumstances must be analyzed by itself. For further information concerning additional cases dealing with public and private highways, streets or alleys, burial grounds, churches, schools and railroads, reference may be made to:

139 ALR 1339. Statement as to use being made or intended to be made of lands excepted or reserved from conveyance as limiting effect of exception or reservation.

FOR FURTHER REFERENCE

| 162 ALR 288 | *Deed or mortgage as affected by uncertainty of description of excepted area* |
| 61 ALR 2d 1390 | *Admissibility of parol evidence with respect to reservations or exceptions upon conveyance of real property* |

REFERENCE

23 American Jurisprudence 2d, DEEDS

26 Corpus Juris Secundum, DEEDS

80 American Jurisprudence 2d, Wills

---

[74] 37 NE 484, 142 NY 555 (1894).

CHAPTER 7

WORDS AND PHRASES

> When an instrument does not, by its terms, clearly and plainly describe
> the land affected, or where it is phrased in language susceptible of more
> than one construction, intention of parties is to be ascertained, not solely
> from the words of the instrument, but from its language when read in light
> of the circumstances surrounding the transaction.

> —*Thomas et al v. Texas Osage Co-op. Royalty Pool, Inc.,*
> 248 S.W.2d 201 (Tex., 1952)

There are several principles which apply to the interpretation of land records regarding words and phrases. Many terms have been defined herein, but a number of others are found with regularity and have been ruled on by various courts. These will mostly be found in a set of legal encyclopedias known as *Words and Phrases*,[1] which contains most of the words and phrases a reader is likely to encounter when dealing with land records. For others, either *Webster's Dictionary*, *Black's Law Dictionary*, or *Ballentine's Law Dictionary* will ordinarily suffice. For unique words, or words

---

[1] West Publishing Company, St. Paul, Minnesota. One hundred volumes.

that don't appear in any of the above, special sources must be consulted, which may consist of almost any imaginable reference.[2]

Words in a deed are presumed to have a purpose;[3] otherwise, they need not be included. When someone originally wrote the document, or its description, they were defining something and employing words to tell a story. Documents (deeds, etc.) are also *to be viewed in light of the surrounding circumstances.*[4] In interpreting a document, the court will place itself as nearly as possible in the situation of the parties at the time of the conveyance.[5] Consider then when the document or description was first written and what the conditions and circumstances were at *that* time.

A common mistake illustrating the significance of this rule occurs in relation to the relocation of a road. Suppose that a chain of title is traced back a hundred years with a consistent description calling for a road as one of its boundaries. The questions now are, *what road was called for a hundred years ago, and where was it located?* Figure 7.1 illustrates conditions that may have changed over time.

Another example concerns the location of a land parcel on a road between two residences. A 1900 description mentions "the road leading from John Jones to William Smith." Tracing the description to its origin in 1875 reveals that the description remains the same, however, it is necessary to determine the residences in 1875, *not in 1900.* The persons mentioned may have moved one or more times during the 25-year period. See Figure 7.2.

---

[2] One peculiar but typical example was the call in a Vermont deed for a sugar plum tree, a species which grows in the midwestern United States, but not in the region where the particular description was located. An unrelated source, a trout fisherman's story book, contained a story about good trout fishing when the shadbush (*Amelanchier spp.*), also known as sugar plum, is in blossom, thus ending another mystery.

[3] Each word in deed is presumed to have been used for some purpose and to have some effect. *Wittmeir v. Leonard*, 122 So. 330, 219 Ala. 314 (1929).

It cannot be presumed that words or terms in a conveyance were used without a meaning or having some effect given them, and a construction will be adopted, if not contrary to law, giving effect to the deed and every word and term employed. *Rhomberg v. Texas Co.*, 40 N.E.2d 526, 379 Ill. 430 (1942).

[4] *State v. Ladd*, 268 A.2d 894, 110 N.H. 381 (1970).

It is time of conveyance which must be considered to establish the intent of the grantor of deed. *Freeman v. Affiliated Property Craftsmen*, 72 Cal. Rptr. 357, 266 C.A.2d 723 (1968).

Generally, land rights are to be determined at the time they are granted. *Wilson v. DeGenaro*, 415 A.2d 1334, 36 Conn. Sup. 200 (1979), adopted 435 A.2d 1021, 181 Conn. 480 (1980).

Intent of parties to deed should be determined in light of circumstances existing at time of execution and purpose sought to be accomplished. *Montgomery v. Central Nat. Bank & Trust Co. of Battle Creek*, 255 N.W. 274, 267 Mich. 142 (1934).

[5] In construing a deed, court has duty to place itself as nearly as possible in situation of parties at time instrument was made. *Sanborn v. Keroack*, 171 A.2d 25, 103 N.H. 297 (1961).

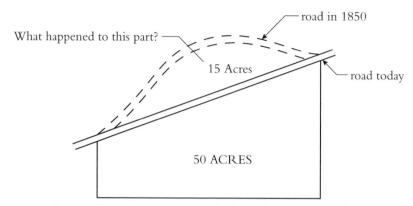

Figure 7.1 Consider conditions at the time a document was written.

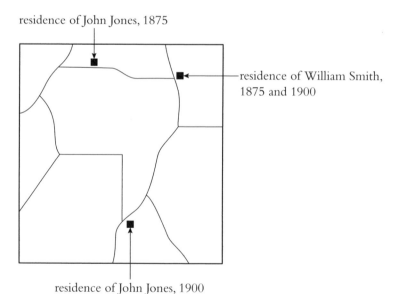

Figure 7.2 Consider circumstances that may have changed between the times descriptions were written.

A deed speaks for itself, and its construction depends upon the language used. Deeds are to be *construed according to their plain terms*,[6] and *words given ordinary meaning*.[7]

A will speaks for the maker's death, and the language used therein must be construed as of the date of its execution and in light of the surrounding circumstances at the time. It is the intention of the testator spoken in the words of his will, and not any intention that may be deduced from speculation as to what he would have done had he anticipated a change of the circumstances surrounding him at the time of the execution of the will, that is to govern in construing the instrument.[8]

In construing a will, the document should be read in the ordinary and grammatical sense of the words employed, unless some obvious absurdity or some repugnance or inconsistency with the declared intentions of the testator, as extracted from the whole instrument, will follow from so reading it.[9]

---

[6] Civ. App. 1930. Deed must be construed according to its plain terms. *Faison v. Faison*, 31 S.W.2d 828 (Tex., 1930), error dismissed.

[7] Nontechnical words in a conveyance are to be given their ordinary and plain meaning unless context shows they are used in a different sense. *Weber v. Graner*, 291 P.2d 173, 137 C.A.2d 771 (1955).

Ordinary meaning must be given words used in deed where meaning is not doubtful. *Raritan State Bank v. Huston*, 161 N.E. 141, 329 Ill. 604 (1928).

In the construction of deeds, the words employed should be given their fair and reasonable meaning. Ind. 1892. *Tinder v. Tinder*, 30 N.E. 1077, 131 Ind. 381, (App., 1913). *Figgins v. Figgins*, 101 N.E. 110, 53 Ind. App. 43 (1912).

Words used to give expression to a deed should be construed according to the common meaning of the language. *G.H. Bass & Co. v. Wilton Lumber Co.*, 104 A. 160, 117 Me. 314 (1918).

Although words may change in meaning, they must be read in deed as parties then understood them and as they were commonly used where deed was written. *Franklin Fluorspar Co. v. Hosick*, 39 S.W.2d 665, 239 Ky. 454 (1931).

In construing deeds, language used must be given its customary meaning. *Ellenwood v. Woodland Beach*, 115 N.W.2d 115, 366 Mich. 367 (1962).

Language in deeds, not technical, must be taken in its ordinary and usual sense. *Bradshaw v. Bradbury*, 64 Mo. 334 (1876).

Words in a deed must be given their ordinary and popular meaning, unless used in a technical sense, or the context shows that they are used in a different sense. *Wood v. Mandrilla*, 140 P. 279, 167 C. 607 (1914).

Courts look to the usual meaning of words in construing a deed. *Wilson v. DeGenaro*, 415 A.2d 1334, 36 Conn. Sup. 200 (1979), adopted 435 A.2d 1021, 181 Conn. 480 (1980).

[8] *Gorham v. Chadwick*, 200 A. 500, 135 Me. 479 (1938), 117 ALR 805, *Carter v. Sunray Mid-Continent Oil Co.*, 94 So. 2d 624, 231 Miss. 8 (1957); *Compton v. Rixey's Ex'rs*, 98 S.E. 651, 124 Va. 548 (1919), see 5 ALR 465.

[9] *Galloway v. Darby*, 151 S.W. 1014, 105 Ark 558 (1912), *McCormick v. Reinberger*, 234 S.W. 300, 192 Ky. 608 (1921), *Bender v. Bender*, 75 A. 859, 226 Pa. 607 (1910), *Love v. Love*, 38 S.E.2d 231, 208 S.C. 363 (1946), see 168 ALR 311.

Words in a will, too, are to be interpreted according to their ordinary and natural meaning unless a different meaning is indicated by the context or by the circumstances of the case, or unless it clearly appears that they were used in a technical or special sense. A testator is presumed to have been familiar with the ordinary and natural meaning of the terms used in his will and to have used the words contained therein in their usual sense and to have intended that such meaning be adopted, unless an intention to use the words of the instrument in some other sense is clearly manifested.[10]

If words have a primary and secondary meaning they will be construed according to their primary signification unless an intention to use them in some other sense manifestly appears.[11]

Law in Existence at the Time. A deed should also be construed in light of the law which existed when the deed was executed.[12] Once again, it is important to know when the document or its description, was first created. Since laws continually change, an interpretation today may differ from an interpretation of the same facts many years ago.

---

[10] *Travers v. Reinhardt*, 205 U.S. 423, 51 L.Ed. 865, 27 S.Ct. 563 (1907); *Smith v. Bell*, 31 U.S. 68, 8 L.Ed. 322 (1832), *Edgerly v. Barker*, 31 A. 900, 66 N.H. 434 (1891); *Phelps v. Harris*, 101 U.S. 370, 25 L.Ed. 855 (1879), *Wilson v. Witt*, 112 So. 222, 215 Ala 685 (1927), see 52 ALR 1095, *Gray v. Francis*, 124 S.E. 446, 139 Va. 350 (1924); In *Re Westlake's Estate*, 266 P. 714, 83 Colo. 540 (1928), *Gildersleeve v. Lee*, 198 P. 246, 100 Or 578 (1921), see 36 ALR 1166; *Northern Trust Co. v. Wheeler*, 177 N.E. 884, 345 Ill. 182 (1931), see 83 ALR 154; *Stanton v. Stanton*, 101 A.2d 789, 140 Conn. 504 (1953); *Gannon v. Albright*, 81 S.W. 1162, 183 Mo. 238 (1904), *Blaisdell v. Coe*, 139 A. 758, 83 N.H. 167 (1927), see 65 ALR 626.

The testator's intention, if clear may not be disregarded on the basis of a literal interpretation of his testamentary words. *Re McFerren's Estate*, 76 A.2d 759, 365 Pa. 490 (1950), see 22 ALR2d 451.

When the ordinary sense of words used in a will would be repugnant or inconsistent with the remainder of the instrument, that ordinary sense is not to be adhered to. *Boston Safe-Deposit & Trust Co. v. Coffin*, 25 N.E. 30, 152 Mass. 95 (1890).

The construction of a will in clear and unmistakable language is to be determined by what the testator said and not what he may have intended. *Palmer v. Crews*, 35 So.2d 430, 203 Miss. 806 (1948), see 4 ALR2d 483.

[11] *Wilson v. Witt*, 112 So. 222, 215 Ala 685 (1927), see 52 ALR 1095; In *Re Hunt's Estate*, 23 N.E. 120, 117 N.Y. 522 (1889),.

[12] *Stuart v. Fox*, 152 A. 413, 129 Me. 407 (1930), certiorari denied 52 S. Ct. 15, 284 U.S. 572, 76 L. Ed. 498 (1931).

A grant of land must be interpreted, understood, limited and restrained, according to the law of the country in force at the time when the grant was made. *Elder v. Delcour*, 269 S.W.2d 17 (Mo. 1954).

The state of the law at the date of the deed may be considered in determining the meaning and operation of its words. *Hamlen v. Keith*, 50 N.E. 462, 171 Mass. 77 (1898).

The law of the state in which land is situated controls its descent, devise, alienation and transfer, and the construction of instruments intended to convey it. *U.S. Trust Co. of N.Y. v. Boshkoff*, 90 A.2d 713, 148 Me. 134 (1952).

Professional or Lay Draftsperson. It is frequently recognized that in ascertaining the meaning of words used in a will, a distinction exists between instruments drawn by skilled testamentary draftsmen and those prepared by persons obviously unlearned in the law; words found in professionally rendered documents are to be construed with some strictness, with emphasis being placed upon their accepted technical meaning, while the language exhibited in wills drawn up by nonprofessionals are to be interpreted liberally with reference to its popular meaning. Thus, the opinion has been expressed that where a will has been drawn by one learned in the law, the words employed will ordinarily be given their accustomed technical meaning, but where it is drawn by a layman the language used may be given the meaning it would commonly have to a person in his situation.[13]

Holographic wills, being usually prepared by one who is unlearned in the law, should be construed more liberally than ones drawn by an expert, and the words and phrases employed in such instruments should be interpreted according to their ordinary acceptation, even though they may have a different technical legal meaning, where the circumstances surrounding the execution of the will indicate that the testator so intended.[14]

Technical Words and Phrases. Technical words when used in a will should be construed according to their technical meaning, as established by reference to the science or art to which they are peculiar, unless it appears from the will that they were used in a different sense, or unless a different construction is necessary in order to give sense to the will. Where an established rule of law has annexed a definite meaning to the language used in a will, it is as if such legal interpretation of the language so used were written therein at length. It is presumed, in the absence of evidence to the contrary, that technical words that have a definite and long-accepted meaning, appearing in a will drafted by a person skilled in the use and meaning of legal terminology, were used by the testator correctly and with the intent that they be interpreted in conformity with the law.[15]

Wills, however, of all the classes of legal instruments, are least to be governed in their construction by their technical terms, especially in this country, where wills are frequently drawn by persons unacquainted with legal phraseology and ignorant of

---

[13] *Re Boutelle's Estate*, 15 N.W.2d 506, 218 Minn. 158 (1944), see 154 ALR 966, *Commonwealth v. Wellford*, 76 S.E. 917, 114 Va. 372 (1913); *Buchwald v. Buchwald*, 199 A. 795, 175 Md. 103 (1938), *Burton v. Kinney*, 231 S.W.2d 356, 191 Tenn. 1 (1950), see 19 ALR2d 366.

[14] In *Re Lanart's Estate*, 9 Alaska 535 (1939), affirmed (C.A.9 Alaska) 111 F.2d 88 (1939); *Anderson v. Gibson*, 157 N.E. 377, 116 Ohio St. 684 (1927), see 54 ALR 92.

[15] *Phelps v. Harris*, 101 U.S. 370, 25 L.Ed. 855 (1879), *Galloway v. Darby*, 151 S.W. 1014, 105 Ark 558 (1912), In *Re Boutelle's Estate*, 15 N.W.2d 506, 218 Minn. 158 (1944), see 154 ALR 966, *Alper v. Alper*, 65 A.2d 737, 2 N.J. 105 (1949), see 7 ALR2d 1350; *Ward v. Stow*, 17 N.C. 509 (1834); *Doe v. Roe*, 2 Del. 103 (1835); *Rothschild v. Weinthel*, 131 N.E. 917, 191 Ind. 85 (1921), see 17 ALR 1377; In *Re Davidson's Will*, 26 N.W.2d 223, 223 Minn. 268 (1947), see 170 ALR 215.

the meaning that the law attaches to the words they use. Not infrequently the testator is like the writer described by Byron as having just enough of learning to misquote, and accordingly, where a will bears earmarks of having been drawn by a layman, and not by a lawyer, the court, in the endeavor to arrive at the intent of the testator, will not view the language technically, but liberally, and with reference to its popular meaning.[16]

If, in describing an estate, the grantor has used words which in law of real property have come to have a definite legal significance, such words must be given their legal effect.[17]

A deed should be construed with reference to the actual state of the land at the time of its execution.[18]

Interpreting Words and Phrases. Every word in a deed, if possible, should have its proper effect and the meaning of one word or phrase may be ascertained by comparing it with other words or phrases in the same instrument.[19]

Language of Instrument. Mere matter of convenience or taste cannot be allowed to overrule the express language of an instrument.[20] And, a grammatical construction should never be allowed to interfere with the intention of an instrument.[21]

Misspellings. Where a word in a deed is misspelled, the court will construe the deed according to the meaning of the word actually used, especially where the latter construction would give no effect to the clause containing the word.[22]

---

[16] *Giles v. Little*, 104 U.S. 291, 26 L.Ed. 745 (1881), (ovrld on other grounds *Roberts v. Lewis*, 153 U.S. 367, 38 L.Ed. 747, 14 S.Ct. 945); *Bosley v. Wyatt*, 55 U.S. 390, 14 L.Ed. 468 (1852), *Wescott v. Binford*, 74 N.W. 18, 104 Iowa 645 (1898), *Kelly v. Reynolds*, 39 Mich. 464 (1878); *Lightfoot v. Mayberry* (Eng) [1914] AC 782, 83 LJ Ch 627.

[17] *Sauls v. Cox*, 67 N.E.2d 187, 394 Ill. 81 (1946).

[18] *Farrow v. Trickey*, 374 S.W.2d 49 (Mo., 1963).

Conn. Super. 1979. Generally, land rights are to be determined at the time they are granted. *Wilson v. DeGenaro*, 415 A.2d 1334, 36 Conn. Sup. 200 (1979), adopted 435 A.2d 1021, 181 Conn. 480 (1980).

[19] *Saulsberry v. Maddix*, 125 F.2d 430 (C.C.A.Ky., 1942), certiorari denied 63 S. Ct. 36 (1942), two cases, 317 U.S. 643, 87 L.Ed 518 (1942).

[20] *Fratt v. Woodward*, 91 Am. Dec. 573, 32 C. 219 (1867).

[21] *Hancock v. Watson*, 18 C. 137 (1861).

[22] *Baustic v. Phillips*, 121 S.W. 629, 134 Ky. 711 (1909).

Errors in spelling in deeds are usually unimportant, and, where it is plain that a word is misspelled, the deed will be construed to meaning of word intended, rather than according to meaning of word actually used. *Anderson & Kerr Drilling Co. v. Bruhlmeyer*, 136 S.W.2d 800, 134 Tex. 574 (1940), see 127 ALR 1217, answers to certified questions conferred to 138 S.W.2d 1118 (1940).

Poorly Written Document. The words and phrases employed in an ineptly drawn deed may be transposed so as to bring them in accord with the evident intention of the parties, and glaringly omitted words may be supplied to make the meaning clear.[23]

Punctuation. Punctuation is ordinarily given slight consideration in construing a deed.[24]

> A comma used in descriptive language referring to "East 1/2, S.W. 2/4" did not mean "of," but was equivalent to an "and." (*U.S. v. 12,918.28 Acres of Land in Webster Parish, D.C.* La. 61 F.Supp. 545, affirmed, C.C.A., *Crichton v. Saucier,* 159 F.2d 303)
>
> A period separating a portion of the description of land in a recorded copy of a deed containing no reference to township number or range from portion of description that did refer to township number and range established that township and range related only to portion of description following the period and not to that preceding it, particularly where the two portions of the description appeared in separate paragraphs in the original deed. (Ibid.)

Words will control punctuation marks rather than the other way around;[25] however, punctuation will be resorted to in order to settle the meaning of an instrument, after all other means fail.[26]

Additional Wording in the Deed. Words that are added in the latter part of a deed, for the sake of greater certainty, may be resorted to explain the preceding parts which are not entirely clear.[27]

---

[23] *Kentucky Real Estate Board v. Smith*, 114 S.W.2d 107, 272 Ky. 313 (1938).

[24] *Franklin Flourspar Co. v. Hosick*, 39 S.W.2d 665, 239 Ky. 454 (1931).

Berry v. Hiawatha Oil & Gas Co., 198 S.W.2d 497, 303 Ky. 629 (1947).

In construing a deed, the words contained in the instrument, and not the punctuation, should be the controlling guide, but courts are not required to disregard all punctuation and look solely to the language of the instrument. *Harris v. Ritter*, 279 S.W.2d 845, 154 Tex. 474 (1955).

In the construing of deeds, no regard shall be given to punctuation. *Vinson v. Vinson*, 4 Ill. App. 138 (1879).

Punctuation will not be permitted to change the meaning of a deed, if reading it altogether, its meaning is clear. *Wilson v. Wilson*, 109 N.E. 36, 268 Ill. 270 (1915).

[25] *Teachers' Retirement Fund Ass'n of School District No. 1 v. Pirie*, 34 P.2d 660, 147 Or. 629 (1934).

[26] *Ewing's Lessee v. Burnet*, 36 U.S. 41, 11 Pet. 41, 9 L.Ed 624, 1 Ohio F. Dec. 574 (1837).

[27] *Wallace v. Crow*, 1 S.W. 372 (Tex., 1886).

Words Cannot Be Added. The effect of a deed cannot be changed by the addition of words not used by the grantor, and the court will not add words to modify a clear meaning of the deed.[28]

Abbreviations. Courts generally take judicial notice of commonly used initials and abbreviations in descriptions.[29] They have held that "NE. SE.," followed by the numbers of a section, township, and range, indicate the northeast quarter of the southeast quarter of the section named. They have also said the same about "NE.-NE." In addition, they have said that courts judicially know that "Sec. 23, 38, 14" means "section 23, township 38, range 14."[30]

Appendix One is devoted to particular words or phrases which are commonly encountered. Most of the definitions are as interpreted by the courts.

REFERENCE

80 American Jurisprudence 2d, Wills

[28] *Triplett v. Triplett*, 60 S.W. 2d 13, 332 Mo. 870 (1933).

Seested v. Applegate, 26 S.W.2d 796 (Mo.App., 1930).

While effect should be given to every word of written instrument whenever possible, it is never permissible to insert meaning to which parties did not agree. *Tubb v. Rolling Ridge, Inc.*, 214 N.Y.S. 2d 607, 28 Misc.2d 532 (1961).

[29] *Ballard v. Roanoke Bank*, 65 So. 356, 187 Ala. 335 (1914). ("T" for township, "R" for range).

Taylor v. Wright, 13 N.E. 529, 121 Ill. 455 (1887). ("W" for west).

Beedy v. Finney, 91 N.W. 1069, 118 Iowa 276 (1902). (character "&"); see Appendix One.

Beggs v. Paine, 109 N.W. 322, 15 N.D. 436 (1906). ("N.W." for northwest).

State ex rel. Wyatt v. Vaile, 26 S.W. 672, 122 Mo. 33 (1894). ("ex,""a," and "cor" for "except,""acres," and "corner").

Ottumwa, C.F. & St. P.R. Co. v. McWilliams, 32 N.W. 513, 71 Iowa 164 (1887). ("tp" for township, and "r" for range).

A description by figures and customary abbreviations is good. *Moseley's Adm'r v. Mastin*, 37 Ala. 216 (1861).

A description by figures is the full equivalent of the spelled-out words for such figures. *Middlebury College v. Cheney*, 1 Vt. 336 (1814).

[30] Patton, § 160.

CHAPTER 8

THE USE OF EXTRINSIC EVIDENCE

No deed or conveyance of land was ever made, however minute and specific the description, that did not require extrinsic evidence to ascertain its location; and this is so whether the description be by metes and bounds, reference to other deeds, to adjoining owners, watercourses, or other description of whatever character.

—Peacher v. Strauss,
47 Miss. 353 (1872)

As discussed in Chapter 5, it is the grantor's intention, expressed within the four corners of the instrument, that governs in determining the title conveyed. When the intention is not clear, or if one or more conflicts arise, there are rules that govern further interpretation. For instance, contemporaneous instruments executed at the same time, relating to the same subject matter and between the same parties may be taken together and construed as one instrument. Also, subsequent actions of the parties in relation to their conveyances may indicate their previous intentions. However, there must be ambiguity within the description before resort is made to extraneous matter.[1]

Extrinsic evidence, also known as extraneous evidence or *evidence aliunde*, is that which is not furnished by the document itself, but is derived from outside sources.[2] The use of such evidence is permissible, and generally essential, in cases where ambiguity exists within the document.

In *Wigram on Wills*, the author states "A written instrument is not ambiguous because an ignorant and uninformed person is unable to interpret it. It is ambiguous only if found to be of uncertain meaning when persons of competent skill and information are unable to do so."

Judge Story, in the case of *Piesch v. Dickson*[3] said that the rules in regard to ambiguities are simple; the difficulty lies in the application of these rules to particular instances.

[1] *Curran v. Maple Island Resort Ass'n.*, 14 N.W.2d 655, 308 Mich. 672 (1944).

[2] *Black's Law Dictionary.*

[3] 1 Mason 9, Fed. Cas. No. 10,911 (1815).

Evidence to Aid Construction. Where an ambiguity exists in an instrument, extrinsic evidence is admissible, not to contradict or vary the terms of the description, but to place all the facts, circumstances, and positions of the parties before the court in order that the true intent of the grantor be determined.[4]

In the case of *Guilmartin v. Wood*,[5] it appeared that a conveyance described the lot in dispute as being on the east side of the street, whereas the lot intended to be conveyed was on the west side of the street. The court held that parol evidence was not admissible to show title in the lot intended to be conveyed. The court said, "Parol evidence is admissible to explain an ambiguity that does not appear on the face of the writing, but arises from some extrinsic, collateral matter—to point out, and connect the writing with the subject-matter, and to identify the object proposed to be described. The oral evidence must not be inconsistent with the writing. In construing a deed, an ambiguity in the description of the premises conveyed may be explained by parol evidence; and, where the description is by metes and bounds, evidence of the situation and locality of the premises, and of their identity, according to the description in the conveyance, is admissible. But such evidence is not admissible to show a mistake in the description, or to alter or vary the boundary, or to substitute another and different boundary for the one expressed in the conveyance.

It may be conceded that this part of the description is a mistake of the draftsman of the deed. If so, a court of law is without power to reform it; and, without having been reformed by a court having jurisdiction, it is inoperative to pass the legal title to land situate on the west side of the street."

Definitions. The term "ambiguity" is interpreted as connoting any doubt, uncertainty, double meaning, or vagueness that is inherent in the descriptive words themselves, or that may arise in the application of the description to its subject, the surface of the earth.

A *latent ambiguity* in the description of land is an uncertainty which does not appear on the face of the instrument, but which is shown to exist for the first time by matter outside the writing, when an attempt is made to apply the language to the ground.[6]

A *patent ambiguity* in the description of land is such an uncertainty appearing on the face of the instrument that the court, reading the language in the light of all the facts and circumstances referred to in the instrument, is unable to derive therefrom the intention of the parties as to what land is to be conveyed.[7]

---

[4] Wide latitude is permitted in the interpretation of evidence to determine the intention of parties to a deed. *Gulf View Courts v. Galveston County*, 150 S.W.2d 872 (Tex. Civ. App., 1941).

[5] 76 Ala. 204 (1884).

[6] *Reed v. Locks & Canals*, 8 How. 274, 12 L.Ed. 1077 (1850); *Atkinson v. Cummins*, 9 How. 479, 13 L.Ed. 223 (1850); *Nichols v. Turney*, 15 Conn. 101 (1842); *Abbott v. Abbott*, 51 Me. 575 (1863); *Dorsey v. Hammond*, 1 Harr. & J. 190 (Md., 1801); *Storer v. Freeman*, 4 Am.Dec. 155, 6 Mass. 435 (1810); *Bell v. Woodward*, 46 N.H. 315 (1865).

[7] Ibid.

Contemporaneous Instruments. In ascertaining the intent of a deed, separate or different instruments may be construed together, where they are executed at the same time, and related to the same subject matter, and are between the same parties, or are parts of the same transaction. Decisions of the courts have included the following:[8]

- Where a deed is made to correct a mistake and supply an omission in a previous deed between the same parties
- Where a separate instrument contains a declaration of the grantor's objects and purposes
- Where a deed has been executed in compliance with a bond previously given by the parties for a deed
- Where one instrument is attached to, and made part of, the other
- Where there is an incorporation by reference to other instruments to orders or judgments

On the other hand, two instruments that are not related to the same subject matter may not be construed together, nor will they be so construed when not part of the same transaction. Even where they are executed at the same time and refer to the same subject matter, they cannot be construed together as one instrument or as constituting parts of one transaction where they are not between the same parties.[9] In construing an unambiguous conveyance, resort cannot be made to prior transactions, except where two parties claim under a common source of title, wherein it has been held that the rights and interests of the parties were governed by the provisions of the deed of the original grantor.[10]

Surrounding Circumstances. Where the language of an instrument is ambiguous, the intention of the parties may be ascertained by a consideration of the surrounding circumstances existing at the time of its execution.[11] In so doing, the

---

[8] *Metzger v. Miller*, D.C.Cal., 291 F. 780 (1923), *Arcari v. Strouch*, 158 A. 222, 114 Conn, 200 (1932).

[9] A plan of other land, acquired by the grantor from a different source than the land conveyed is not admissible as an aid to the construction of a deed conveying a right of way with the land. *Hurd v. General Electric Co.*, 102 N.E. 444, 215 Mass. 358 (1913).

[10] Ibid.

Where ambiguity in a chain of title exists, the intention of the parties may be ascertained from a previous deed. *Weniger v. Ripley*, 293 P. 425, 134 Or. 265 (1930).

[11] In the construction of a valid deed ambiguous on its face there are but two possible sources of information, namely, the facts and circumstances of the case and the intention of the parties as declared by them before, at or after the making of the deed. *Arab Corp. v. Bruce*, C.C.A. La., 142 F.2d 604 (1944).

Deed must be given fair and reasonable meaning to harmonize with circumstances under which it was given. *Brito v. Slack*, Civ. App., 25 S.W.2d 881.

court will place itself as nearly as possible in the position of the parties when the instrument was executed and interpret the document in the light of the facts then known by the parties. The object of the admission and consideration of evidence of surrounding circumstances is to aid the court in construing the language of the instrument and ascertaining the grantor's intent therefrom, and the surrounding circumstances will not be permitted to place a construction on the deed inconsistent with the words used so as to add to or detract from or alter the intent; hence, where the language of the deed is plain, certain, and unambiguous, the surrounding facts and circumstances will not be considered.[12]

Where the language of the deed is ambiguous, the court may consider its origin and the sources of its derivation, preliminary negotiations of the parties, all the surrounding circumstances or the existing state of facts, the situation of the parties and of the property,[13] or the condition or state of things granted at the time, the state of the country, the relationship of the parties, the state of the law at the date of the deed,[14] previous agreements that the deed is to carry into effect, the object or purpose to be subserved, and the person by whom the deed is drawn.[15] Generally, all sources of inquiry naturally suggested by the description, or that may have acted on the minds of the parties, are open to examination within the limits of the rules relating to parol evidence in such cases. In this connection, it may be noted that a number of facts all pointing the same way may have an effect that no one of them would have had alone. The deed, however, must receive its construction as of its date and the date of its delivery, and not in the light of subsequent events.[16]

---

[12] *Lassiter v. Goldblatt Bros.*, 41 N.E.2d 803, 220 Ind. 215 (1942), *Ruprecht v. Nicholson*, 264 P. 332, 88 Cal.App. 762 (1928), *Duffield v. Duffield*, 127 N.E. 709, 293 Ill. 300 (1920), *Sword v. Sword*, 252 S.W. 2d 869 (Ky, 1952).

[13] Intention of the parties to deed may be ascertained from language as a whole and surrounding facts, such as continued possession of grantor after execution and delivery of deed. *Mitchell v. Spillers*, 47 S.E.2d 564, 203 Ga. 565 (1948).

In interpreting grantor's intent from a deed, trial court is permitted to take into account material circumstances and pertinent facts known to the parties at the time of execution. *Darman v. Dunderdale*, 289 N.E.2d 847, 362 Mass. 633 (1972).

In the construction of a grant, the court will take into consideration the circumstances attending the transaction, and the particular situation of the parties, the state of the country, and of the thing granted, at the time, in order to ascertain the intent of the parties. *Adams v. Frothingham*, 3 Mass. 352, 3 Am. Dec. 151 (1807).

[14] Widespread opinion as to law, justified by judicial opinions, must be given weight in ascertaining intent of parties to particular instrument. *Erickson v. Ames*, 163 N.E. 70, 264 Mass. 436 (1928).

[15] Estate created by deed held to be determined, not alone from the words, but from the situation, circumstances, and the context and the facts as to whether the scrivener knew the use of legal technical words. *Smith v. Bachus*, 70 So. 261, 195 Ala. 8 (1915).

[16] *Valdez Bank v. Von Gunther*, 3 Alaska 657 (1909), *Los Angeles County v. Hannon*, 112 P. 878, 159 Cal. 37 (1910), Ann. Cas. 1912B 1065, *Crocker v. Cotting*, 44 N.E. 214, 166 Mass. 183 (1896), 33 L.R.A. 245, *Harvey v. Inhabitants of Sandwich*, 152 N.E. 625, 256 Mass. 379 (1926).

The case of *Thomas et al. v. Texas Osage Co-op Royalty Pool, Inc., et al.*[17] is an example where the court took into consideration many of the aforementioned items in arriving at its decision, wherein it stated:

> Where, in addition to describing eight tracts by metes and bounds, the deed provided that the conveyance was to cover all lands now owned by the grantor in stipulated surveys, whether therein "properly" described or not, "containing 276.5 acres of land more or less," and the grantor in fact owned 434.75 acres of land in the surveys, the general description had reference only to lands actually described or attempted to be described by metes and bounds and did not enlarge the grant.

> Where, after describing the eight tracts by metes and bounds, the deed provided that the conveyance was to cover all the lands owned by the grantors in stipulated surveys, whether therein "properly" described or not, "containing 276.5 acres of land more or less," but it was twice recited in the deed that eight tracts, when added together, contained 276.5 acres of land, more or less, and, at the time of the conveyance, the grantor owned 434.75 acres of land in the surveys in question, the deed was ambiguous and susceptible of more than one construction.

> A particular description in a deed does not override a general description where it appears that the property covered by words of particular description is not the whole of the property intended to be conveyed and that words of general description are intended to have an enlarging effect; and where both general and particular description refer to the same land, and the two cannot be reconciled, the particular description controls the general one.

> When an instrument does not, by its terms, clearly and plainly describe the land affected, or where it is phrased in language susceptible of more than one construction, the intention of the parties is to be ascertained, not solely from the words of the instrument, but from its language when read in the light of the circumstances surrounding the transaction.

> Where there is ambiguity in the deed, the court may consider extrinsic facts and circumstances in aid of the description therein, and this rule is applicable either when uncertainty appears on the face of the deed or when an effort to apply the description to the ground gives rise to an ambiguity.

A general description may be looked to in aid of a particular description that is defective or doubtful, but not to control or overrule a particular description about which there can be no doubt.

> As heretofore stated, after describing the eight tracts by metes and bounds, the deed provides that the conveyance is to cover all lands now owned by grantors in the above stipulated surveys whether herein "properly" described or not, "containing 276 1/2 acres, more or less." The grantors did not say that they intended to convey all the lands in the mentioned surveys, whether described therein or not, but said whether "properly" described therein or not. To our minds this means that the land intended to be conveyed

---

[17] 248 S.W.2d 201 (1952).

had been attempted to be described. The McKnight tract was not described, properly or otherwise. If the grantee had said all land in the mentioned survey or surveys whether described therein or not, it would be more plausible to say that the clause was inserted in the deed to enlarge the particular description therein. The language used in our opinion, does not enlarge the grant but refers to the land described by metes and bounds. The rule is that a particular description in a deed does not override a general description of the land where it appears that the property covered by the words of particular description is not the whole of the property intended to be conveyed and that the words of general description are intended to have an enlarging effect. Reference to 41 Tex. Jur., 1040, the court stated, "where a general and a particular description refer to the same land and the two cannot be reconciled, the particular description controls the general one."

It will be noted that the first tract in the deed described the land as being a part of the original Josiah Allen Survey but contains the further provisions that "the part herein conveyed described as follows," then the land conveyed out of the Josiah Allen Survey was described by metes and bounds.

It is not clear that it was the intentions of the grantors to convey not only the land described by metes and bounds, but all of the land owned by them in the surveys mentioned. To our minds, the deed is susceptible to the construction that the grantors only intended to convey the land described or attempted to be described by metes and bounds.

In the second deed the property is described as a part of the Josiah Allen Survey and described by metes and bounds as containing 83 1/2 acres, more or less. This deed contains the provision, it being mutually understood and agreed that this conveyance is to cover all lands now owned by the grantors in the above stipulated surveys whether herein properly described or not and containing 83 1/2 acres more or less." There the grantor and the grantee had an opportunity to again describe the McKnight tract by metes and bounds. This was not done and is a strong circumstance tending to show that it was not the intention of the parties that the McKnight land be conveyed by the deed. The parties, in failing to describe the McKnight land by metes and bounds, in the second deed, placed their construction upon the first deed.

Practical Construction: Subsequent Action of the Parties.

Where a deed is of doubtful meaning, or the language therein is ambiguous, subsequent acts of the parties, showing the construction they put on the instrument, are entitled to great weight determining what the parties intended.[18] However, the rules as to practical construction of an instrument only apply when the language on its face is doubtful, uncertain, or ambiguous. Further, in order that it may apply in a given case, it must appear that the particular construction was participated in by all the parties in interest.[19]

---

[18] Where parties put a reasonable practical construction on a deed and adhere to the construction for a period of years, that construction is binding on subsequent purchasers who have no notice of any different oral arrangement. *Mulder v. Stands*, 226 P.2d 463, 71 Idaho 22 (1956).

[19] Where deed could be construed within its four corners, evidence of acts of grantees after conveyance had been accomplished had no probative value. *Schwab v. Schwab*, 112 N.Y.S.2d 354, 280 App. Div. 139 (1952).

In order to constitute a "practical construction" of a deed, it must reasonably appear that acts were done with knowledge and in view of a purpose at least consistent with that to which they are sought to be applied. So, the mere statement by the grantor as to the meaning of the deed, not carried into effect by any act, does not show such a "practical construction" as should govern the interpretation of the deed. A party failing to object cannot be held to have placed a practical construction on a deed that will enlarge his rights. In addition, a grantee in a deed who attempts to procure a new deed from the grantor because of a doubt as to the meaning of the original deed does not lose any rights granted by the original deed.[20]

USE OF PAROL EVIDENCE

Whenever the terms of an instrument are susceptible of more than one interpretation, or an ambiguity arises, or the intent and object of the instrument cannot be ascertained from the language employed therein, parol evidence may be introduced to demonstrate what the parties had in mind at the time of making the contract or executing the instrument, and to determine the object for or on which it was designed to operate. Written words may have more than one meaning, and while parol evidence will not be allowed to change a "plain meaning," it may be used to eliminate a doubtful one.[21] If a written contract is so ambiguous or obscure in its terms that the contractual intention of the parties cannot be fully understood from a mere inspection of the instrument, extrinsic evidence may be received to enable a court to make a proper interpretation of the instrument, and in such a case the extrinsic evidence is considered to be an aid to the interpretation.[22]

In determining, so far as the parol evidence rule is concerned, whether or not an ambiguity exists in a document, the test lies, not necessarily in the presence of particular ambiguous words or phrases, but rather in the meaning of the document itself, whether or not particular words or phrases in themselves are uncertain or doubtful in meaning. Accordingly, a document may be ambiguous so as to warrant the admission of parol evidence notwithstanding the fact that it contains no words or phrases which are ambiguous in themselves. The ambiguity in the document may arise solely from the unusual use of otherwise unambiguous words or phrases. An ambiguity may arise from words which are plain in themselves, but uncertain when

---

[20] *Arab Corp. v. Bruce*, D.C. La., 50 F.Supp 350 (1943), affirmed, C.C.A., 142 F.2d 604 (1944), *Fullagar v. Stockdale*, 101 N.W. 576, 138 Mich. 363 (1904), *Kentucky Diamond Min. & Dev. Co. v. Kentucky Transvaal Diamond Co.*, 132 S.W. 397, 141 Ky. 97 (1910), Ann.Cas. 1912C 417.

[21] Words on a plat indicating a conditional dedication of a road were ambiguous as regards dedication and, being subject to more than one meaning, parol evidence was properly admitted to ascertain the intent of the parties with respect thereto. *Houston v. McCarthy*, 340 S.W.2d 559 (Tex. Civ. App., 1960).

[22] *Schmittler v. Simon*, 21 N.E. 162, 114 N.Y. 176 (1889); *Lossing v. Cusman*, 88 N.E. 649, 195 N.Y. 386 (1909).

applied to the subject matter of the instrument, thereby warranting the admission of parol evidence.[23] Where the subject matter of a written contract or other instrument is not properly or clearly identified or described, or where the description or specification of the subject matter is ambiguous or uncertain, parol, or extrinsic evidence is generally held admissible for the purpose of clarifying the instrument in this respect. In short, an ambiguity, so far as the parol evidence rule is concerned, may arise from the use of words if either their meaning or their application is doubtful or uncertain, and parol evidence is admissible to explain the meaning of words used in a writing which are ambiguous when applied to the subject matter, as well as when the meaning of the writing, looking only at the language thereof, is uncertain.[24]

To be admissible under the ambiguity exception to the parol evidence rule, the oral testimony must clarify an existing ambiguity, and cannot establish an understanding at variance with the plain terms of the written instrument. The evidence sought to be introduced for explanatory purposes must not be inconsistent with the written terms of the instrument, and a word or term in a contract, to be ambiguous, must have some "stretch" in it—some capacity to connote more than one meaning—before parol evidence is admissible. The mere fact that there is a dispute between the parties as to the interpretation of a document does not mean that there is an ambiguity justifying the admission of parol evidence for explanatory purposes.[25]

Construction of Instrument. The principles governing the admissibility of parol evidence as an aid to the interpretation or construction of written instruments *follow the general rules with regard to the construction of such instruments.* For instance, where writing is such that no construction is called for, such as where the language is plain and unambiguous and the intent of the parties is to be gathered from the four corners of the instrument, it follows that parol evidence is not admissible for the purpose of showing the meaning of the language used in the writing. On the other hand, where the language in the writing is ambiguous or uncertain to the extent that its meaning cannot be determined from the application of the general rules of construction to the instrument itself, so that the surrounding circumstances must be considered to ascertain the intention of the parties, parol evidence is admissible to explain, rather than to vary or contradict, the meaning of

---

[23] *Evidence aliunde* is admissible in all cases where there is a doubt as to the true location of a survey or a question as to the application of a grant to its proper subject matter. *Asberry v. Mitchell,* 93 S.E. 638, 121 Va. 276 (1917).

[24] *Midkiff v. Castle & Cooke, Inc.* 368 P.2d 887, 45 Hawaii 409 (1962); *Harten v. Loffler,* 212 U.S. 397, 53 L.Ed. 568, 29 S.Ct. 351 (1908), *Kramer v. Gardner,* 116 N.W. 925, 104 Minn. 370 (1908), *Arnold's Estate,* 87 A. 590, 240 Pa. 261 (1913), *Klueter v. Joseph Schlitz Brewing Co.,* 128 N.W. 43, 143 Wis. 347 (1910).

[25] *Conrad Milwaukee Corp. v. Wasilewski,* 141 N.W.2d 240, 39 Wis.2d 481(1968), *Buckbee v. P. Hohenadel Jr. Co.* (C.A.7) 224 F. 14, *Kilbourne-Park Corp v. Buckingham,* 404 P.2d 244 (Wyo, 1965); *O'Connor Oil Corp. v. Warber,* 141 N.W.2d 881, 30 Wis.2d 638 (1966), *Midkiff v. Castle & Cooke, Inc.,* 368 P.2d 887, 45 Hawaii 409 (1962).

the language used and thus resolve the ambiguity or uncertainty. Accordingly, in considering the admissibility of parol evidence to ascertain the meaning of the language used in a written instrument, the general rules of construction of such instruments, as developed in other chapters, should also be considered. As an example, even though a written instrument may be ambiguous or its terms inconsistent or conflicting, parol evidence may nevertheless be inadmissible where the ambiguity, inconsistency, or conflict may be resolved, without going outside the instrument, by applying the principles of construction concerning the controlling effect among written, typewritten, and printed matter.[26]

In a number of cases involving written instruments, the courts have stated that parol evidence is admissible to explain the writing or to show the purpose and character of the transaction,[27] or to show the object of the parties in executing the instrument. An examination of the facts in these instances usually discloses that the basis of the admission is either ambiguity in the writing or the fact that the evidence will aid in the interpretation of the writing by placing the court in the position of the parties at the time they entered into the contract. Where the terms of a writing are ambiguous, parol evidence is generally held to be admissible to clarify the ambiguity. Where the language of the contract is plain and unambiguous, however, parol evidence as to its meaning is generally held to be inadmissible.[28]

In the admission of parol or extrinsic evidence, the line that separates evidence that aids the interpretation of what is in an instrument from direct evidence of intention independent of the instrument must be kept in view, since it is the duty of the court to declare, not what was intended to be written, but the meaning of that which was actually written, in the instrument. In other words, the extrinsic evidence is

---

[26] There is an abstract distinction between "construction" and "interpretation," in that "construction" is the drawing of conclusions from elements known from, given in, and indicated by, the language used, while "interpretation" is the art of finding the true sense of the language itself, or of any form of words or symbols. In other words, "interpretation" is used with respect to language or symbols themselves, while "construction" is used to determine, not the sense of the words or symbols, but the legal meaning of the entire contract. However, in many contract cases it would be difficult, as well as unrewarding, to determine where "interpretation" ends and "construction" takes over. It is doubtful if the abstract distinction between the terms is of any vital significance, and certainly, in common and legal usage, "construction" and "interpretation" are used interchangeably. See 17 Am Jur2d, Contracts, § 240; 30 Am Jur 2d Evidence, § 1065.

For references, see cases under *Bills and Notes, Building and Construction Contracts, Contracts, Deeds, Insurance, Landlord & Tenant, Vendor & Purchaser,* and *Wills.*

[27] Testimony that a survey was made for the purpose of running a line between swamp and upland is not admissible as tending to vary a signed certificate on a plat showing such line, that it is the plat referred to in a certain deed and mortgage, and that the line shown as separating the upland from the lowland is the line agreed on in such deed and mortgage. *Halsey v. Minnesota-South Carolina Land & Timber Co.,* 177 S.E. 29, 174 S.C. 97 (1934), see 100 ALR 1.

[28] *Cooper v. Berry,* 21 Ga. 526 (1857), *Wilson v. Cochran,* 48 Pa. 107 (1864), *Baker v. Gregory,* 28 Ala. 544 (1856), *Smith v. Vose & Sons Piano Co.,* 80 N.E. 527, 194 Mass. 193 (1907), *Whitaker v. Lane,* 104 S.E. 252, 128 Va. 317 (1920), see 11 ALR 1157; *Brick v. Brick,* 98 U.S. 514, 25 L.Ed. 256, *Murray v. Gadsden,* 91 App. DC 38, 197 F.2d 194 (1952), see 33 ALR2d 554.

admissible, not to show that the parties meant to say something other than what they stated in the instrument, but to show what they meant by what they stated in the instrument. To be admissible under the ambiguity exception to the parol evidence rule, the oral testimony must clarify an existing ambiguity, and cannot establish an understanding at variance with the plain terms of the written instrument.[29]

Patent and Latent Ambiguities. In the past, ambiguities were classified as *patent* and *latent*, and this classification may still exist in some jurisdictions. Although the prevailing general rule is now otherwise, some courts, mainly in earlier decisions, have followed the view that the admissibility of parol or extrinsic evidence to aid in the description of property based upon whether the ambiguity is a patent or latent one, parol, or extrinsic evidence being admissible in the latter case but not in the former.[30]

Both in jurisdictions where the distinction between patent and latent ambiguities is recognized and in jurisdictions where it is not, it is generally held that a "latent" ambiguity—that is, an uncertainty that does not appear on the face of the instrument but that is shown to exist for the first time by matter outside the writing—may be explained or clarified by parol evidence. This does not vary or contradict the terms of the instrument but merely aids the court in ascertaining the true intention of the parties. The theory is that since a latent ambiguity is disclosed only by extrinsic evidence, it may be removed by extrinsic evidence. An ambiguity is properly latent, in the sense of the law, when the questionable expression or obscure intention does not arise from the words themselves but from the ambiguous or obscure state of extrinsic circumstances to which the words of the instrument refer, and that is susceptible of explanation by a mere development of extraneous facts without altering or adding to the written language or requiring more to be understood thereby than will fairly agree with the ordinary or legal sense of the words made use of.[31]

There is also the view that there is an intermediate class of ambiguities, sharing the nature of both patent and latent ambiguities, which may be referred to as "mixed" ambiguities. This intermediate class has been considered to exist when the words are all sensible and have a settled meaning, but at the same time consistently

---

[29] *Payne v. Commercial Nat. Bank*, 169 P. 1007, 177 Cal. 68 (1917), *Barnhart Aircraft, Inc. v. Preston*, 297 P. 20, 212 Cal. 19 (1931), *Bauer v. Taylor*, 118 S.W.2d 826 (Tex.Civ.App.) error refused.

Parol evidence is not admissible to prove that parties intended something different from that which the written language of a deed expresses or which may be the legal inference and conclusion drawn from that language, but, where words are doubtful or ambiguous, it is competent to give in evidence existing circumstances to give definite meaning to language used and to show the sense in which particular words were probably used, especially in matters of description. *Oldfield v. Smith*, 24 N.E.2d 544, 304 Mass. 590 (1939).

[30] See 102 ALR 287.

[31] *Norton v. Larney*, 266 U.S. 511, 69 L.Ed. 413, 45 S.Ct. 145 (1924), *Claremont v. Carleton*, 2 N.H. 369 (1821), *Mumford v. Memphis & C.R.Co.*, 2 Lea (81 Tenn) 393 (1879), *Hammond v. Capital City Mut. F Ins. Co.*, 138 N.W. 92, 151 Wis. 62 (1912); *Cordas v. Wright*, 277 P.2d 520, 129 Cal.App.2d Supp. 867

allow two possible interpretations, according to the subject matter in the considera-tion of the parties. In such a class of cases, parol evidence is admissible to explain the ambiguities.[32]

The distinction between patent and latent ambiguities is gradually disappearing, and a tendency is manifested in the more recent cases to admit parol evidence even in cases of patent ambiguities, that is, parol evidence is held admissible to explain an ambiguity regardless of whether it is latent or patent.[33] In other words, the distinction between patent and latent ambiguities in determining the admissibility of parol evidence has been discarded or disregarded in a number of jurisdictions.[34]

Particular Situations. Parol evidence has been admitted to explain a wide variety of particular words and phrases that, in the instruments containing them, were of ambiguous or uncertain meaning, particularly where the words or phrases have technical or local meanings not commonly known, and where abbreviations, symbols, or figures and the like are used in substitution for words and phrases. However, where the particular words and phrases used in a written agreement have a well-understood general meaning, parol evidence is not admissible for the purpose of showing a meaning other than the generally accepted meaning.[35]

(1954), *Shelley v. Nichols*, 261 P.2d 771, 120 Cal.App.2d 602 (1953), *Taylor v. McCowen*, 99 P. 351, 154 Cal. 798 (1908); *Forsyth Mfg. Co. Castlen*, 37 S.E. 485, 112 Ga. 199 (1900).

Even under the latent ambiguity exception to the parol evidence rule, parol or extrinsic evidence is admis-sible only to explain an ambiguity, not to change or contradict the language of the written instrument. *Hardin v. Ray*, 404 S.W.2d 764 (Mo.App., 1966).

For the purposes of the admission of parol evidence, an ambiguity is latent where the written language is apparently clear and certain but becomes doubtful in the light of something extrinsic or collateral, and is patent when the language itself is doubtful or susceptible of more than one meaning. *Stoffel v. Stoffel*, 41 N.W.2d 16, 241 Iowa 427 (1950), see 14 ALR2d 891.

[32] *Hall v. Equitable Life Assur. Soc.*, 295 N.W. 204, 295 Mich. 404 (1940), *Ganson v. Madigan*, 15 Wis. 144 (1862); *Blair v. Wessinger*, 178 P. 545, 39 Cal.App. 269 (1918); *Schlottman v. Hoffman*, 18 So. 893, 73 Miss. 188 (1895).

[33] Under the enlightened modern view, there is no real difference in the rules of evidence applicable to patent and latent ambiguities. *Haupt v. Hichaelis*, 231 S.W. 706 (Tex.Com.App., 1921).

[34] *Stoffel v. Stoffel*, 41 N.W.2d 16, 241 Iowa 427 (1950), see 14 ALR2d 891; *Lambdin v. Dantzebecker*, 181 A. 353, 169 Md. 240 (1935), see 102 ALR 277; *Forsyth Mfg. Co. v. Castlen*, 37 S.E. 485, 112 Ga. 199, *Shore v. Miller*, 4 S.E. 561, 80 Ga. 93 (1887).

[35] *O'Hear v. DeGoesbriand*, 33 Vt. 593 (1861); *Johnston v. Cox*, 154 So. 206, 114 Fla. 243 (1934).

The terms "about," "more or less," and terms of similar import in contracts of sale, indicating the amount of the subject matter, have been held not to produce ambiguities, and therefore, parol evidence thereof is held not to be admissible. *Brawley v. United States*, 96 U.S. 168, 24 L.Ed. 622 (1877). On the contrary, parol evidence of the circumstances existing at the time the contract for the sale of real property was made has been held admissible for the purpose of identifying the premises and of removing the ambiguity cre-ated by the use of the word "about." *Harten v. Loffler*, 212 U.S. 397, 53 L.Ed. 568, 29 S.Ct. 351 (1908).

Land Identified by Descriptive Name. Courts have been liberal in allowing parol evidence to identify land conveyed in a deed describing it by a descriptive name by which it is known—such as the farm, in a certain place, on which the grantor then lives, land on which the grantee lives, land owned by the grantor in a certain locality, land inherited by the grantor, land conveyed to the grantor by a certain person, all in the grantor's possession, land purchased by a named person from another, the home farm, homestead, or similar terms.[36]

Explanation of Lot or Section Numbers. Parol evidence is usually allowed to identify the land intended to be conveyed where it is described merely by a lot number or a lot and block number, and it appears that there are several lots of that number or no lot of that number. Similarly, an ambiguity created by describing the property by the wrong section number may be clarified by evidence as to the surrounding circumstances, enabling the court to determine what property the deed was intended to convey. However, if the description is too indefinite and uncertain to identify the land or to furnish means of identifying it by parol evidence, such evidence is, of course, inadmissible.[37]

In the Alabama case of *Reynolds v. Lawrence*,[38] it appeared that a deed conveying land described it as the "S. 1/2 and N.E. 1/4 of N.W. 1/4 of Sec. 29," and so on. It was sought to explain this description by parol evidence alleging that the lands conveyed were the S. 1/2 of the N.W. 1/4 and the N.E. 1/4 of the N.W. 1/4 of section 29. The court held that the evidence was admissible, saying "While it is a correct general principle of law that, if an ambiguity is patent on the face of the deed, it cannot be made certain by parol proof as to what was the intention of the parties, but the instrument must be construed by the court, yet the court is entitled to the light of all the circumstances surrounding the parties, in order to enable it to determine the property intended to be conveyed by the deed. This has been called an intermediate class, partaking of the nature of both patent and latent ambiguities; and this court, speaking through Justice Stone, has clearly expressed this distinction in a case where lands were described by government numbers, yet failed to state in what county or state they were situated, and proof was permitted to be made of the fact that the party making the deed was living in a certain county in Alabama, on lands answering to said description."[39]

---

[36] *Reeves v. Whittle*, 153 S.E. 53, 170 Ga. 408 (1930); *Doolittle v. Blakesley*, 4 Day 265 (Conn., 1810); *Robeson v. Lewis*, 64 N.C. 734 (1870); *Clements v. Pearce*, 63 Ala. 284 (1879), *Hinton v. Moore*, 51 S.E. 787, 139 N.C. 44 (1905); *Robbins v. Harris*, 2 S.E. 70, 96 N.C. 557 (1887); *Marshall v. Carter*, 85 S.E. 691, 143 Ga 526 (1915); *Brusseau v. Hill*, 256 P. 419, 201 Cal. 225 (1927); *Derrick v. Sams*, 25 S.E. 509, 98 Ga 397 (1896); *Carson v. McCaskill*, 99 S.E. 108, 111 S.C. 516 (1918); *Powers v. Scharling*, 67 P. 820, 64 Kan. 339 (1902); *Abercrombie v. Simmons*, 81 P. 208, 71 Kan. 538 (1905), *Emery v. Webster*, 42 Me. 204 (1856), *Woods v. Swain*, 70 Mass. 322 (1855).

[37] *Wetzler v. Nichols*, 101 P. 867, 53 Wash. 285 (1909); *Jordan v. Tinnin*, 342 So.2d 748 (Ala., 1977); *Powell & Kendall v. Lawson*, 77 S.E. 183, 12 Ga.App. 350 (1912).

[38] 119 Am. St. Rep. 78, 40 So. 57, 147 Ala 216 (1906).

[39] *Chambers v. Ringstaff*, 69 Ala 140 (1881).

Explanations of Parts or Fractions of a Tract. Parol evidence is admissible to identify the land described in a deed as a part, one-half or other fraction, or a certain number of acres out of a particular tract of land, where there is a further description sufficient to serve as a guide to the location of the land intended to be conveyed. In the case of *Light v. Crowson Well Service, Inc.*,[40] the Louisiana court stated that parol evidence should have been considered by the trial court to aid in the construction of a mineral deed that conveyed 1/116th fractional interest in a mineral estate, but then recited that it was the intention of the parties to convey 61 "mineral acres" out of a 366-acre tract that would have produced a one-sixth undivided interest.

Land described as the "south half" of a certain quarter section may be shown by extrinsic evidence to be one-half in area of said section and not one-half according to the government survey. Where the further description is not sufficient to serve as a guide to the location of the land intended to be conveyed, parol evidence is inadmissible; the land must be pointed out with some certainty.

The admission of parol evidence in such a case has been denied on the ground that such a description is unambiguous. A deed to the "north half" of an irregular tract of land conveys one of two equal parts comprising the tract and is unambiguous. Hence, parol evidence of transactions and conversations prior to the execution of the deed is inadmissible. In the Texas case of *Rutherford v. Randall*,[41] the court stated that there was no ambiguity presented in a mineral deed that conveyed a 1/240th interest, even though the same clause provided that the interest conveyed was that which had been received from the grantor's parents, a 1/24th interest, where the recital that the conveyance was the portion previously conveyed by the grantor's parents immediately followed the language "for a better description of said land," and was thus no more than a description of the lands where the interests were located.

If a given quantity of land is excepted from a corner of a tract, it must, as a general rule, be laid off in a square form. But parol evidence is admissible to show that this was not the intention of the parties, and their intention, when shown, must be respected.[42]

Where land is conveyed in sections or subdivisions thereof, it is presumed that reference is to the public surveys of the United States. Where, however, land is conveyed in fractions of sections, as the south half of a certain section, extrinsic evidence is admissible to show that one-half in area was meant.[43]

Common or Uncommon Abbreviations. Abbreviations used in the description of property in a deed may be explained by extrinsic evidence, and some courts take judicial notice of the fact that in modern usage the terms "Township" and "Range" are frequently abbreviated.[44]

---

[40] 313 So.2d 803 (La., 1975).

[41] 593 S.W.2d 949, see 12 ALR4th 788.

[42] *Robinson v. Taylor*, 123 P. 444, 68 Wash. 351 (1912); *Lego v. Medley*, 48 N.W. 375, 79 Wis. 211 (1891).

[43] *Richards v. Renehan*, 253 P.2d 1046, 57 N.M. 76 (1953), *Prentiss v. Brewer*, 17 Wis. 635 (1864).

[44] *Hull v. Croft*, 132 Ill.App. 509 (1907), *Douglass v. Byers*, 76 P. 432, 69 Kan. 59 (1904); *Richards v. Renehan*, 253 P.2d 1046, 57 N.M. 76 (1953).

Abbreviation generally means the contraction, rather than the omission, of words, and does not embrace ambiguities arising from incomplete or indefinite statements in the writing.[45]

Parol or extrinsic evidence as to the meaning of abbreviations, signs, symbols, figures, or words which have no apparent meaning in contracts or other writings has quite generally been admitted. However, even though parol evidence is admissible to show the meaning of certain abbreviations, it has generally been held not be admissible to show the intention of the parties in using them.[46]

It was held in *Barry v. Coombe*[47] in 1828 that extrinsic evidence was admissible to show that the letters "E.B.," in the description of property in a memorandum of sale (required to be in writing under the Statute of Frauds), "Your 1/2 E.B. wharf and premises," meant eastern branch, the court being of the opinion that this was more in the nature of a latent ambiguity, as to which extrinsic evidence is admissible, or at least a mixed ambiguity, both patent and latent.

In the Illinois case of *McChesney v. Chicago*,[48] parol evidence was held to be admissible to show that the expression "sec. 23, 38, 14," as used in a description of property, meant section 23, township 38, and range 14, because such descriptions have a well-defined meaning among surveyors.

In *Hull v. Croft*,[49] it was said that courts take notice without proof of the meaning of the initials and abbreviations ordinarily employed in descriptions of land, and, if necessary, extrinsic evidence may be heard to explain the meaning of such abbreviations.

The Kansas court said in *Douglass v. Byers*,[50] that while a description of land cannot be supplied by parol evidence, such evidence is competent to explain abbreviations in a deed and clear up the ambiguities.

Parol evidence was also held to be admissible in the case of *Converse v. Wead*[51] to show that "Ill. C. R.R. Co.," "Jas. S. Allen," "D," "Mar. 5, 69," "E1/2 S.W. 1/4," in extracts and minutes of records of deed, meant respectively "Illinois Central Railway Company," "James S. Allen" grantee, "Deed," "March 5, 1869," the date of the deed, and the description of the property. The court said that parol evidence may be received to explain the meaning of abbreviations in written instruments and show the words for which they stand.

On the ground, that parol evidence is admissible to explain abbreviations in writings which in themselves are unintelligible, where such explanation is not inconsistent

---

[45] *Carland v. Western U. Telegraph Co.*, 76 N.W. 762, 118 Mich. 369 (1898), 43 L.R.A. 280, 745 Am.St.Rep. 394; *DeLavalette v. Wendt*, 75 N.Y. 579, 31 Am.Rep. 494 (1879). see 100 ALR 1465, "Admissibility of parol evidence as to meaning of cryptic words, abbreviations, signs, symbols, or figures appearing in written contracts or other writings."

[46] Ibid.

[47] 1 Pet. (U.S.) 640, 7 L.Ed. 295 (1828).

[48] 173 Ill. 75, 50 N.E. 191 (1898).

[49] 132 Ill. App. 509 (1907).

[50] 69 Kan. 59, 76 P. 432 (1904).

[51] 142 Ill. 132, 31 N.E. 514 (1892).

with the written terms, parol evidence was held admissible in the Indiana case of *Barton v. Anderson*[52] to show that "120 ft. Wash. St. S.W. Cor. out 66," as used in a tax duplicate in describing a piece of land, meant "120 feet on Washington Street, south-west corner, out-lot 66."

Parol evidence was again held to be admissible to explain the meaning of figures and lines appearing on a plan or diagram for the improvement of a street, in *Hyde Park v. Andrews*,[53] where the court said: "Such diagrams are seldom fully explained by notes or references attached, so as to be fully understood except by persons having skill as surveyors. The field notes of the surveys would not be understood by a majority of the community unaided by a person who is familiar with their meaning. They consist of contractions and technical words peculiar to the science, that are not intelligible to persons generally. Hence, in most disputed questions of these surveys, to enable a jury to determine true lines, courses, and distances, resort must be had to oral evidence explaining the field notes and plats."

Illegible, Canceled, and Obliterated Words. Sound reason as well as authority favors the admissibility of extrinsic evidence when offered for the purpose of reproducing words that are illegible or that have been canceled or erased. In the case of *Duffin v. People*,[54] the court stated that parol evidence may be given in the case of a note written in ink that faded quickly, where a photographic copy was made before the writing disappeared. In addition, the Maine court stated in *Fenderson v. Owen*[55] that where it was necessary to determine the date of a paper offered in evidence, and the name of the month was so inartificially written that upon inspection the presiding judge was unable to determine whether it should read June or January, extraneous evidence was admissible to show the true date.

Under such circumstances, such evidence does not vary or contradict the writing, but is necessary in discovering what the writing actually is.[56]

Technical and Trade Terms. Parol evidence is always receivable to define and explain the meaning of words or phrases in a written instrument which are technical and not commonly known, or which have two meanings—the one common and universal and the other technical. Additionally, where a new and unusual word or phrase is used in a written instrument, or where a word or phrase is used in a particular sense as applicable to a particular trade, business, or calling or to any

---

[52] 104 Ind. 578, 4 N.E. 314 (1892).

[53] 87 Ill. 229 (1877).

[54] 107 Ill. 113 (1883).

[55] 54 Me. 372 (1867).

[56] *Goldsmith v. Picard*, 27 Ala 142 (1855), *Robinson v. Cutter*, 40 N.E. 112, 163 Mass. 377 (1895), *Thomas v. Thomas*, 19 N.W. 104, 76 Minn. 237 (1899); *Duffin v. People*, 107 Ill. 113 (1883), *Fenderson v. Owen*, 54 Me. 372 (1867).

particular class of people, it is proper to receive extrinsic evidence to explain or illustrate the meaning of that word or phrase. For example, in the case of *Miller v. Wiggins*[57] the Pennsylvania court decided that a contract to pay for masonry at a certain rate per perch "measured in wall" is not so plain that only one conclusion can be drawn as to the method of measurement, and therefore, parol evidence was admissible to explain the trade meaning of the words quoted.

Such evidence neither varies nor adds to the written memorandum, but merely translates it from the language of trade into the ordinary language of people in general. Under this rule, parol evidence is admissible to show that seemingly ambiguous statements of description have a recognized meaning in the trade or business to which the contract relates.[58]

A well-recognized technical meaning of a term does not necessarily preclude oral evidence of an intended or understood modified meaning, where the circumstances and language in connection with which it is used tend to obscure it and leave in doubt the light in which the parties to the agreement regarded it.[59]

Local Terminology. Although extrinsic evidence is not admissible to change the meaning of a word having a general well-defined signification, if a word is used which has no definite and specific general meaning, its local meaning may be proved. But a party to a written contract may not introduce parol evidence of the interpretation put upon the language of such contracts in states other than that according to the laws of which the contract is to be interpreted, to show that he intended to use the language in accordance with the interpretation given to it in such other states, and that the other party had reason to know that he was doing so.[60]

Figures, Abbreviations, and Characters; Signs and Symbols. Where an abbreviation, symbol, or figure used in a contract has a plain unambiguous meaning, parol evidence is not admissible to show a meaning different from that called for by its terms. Moreover, parol evidence is not admissible to show the intention of the parties in using abbreviations, signs, symbols, or figures, or to show what was said about their use. But where their meaning is ambiguous, uncertain, or obscure, parol evidence may be received to show in what sense figures or abbreviations were used in a contract, to explain abbreviations, characters, or marks, as used in a particular

---

[57] 76 A. 711, 227 Pa. 564 (1910).

[58] *Salmon Falls Mfg. Co. v. Goddard*, 14 How (U.S.) 446, 14 L.Ed. 493, *Re Curtis*, 30 A. 769, 64 Conn. 501 (1894), *Levi v. Schwarz*, 95 A.2d 322, 201 Md. 575 (1952), see 36 ALR2d 1241, *Ganson v. Madigan*, 15 Wis. 144 (1862); *Coughlin v. Bair*, 262 P.2d 305, 41 Cal.2d 587 (1953), *Hatch v. Douglas*, 48 Conn. 116 (1880), *Fairly v. Wappoo Mills*, 22 S.E. 108, 44 S.C. 227 (1894); *Maurin v. Lyon*, 72 N.W. 72, 69 Minn. 257 (1897).

[59] *Brown v. A.F. Bartlett & Co.*, 167 N.W. 847, 201 Mich. 268 (1918).

[60] *Galena Ins. Co. v. Kupfer*, 28 Ill. 332 (1862), *Collender v. Dinsmore*, 55 N.Y. 200 (1873), *Inman Mfg. Co. v. American Cereal Co.*, 110 N.W. 287, 133 Iowa 71 (1907).

business, which are unintelligible to persons unacquainted with such business, and to show the meaning of abbreviations, signs, symbols, figures, and words, in contracts or other writings, which have no apparent meaning, provided such explanation is consistent with other terms of the contract. In some such cases, parol evidence has been admissible on the theory that it merely identified the subject matter of the contract. In other cases, it has been admitted on the theory that the abbreviations, and so forth, have a recognized meaning in a given trade, and that the evidence merely translates the abbreviation from the language of the trade into ordinary language. If the language of a writing has a secret meaning, it is proper to show the fact. Codes or ciphers play an important part in business affairs, and there seems to be no reason why they are not as much proper subjects for translation as foreign languages.[61]

Practical Construction by the Parties. Whenever the language of a description renders the location of land doubtful, the construction put on the deed by the parties in locating the premises may be resorted to in order to determine their intention, especially with respect to boundaries. In the Maine case of *Moore v. Fletcher*,[62] the court stated that in the determination of what passes under a grant of a mill privilege or the privilege of a mill, the use that was made of the mill by the grantor is an aid in ascertaining the property that passes; the property passing cannot be curtailed by proof that less land might be used during the operation of the mill than was in fact used.

Statements of the parties are sufficient for this purpose, but unless there is an ambiguity such evidence is not admissible. In addition, a practical construction, when admissible at all, must be participated in by all the parties.[63]

Exceptions and Reservations. Parol evidence is held by most courts to be admissible for the purpose of aiding in the construction of the provisions of deeds as to reservations or exceptions by the grantor where the meaning or application of the provision in question is doubtful or ambiguous. However, such evidence is generally inadmissible where the meaning of the provision in question is clear and

---

[61] *White v. Oliver*, 49 P.2d 147, 173 Okla 559 (1935), see 100 ALR 1461; *Jaqua v. Witham & Anderson Co.*, 7 N.E. 314, 106 Ind. 545 (1886), *Vogt v. Schienebeck*, 100 N.W. 820, 122 Wis. 491(1904), *National Spun Silk Co. v. Peerless Silk Mills Corp.*, 116 A. 711, 97 N.J.L. 190 (1922); *Brewer v. Horst & L. Co.*, 60 P. 418, 127 Cal. 643 (1900), *Fenderson v. Owen*, 54 Me. 372 (1867); *Berry v. Kowalsky*, 30 P. 202, 95 Cal. 134 (1892), *Dages v. Brake*, 83 N.W. 1039, 125 Mich. 64 (1900), *Maurin v. Lyon*, 72 N.W. 72, 69 Minn. 257 (1897); *United States v. Hardyman*, 13 Pet. (U.S.) 176, 10 L.Ed. 113 (1839), *Griffin v. Eerskine*, 118 N.W. 906, 131 Iowa 444 (1906); *Kirby Planing Mill Co. v. Hughes*, 75 S.E. 1059, 11 Ga.App. 645 (1912); *Carland v. Western U. Tel. Co.*, 76 N.W. 762, 118 Mich. 369 (1898).

[62] 16 Me. 63 (1839).

[63] *French v. Hayes*, 43 N.H. 30 (1861); *Chapman v. Crooks*, 2 N.W. 924, 41 Mich. 595 (1879), *Jackson v. Perrine*, 35 N.J.L. 137 (1871); *Snadon v. Gayer*, 566 S.W.2d 483 (Mo.App., 1978).

unambiguous, either intrinsically or from other provisions of the instrument, or where it would tend to vary or contradict the terms of the provision. In the case of *Holland v. State*,[64] the Florida court decided that a warranty deed, in the form prescribed by statute, and conveying a strip of land to the state, operated to convey the fee simple title of the land in the absence of words of limitation or other expressions of a contrary intention in the instrument, despite the grantors' claim that the deed was ambiguous on its face concerning the extent of the estate conveyed and thus required clarification by parol evidence.

In some instances the admission of parol evidence to aid in the construction of the instrument has been expressly limited to the showing of the surrounding circumstances, or, although not so expressly limited, has not gone beyond such showing, and in some instances in which the evidence has been held to be inadmissible it has consisted merely of oral statements or conversations of the parties, or of the unexpressed intention of the grantor, with respect to the claimed reservation or exception.[65]

While physical conditions or objects in existence at the time of a grant are usually referred to for the purpose of identifying the land excepted from a grant, references to acts or agreements of parties relating to the excepted part, or to identifying occurrences, are recognized so long as they serve as means of determining the location and boundaries of the parcel being excepted. Assuming competency, it is not material whether the extrinsic evidence consists of official records, written contracts, or parol evidence of agreements.[66] In construing reservations or exceptions in deeds, courts endeavor, if possible, to ascertain the intention of the parties, particularly that of the grantor, from the language contained in the instrument, and give that intention effect so long as it does not conflict with any rule of law. Emphasis is given to a determination of what the grantor meant by the language of the reservation, and a reasonable construction should be given. The intention of the parties in the use of words indicating an exception in the description of real estate is a question for the court.[67]

Ordinarily, a grantor is presumed to have made all the reservations and exceptions he intended to make and is not permitted to detract from his grant by showing that some reservation was intended but not expressed. If no reservations or exceptions are found in the deed, none should be presumed, and ordinarily a parol reservation or exception is ineffective.[68]

---

[64] 388 So.2d 1080 (Fla. App. DI., 1980).

[65] *Middle Creek Coal Co. v. Harris*, 265 S.W. 465, 205 Ky. 119 (1924), later app 290 S.W. 468, 217 Ky 620 (1927); *Beury v. Shelton*, 144 S.E. 629, 151 Va. 28 (1928); *Witherspoon v. Campbell*, 69 So.2d 384, 219 Miss. 640 (1954), *Countiss v. Baldwin*, 151 S.W.2d 235 (Tex.Civ.App., 1941); *Read v. Gould*, 77 S.E. 642, 139 Ga. 499 (1912).

[66] *Texas Co. v. Wall*, (C.A.7, Ill.), 107 F.2d 45.

[67] *Phoenix Title & Trust Co.. v. Smith*, 416 P.2d 425, 101 Ariz. 101(1966), *Kuhn v. Kruger*, 226 N.E.2d 902, 141 Ind.App. 161 (1967), *McDonald v. Antelope Land & Cattle Co.*, 294 N.W.2d 391 (N.D., 1980); *Bauer v. Bauer*, 141 N.W.2d 837, 180 Neb. 177 (1966).

[68] *Tong v. Feldman*, 136 A. 822, 152 Md. 398 (1927); *Cutright v. Richey*, 257 P.2d 286, 208 Okla. 413 (1953).

If the language used in a purported reservation or exception is ambiguous or obscure in its meaning, the subject matter and the attendant circumstances will be considered by the court, including the purpose for which the property or right reserved is intended to be used. The court may also look to the subsequent conduct of the parties to a deed as a strong indication of the construction which they placed on the exception or reservation contained in the instrument.[69]

PARTICULAR KINDS OF AMBIGUITIES

Through research and cursory reading, it may be seen that, under every conceivable state of facts and in every imaginable circumstance, the cases may be counted by the hundreds, if not by the thousands, where contracts, wills, and deeds are made effective by the identification, through extrinsic evidence, of the person or subject intended, and in no way violating the rule that such evidence "cannot be admitted to contradict, add to, subtract from, or vary the terms of a written instrument."

It has been said that "it is nevertheless certain that some evidence from without must be admissible in the explanation or interpretation of every contract. If the agreement is that one party shall convey to the other, for a certain price, a certain parcel of land, it is only by extrinsic evidence that the persons can be identified who claim or are alleged to be parties, and that the parcel of land can be ascertained. It may be described by bounds, but the question then becomes where are the streets, or roads, or neighbors, or monuments referred to in the description? And it may sometimes happen that such evidence is necessary to identify these persons or things. Hence, we may say as the general rule that, as to the parties or the subject matter of a contract, extrinsic evidence may and must be received and used to make them certain, if necessary for that purpose."[70]

"If the meaning of the instrument, by itself, is intelligible and certain, extrinsic evidence is admissible to identify its subjects or its objects, or to explain its recitals, or its premises, so far, and only so far, as this can be done without any contradiction of, or any departure from, the meaning which is given by a fair and rational interpretation of the words actually used."[71]

[69] *Guido v. Baldwin*, 360 N.E.2d 842, 172 Ind.App. 445 (1977), *Cravens v. Jolly*, 623 S.W.2d 569 (Mo.App., 1981), *Bauer v. Bauer*, 141 N.W.2d 837, 180 Neb. 177 (1966).

[70] *Peacher v. Strauss*, 47 Miss. 353 (1872); *Boyd v. Miller*, 117 N.E. 559, 68 Ind. App. 454 (1917).

[71] Ibid.

DEEDS, MORTGAGES, AND RELATED INSTRUMENTS

Where a mortgage described a small parallelogram of land and gave the "beginning corner" and the several courses and distances so that it might be easily identified, it was held in the Missouri case of *Orr v. How*[72] that parol evidence might be resorted to for the purpose of identifying the land as existing on the ground. Parol evidence, said the court, was not admissible to control the description of the premises in a deed.

In a 1916 Vermont case[73] the court said, "Parol evidence of intention is not received in aid of construction to make a new contract. It is permissible only when the language is capable of two or more constructions either of which preserves the integrity of the written contract. Then, by the aid of extrinsic facts, the court may determine which interpretation should be given. Within these limits the writing is not altered nor varied, but its language speaks the intention of the parties. But to carry construction beyond that point, and give a meaning to the language used of which it is not fairly capable, though found to accord with the intention of the parties, would be to set aside the writing, and substitute another and different one. This is never permissible."

The Pennsylvania court stated in 1908,[74] "It is quite impossible in most cases so to describe land as to avoid the necessity of parol proof for its identification; for whether it be described by metes and bounds, by monuments erected upon the ground, or by adjoiners, its identification becomes a subject of parol proof."

In Georgia, a description was "All of that tract or parcel of land lying and being in the town of Lavonia, Georgia, one improved lot known as lot No. 1, of the J. & P. survey, bounded as follows: On east by Red Hollow road, on south by 25-foot alley, on the west by above-named and on the north by Mrs. R. M., the said lot measuring 100 feet front and running back 300 feet." The court in this case[75] held that parol evidence was admissible to apply the description to the property conveyed, and that, for that purpose, it was competent to show where the rear line of the lot under the survey named actually was located, and that, at the time of the sale by the grantor who made the deed, there was a fence line standing, which was pointed out, recognized, and agreed on as the west line of the lot.

In an earlier Oregon case,[76] a deed conveyed a right of way by the following language: "A strip of land 40 feet wide along the bank of the Willamette river, beginning at the southwest corner of the donation land claim of John McCoy and wife in T. 10 S.R. 4 W. of Willamette meridian, etc., and extending down said river to the northwest corner of said claim, or as far as the land of F. D. Leverich extends down

[72] 55 Mo. 328 (1874).

[73] *Whittier v. Parmenter*, 90 Vt. 16, 96 A. 378 (1916).

[74] *Ranney v. Byers*, 219 Pa. 332, 123 Am. St. Rep. 660, 68 A. 971 (1908).

[75] *Haley v. Ray*, 142 Ga. 390, 82 S.E. 1058 (1914).

[76] *Wills v. Leverich*, 20 Or. 168, 25 Pac. 398 (1890).

said river." From the description, it was uncertain as to where the west line of the strip was located, since neither the east nor the west line was designated. The court held that, since the description was not clear and intelligible, parol evidence was admissible to show the situation of the parties and the circumstances surrounding the transaction, in order to determine the intention of the parties to the deed, and to give it practical effect.

In the Alabama case of *Sikes v. Shows*,[77] land conveyed by a deed was described as "lot No. 2, of square No. 8, in the town of Rutledge, being 20 feet in front, and running 110 feet back." It appeared in fact that lot No. 2 contained a 30-foot front. The court held that, of the two descriptions, the general and the particular, the general must give way to the particular. Parol evidence was then admitted to show that the 20-foot front specified in the deed was intended to mean 20 feet on the east side of lot No. 2, since that was the side that the grantee was placed in possession and that he had ever since occupied. The court stated: "Where descriptions in deeds are ambiguous or doubtful, and even void on their face for uncertainty, the courts often admit, in aid of the identification of the subject-matter, proof of the situation of the parties and the circumstances surrounding them. This embraces the facts of ownership, possession, change of occupancy, and other circumstances showing the relation of the contracting parties to each other, and to the property at the time the negotiations transpired and the writing was executed. The intention of the parties is thus elicited by showing the practical construction which they themselves placed upon their own contract."

And where it appeared that a partition deed described the boundary line as "line commencing at rock on corner of Ely and Smith streets, or better known as street south of race track, due west along street to Little Buffalo river," and it was sought, in an action to determine the true boundary, to admit evidence of the county surveyor, who made the plat and survey that was incorporated into the deed, that he surveyed the line along Eighth street, which by mistake he called Ninth street on the plat; that he was not sure at the time of the name of the street, and to make sure what street he did mean he added on the plat that it was the street better known as the street south of the race track, it was held in the Virginia case of *Edmunds v. Barrow*,[78] that the evidence was admissible, in order to apply the subject-matter to the ground, and then locate the land. Such evidence, according to the court, did not tend to vary or contradict the deed, but was introduced to determine on which street, Ninth street or the street better known as the street south of the race track, the line was then located. The court said: "There were two descriptions of the line between lots numbered 1 and 2, equally explicit, but repugnant. In locating that line, that repugnancy did appear. Ninth street and the street better known as the street south of the racetrack were not the same. The question then was upon which of those streets the line or originally located, if run."

---

[77] 74 Ala. 382 (1883).

[78] 71 S.E. 544, 112 Va. 330 (1911).

The Supreme Court of New Hampshire, in *Hall v. Davis*,[79] held that the words "Derry old line" used in a deed to define a boundary was susceptible of various meanings, as the original line of Londonderry, or any other line marked by monuments and called by that name; and that evidence that there was such another line showed such an ambiguity in the words of the deed as to admit of parol testimony to prove which line was intended to be designated thereby. The court said: "Thus, if the language of the instrument is applicable to several persons, to several parcels of land, to several species of goods, to several monuments, boundaries, or lines, to several writings, or the terms be vague and general, or have diverse meanings, in all these and the like cases parol evidence is admissible of any extrinsic circumstances tending to show what person or persons, or what things, were intended by the party, or to ascertain his meaning in any other respect; and this without any infringement of the general rule, which only excludes parol evidence of other language declaring the meaning of the parties, than that which is contained in the instrument itself."

General Description. A general description exists where the language of the instrument is concealed in broad terms, and fails to specify or point out in detail the land intended to be conveyed. In such a description, there is an absence of definite monuments, or course and distance, or even an acreage. One of the better examples is a description conveying property to a railroad.

For instance, in the Indiana case of *Indianapolis & V.R. Co. v. Reynolds*,[80] it appeared that a certain instrument, apparently a deed, granting a right of way to the defendant railroad company, providing that the grantor did "forever quitclaim to the Indianapolis & V. Railroad Company the right of way for so much of said railroad as may pass through the following described piece, parcel or lot of land in the county of Marion," and so forth. The instrument gave no further description of the right of way, nor did it define the width of the strip of land granted or to be occupied by the railroad company. The court held that the fact that a railroad corporation was empowered to lay out its road not exceeding 6 rods wide did not raise a legal presumption that a right of way of 6 rods was intended by the grant, and that the grant being ambiguous, parol evidence was admissible to show the acts and declarations of the parties at the time of the execution of the instrument, in order to determine their intention.

Also, the case of *Gowdy v. Cordts*[81] contains a typical example of a general description. In that case the facts showed that one B. owned, in the town of E., several parcels of land, among which was a 10-acre tract. He had purchased an undivided half of this 10-acre tract from T., and the other half from one S. Subsequently B. conveyed certain parcels of his land in the town of E. by deed, which first described four parcels and then "also all the undivided moiety or half part of all that

---

[79] 36 N.H. 569 (1858).
[80] 116 Ind. 356, 19 N.E. 141 (1888).
[81] 40 Hun. (NY) 469 (1886).

certain tract of land situated in the town of E.," describing the 10-acre lot, and then continuing, "being the same premises conveyed to said B. by T.," and so forth. The deed then described another lot and continued, "And also all other real estate and water front on R. creek, situate in the town of E., U. county, and state of N.Y., belonging to me or which I have any interest in." The question was whether these last words conveyed the other undivided half of the 10-acre parcel. In order to answer this question, there was an attempt to bring in parol evidence to show that B. owned lands in the town of E. other than the undivided half of the lot in dispute, which might be conveyed by the general words used in the deed. The court held such evidence was admissible to show the circumstances under which the deed was executed, but that evidence of E.'s intention was not admissible.

Land of Grantor or Mortgagor as a Boundary. Where a deed or mortgage describes a parcel of land as bounded by the land of the grantor or the mortgagor, an ambiguity arises as to the boundary thus defined. Parol evidence has uniformly been allowed in these cases.

In the Massachusetts case of *Hooten v. Comerford*,[82] it appeared that a deed described the land intended to be conveyed as "bounded northerly by a contemplated street called 'Snell street,' easterly by land of the grantor, southerly by land of B., westerly by land of the said C., the grantor." There was nothing in the deed to fix either the east or west boundary line. The court held that parol evidence was admissible to show that bounds or monuments existed at the date of the deed, which were agreed on orally by the parties as showing the true lines of the land conveyed, or were soon afterwards erected or fixed by the parties for that purpose.

And in the Connecticut case of *Carney v. Hennessey*,[83] the grantor owned, at the time of conveyance, not only the land conveyed, but also the land to the north and south of it, and described the land conveyed as bounded on the north and south by his own land. In this description, there was no point of beginning or point of ending given of any of the boundary lines; the only certainty as to the location of the boundaries was that they must include the building he had recently erected, which was expressly conveyed by the deed, and presumably must exclude his own residence. The court held that, since it was clear that a location in substantial accord with the description might exclude all the land in dispute or might include any part of it, and since there was nothing in the deed, either by way of direct description or reference to monument, course, or distance by which to determine the real location within this limitation of land conveyed, the resort must be had to the acts and conduct of the parties at the time and following the time when possession under the deed took place.

---

[82] 152 Mass. 591, 23 Am. St. Rep. 861, 26 N.E. 407 (1891).
[83] 77 Conn. 577, 60 A. 129 (1905).

Description of Land as Owned or Occupied by Grantor or Mortgagor, or as Conveyed to or Inherited by Them. Land described in a deed or mortgage as the land on which a grantor or mortgagor, or a grantee or mortgagee, lived, or as the land inherited or purchased by a grantor or mortgagor from a certain person, or as the land owned by a grantor or mortgagor, and so forth, it is practically uniformly held that the description is not so indefinite or vague as to preclude the admission of parol evidence. In such cases, parol evidence is admissible to identify the land described.

In the case of *Birchfield v. Bonham*,[84] the description in a deed was as follows: "A tract or parcel of land where he (W. Bullein)[the grantor] now lives, containing 50 acres, bounded by lands belonging to Wm. Bonham on one side, and Fall branch on the other." It appeared that the grantor had previously conveyed all the land that he had owned on Fall branch, thus leaving the land of Wm. Bonham as the only boundary for the land described in the deed. The court held that parol evidence was admissible to determine the true limits of the tract whereon W. Bullein lived at the time of the execution of the deed, and thus to identify the land intended to be conveyed by the deed.

Also, in the case of *Hetherington v. Clark*,[85] the Pennsylvania court was presented with a recorded deed which conveyed "all lots, tracts, or pieces of land and reservations, situate in the borough of P., and county of S., N. township, state of P., which we [the grantors] now possess jointly and separately, and are entitled to, and all and every part thereof," and so forth, and it appeared that the grantors had by a prior deed, which was not recorded, conveyed a certain piece of land in the same district, and held that the recorded deed contained a latent ambiguity, and accordingly parol evidence was admissible to show whether the description in that deed conveyed the land conveyed by the prior unrecorded deed. The court stated that written instruments were not to be altered by parol, but that latent ambiguities may be explained, in order that the court might give the instrument its intended effect.

Description by a Particular Name, as "Home Farm," "Homestead," "Mill Spot," and so on. In such cases it seems to be uniformly held that parol evidence is admissible to show what land is so designated, and thus to identify the tract intended to be conveyed.

In the Maine case of *Morrell v. Cook*,[86] parol evidence was held to be admissible to explain whether the description in a deed conveying "the farm," and so on, included the building that constituted the locus in question.

And where a deed described a tract of land, and concluded with the statement that "the above-described tract comprises all my [the grantor's] interest in the home tract of my deceased father," it was decided by the South Carolina court in *Carson v.*

---

[84] 22 S.C.L. (2 Spears) 62 (1843).

[85] 30 Pa. 393 (1858).

[86] 35 Me. 207 (1853).

McCaskill[87] that parol evidence was admissible to show what land was included in the home tract of the grantor's father. The court held that parol evidence to show that the grantor intended to convey all his interest in his land that he inherited from his father was not admissible, since it tended to vary the written language of the deed. The court said: "But a paramount rule is that parol testimony, whether of declarations, acts, or of the res gestae, is inadmissible to vary the terms of a written instrument. When an instrument is ambiguous, parol testimony is admissible to remove the ambiguity; but, except in cases of fraud, accident, or mistake, it is always admitted for that purpose, subject to the limitation that it must be consistent with the instrument, and, therefore, that it must not tend to contradict or vary its terms."

Similarly, in New Hampshire,[88] a description of land designated one of the boundaries as "thence by said S.'s land to the A. H. farm, so-called, thence easterly by said A. H. farm to land owned by G. H.," which was followed by a statement, "Intending to convey the homestead farm of Asa Gee." It appeared that it was impossible to run "by" S.'s land to the A. H. farm, and thence easterly by that farm to land of G. H., and the court held that the last clause in the description, "intending to convey the homestead farm of Asa Gee," should control, and the particular description be rejected for indefiniteness. The court further held that parol evidence was not admissible to show an intent of the parties to the deed different from the intent which appeared on the face of the instrument.

And, where a deed described the land intended to be conveyed as the "R. H. W. homestead," containing 200 acres, more or less," bounded by lands of certain named individuals, and "of others," whose names were not given, it was held by the South Carolina court in *Rapley v. Klugh*[89] that, since the description in the deed was expressed in incomplete terms, parol evidence was admissible to remove all uncertainty, and to identify the land conveyed. This evidence was admissible to explain and make clear that which was not clear and was ambiguous, and not to vary or contradict, or add to, that which was contained in the written instrument.

Where a tract of land intended to be conveyed by a deed was described by section numbers and also as being the "Rush Point Plantation" of the mortgagor, the Louisiana court stated in *Dickson v. Dickson*[90] that parol evidence was admissible to locate the property known as the "Rush Point Plantation," and to show that part of the description by section numbers was erroneous, in that some of the property embraced in such sections did not belong on the mortgagor, and was not a part of the "Rush Point Plantation."

In the Connecticut case of *Brooks Bank & T. Co. v. Dineen*,[91] a description in a mortgage was "About 10 acres of land more or less situated in Cornwall aforesaid known as the Tucker lot." This mortgage also referred to another deed that gave an

[87] 111 S.C. 516, 99 S.E. 108 (1919).

[88] *Peaslee v. Gee*, 19 N.H. 273 (1848).

[89] 40 S.C. 134, 18 S.E. 680 (1893).

[90] 30 La. Ann. 870 (1884).

[91] 97 Conn. 536, 117 A. 551 (1922).

imperfect description of the land. The court held that parol evidence was admissible to show the land intended to be conveyed—what particular piece of land was called the "Tucker lot."

Also, in *John L. Roper Lumber Co. v. Hinton*,[92] it was held that the description of land as "the Old Lebanon juniper swamp" was sufficiently definite to admit parol evidence to "fit the description to the thing"—the description being a "latent ambiguity"—and open to parol evidence to make it more definite.

And where a deed, in describing the land, specified no metes, bounds, or measurements, but purported to convey land by a name or designation as "the mill spot," the Massachusetts court held in *Woods v. Sawin*[93] that parol evidence was admissible to show what was intended by the name used, that is, what particular piece of ground had acquired the name of the "mill spot," by a long-established reputation, before and at the time the deed was executed. The court said that conveying land by designation was a proper mode of conveying land that by reputation had acquired a proper name, but that it necessarily called for *evidence aliunde* to show where the place so called was, and that the same evidence that proved the existence of such a place was competent to prove its limits.

The New York case of *People v. Call*[94] involved a deed that contained a reservation and exception of a piece of land described as "a parcel of fifty (50) acres in the said lot sixteen (16) and lot seventeen (17) known as Pine Point on Sacandaga lake." There was a dispute as to just what land was included in the "Point." The court stated: "Parol evidence is always admissible to apply a writing to its subject. It is a rule that, where words of general description are used in a deed, oral evidence may be resorted to for the purpose of ascertaining and identifying the premises intended to be conveyed. Such evidence is always received where doubt arises upon the face of the instrument as to its meaning, not to enable the court to hear what the parties said, but to enable it to understand what they wrote, as they understood it at the time. . . . The court, therefore, has not merely a right, but it is its duty, to inquire into all the surrounding circumstances that may have acted upon the minds of the parties, before it can approach the construction of the instrument itself. It is common experience that descriptions of property in rural communities are somewhat indefinite, and even inaccurate, and courts deal leniently with such cases in seeking to ascertain the intention of the parties. The identity of the lands conveyed under those circumstances must, generally speaking, partake more or less of a latent ambiguity explainable by testimony dehors the grant. The antiquated doctrine that a document must be construed solely within its four corners, no matter how puzzling the problem, is no longer the law of this state."

Description Only by Lot Number. Where land is described merely by a lot number, or a lot and block number, an ambiguity often arises from the fact that, on the application of the description to the ground, there appear to be several lots of the

---

[92] 260 Fed. 996 (1920).

[93] 4 Gray (Mass.) 322 (1855).

[94] 129 Misc. 862, 223 N.Y. Supp. 257 (1927).

number specified in the instrument, or there appears to be no lot as numbered in the instrument, or that there is some uncertainty as to just what land is included in the lot, and so forth parol evidence is usually allowed in such cases.

Where land conveyed was described as "lots seven (7) and eight (8) in Cove addition to the city of S.," and it appeared that Cove addition to the city of S. consisted of twelve blocks, each block containing sixteen lots, numbered from 1 to 16, inclusive, it was held by the Washington court in *Wetzler v. Nichols*[95] that parol evidence was admissible to remove the latent ambiguity in the deed, to show that the grantor owned lots 7 and 8 in block 5 of Cove addition, and not other lots 7 and 8 in that addition. When an ambiguity, said the court, was made to appear by the introduction of proof outside the deed, it was a latent ambiguity, and might be explained in the same way that it was shown.

In *Haynor v. Excelsior Springs Light, Power, Heat & Water Co.*[96] the Missouri court decided on a description of "lot six (6), block one (1), C.'s addition to Excelsior Springs." The question was whether a certain well was included in the property conveyed, plaintiff having contended that the well was a few feet east of the east line of lot 6, while the defendant contended that it was a few feet west of the same line. The court held that parol evidence was admissible, not for the purpose of contradicting the description contained in the deed, but for the purpose of showing the position of the well with reference to the boundary line defined in the deed.

In the Ohio case of *Eggleston v. Bradford*[97] it was shown that a deed purported to convey lot "No. 16, township 12, range 16, containing 640 acres," and it appeared that the number of the lot "16" was erroneously inserted in the deed. The court held that parol evidence was admissible to show what lot was intended, not to contradict the description, but to locate the deed on the land.

But in *Brandon v. Leddy*[98] the California court examined a deed under which the plaintiff claimed purported to convey "thirty varas of lot six, it being a part of block four, range eight, as shown by the plot of said city of San Jose." The map referred to in the deed, and hence incorporated therein, showed that there were two lots in the city of San Jose answering to the description in the deed—a lot 6 in block 4, range 8, north of the base line (San Fernando street) and a lot 6 in block 4, range 8, south of the base line. The grantor in the deed was the owner of the former lot but not of the latter. The court held that there was a patent ambiguity on the face of the deed, and that parol evidence was not admissible to explain this ambiguity, hence the deed was void for uncertainty.

Also, in the Illinois case of *Ritchie v. Pease*,[99] wherein it appeared that there was no lot answering the description in the deed, parol evidence was held inadmissible. There the deed described the land as "lot 9, in block 28, of Fort Dearborn addition to the city of Chicago," and it was shown there was in fact no property anywhere

[95] 53 Wash. 285, 132 Am. St. Rep. 1075, 101 Pac. 867 (1909).

[96] 129 Mo. App. 691, 108 S.W. 580 (1908).

[97] 10 Ohio 312 (1840).

[98] 67 Cal. 43, 7 Pac. 33 (1885).

[99] 114 Ill. 353, 3 N.E. 897 (1885).

known as "lot 9 in block 28, of Fort Dearborn addition to Chicago." The court said that this was not a case of latent ambiguity, for there was no property answering the description in the deed, and that it was a case requiring the objects or the calls in a deed to be identified and established, for, the words "lot 9 in block 28" being rejected, there were no objects or calls left that could be applied to any property. The court further stated that, had this property, in addition to the descriptive words used, been described as the property of Pease (the grantor) in a given locality, the impossible description might have been rejected, and parol evidence admitted to show what property answered to the description, but that "all of a person's right, title, interest, or claim" in a piece of property that, from the terms used to describe it, could have no existence, was an assertion of no ownership to any specific property.

Reference to a Map or Plat. Where a map or a plat is referred to in an instrument for a further description, there often arises some doubt as to the identity of the particular plat, when there are two or more plats of the same land. It appears that the courts have uniformly allowed the admission of parol evidence to identify which map or which plat it referred to.

A deed referred to a plat in the Illinois case of *Mann v. Bergmann*[100], and the court held that it was proper to identify the plat referred to by parol evidence, and to show that the purchase of the land was made with reference thereto, not with a view to changing the deed, but to show the circumstances under which the property was purchased.

In *Penry v. Richards*,[101] the California court held that, where a deed described the premises conveyed as a piece of land within the limits of a certain town, "known and described on the official map of said town as block No. 6, and containing 150 yards square," parol evidence was admissible to identify the map referred to, and that, when identified, it became a portion of the deed.

Also, where a deed conveying certain lots in the grantor's addition to the city of X contained a reference to a plat of the addition, and it appeared that there were two plats of this addition, drawn up at different times, the Missouri court held, in *Schreiber v. Osten*,[102] that parol evidence was admissible to show to which plat the deed had referenced, the court treating the ambiguity as latent. "Surely," said the court, "parol evidence may be given, and is always given, to locate and identify lands as described in deeds."

Description as "Part," "Half," "Fraction," and so forth, of a Particular Tract. Land described as "a part," "half," "fraction," or as a certain number of acres out of a particular tract of land, with further description sufficient to serve as a guide to the location of the land intended to be conveyed, parol evidence is admissible. But where the instrument does not contain a description sufficient to point out the way

[100] 203 Ill. 406, 67 N.E. 814 (1903).
[101] 52 Cal. 496 (1877).
[102] 50 Mo. 513 (1872).

for the identification of the land, then parol evidence is not admissible, for a conveyance must set forth a subject-matter, either certain within itself or capable of being made certain through extrinsic matter, to which the instrument refers.

In the Alabama case of *Cottingham v. Hill*,[103] a deed contained the following description of the land conveyed: "One acre of land situated on the old C. & C. road on which the schoolhouse is to be built, and more particularly described as part of N.W. 1/4 of N.W. 1/4 of section 9, township 22, range 6 west, in B. county, Ala." The court held that such a description was sufficient to admit parol evidence to identify the land intended to be conveyed by the deed.

Where a deed purported to convey land described as "the southwest quarter of section eleven, containing 40 acres," and it appeared that there were in fact four 40-acre tracts in this southwest quarter, it was held by the Missouri court in *Campbell v. Johnson*,[104] that an ambiguity existed on the face of the instrument, which was patent, since it was impossible to know from the deed what 40 acres was intended to be conveyed, and that extrinsic and parol evidence could not be admitted to explain or remove the ambiguity. The deed was, therefore, held void for uncertainty.

Also, in *Diffie v. White*,[105] the Texas court, in holding that a deed describing the land conveyed as "420 acres out of the northeast portion of the R. H. survey said that the uncertainty of description, if any, appeared on the face of the deed, and extrinsic evidence was not admissible to supply that which was lacking.

In addition, the West Virginia court, in *Davis Colliery Co. v. King*,[106] was faced with certain land excepted from a deed as "excepting, also, two other small tracts of land not to exceed in all 300 acres." The court stated that parol and other extrinsic evidence was not admissible to identify the land, since there were no words in the deed by which the land excepted could be identified; such absent words could not be supplied by parol evidence.

Where land was described in a deed as a "lot of land . . . known as the east half of the southwest division of section 17," it was held by the Ohio court in *Schlief v. Hart*[107] that parol evidence of extrinsic facts and circumstances was admissible to show that the tract of land "known as the east half" was in fact less than the mathematical half of the division. Such evidence, when admitted, according to the court, did not vary or contradict the terms in the deed.

Again, where land was described in a deed as the south half of a particular quarter section, and it appeared that the deed might be construed as referring to the south half according to the government survey, which was a prima facie presumption, or that it might be construed to refer to one half in area, or quantity, it was held by the Wisconsin court in *Prentiss v. Brewer*[108] that parol evidence was admissible to show which was the true meaning as intended by the parties to the deed.

---

[103] 119 Ala. 353, 72 Am. St. Rep. 923, 24 So. 552 (1898).

[104] 44 Mo. 247 (1869).

[105] Tex. Civ. App., 184 S.W. 1065 (1916).

[106] 80 W. Va. 371, 92 S.E. 657 (1917).

[107] 29 Ohio St. 150 (1876).

[108] 17 Wis. 635 86 Am. Dec. 730 (1862).

Description as "All the Grantor's Land" in a Certain Locality. In the case of *Euliss v. McAdams*,[109] the North Carolina court held that a description of land in a deed, as "all their [the grantor's] lands lying between Haw River and Stony Creek up to the line of" persons named, was definite enough to warrant the admission of parol evidence to identify the land.

Call for a Natural Monument. Parol evidence seems to have always been admitted to locate natural monuments, such as trees, paths, fords, and so forth, called for in a document, for it is only through such evidence that these monuments can be identified. Frequently an ambiguity arises in the application of the description to the ground, from the fact that there appear to be two or more monuments answering the description in the instrument. In such cases, parol evidence is practically unanimously permitted.

In one case, a deed gave the point of beginning as "at a pine on the east side of Gum Swamp." It was held by the North Carolina court in *Broadwell v. Morgan*,[110] that it was sufficiently definite so as to render parol evidence admissible to locate it. While parol evidence is generally not admissible to locate monuments considered temporary, it was admissible to locate a permanent monument; a pine was a natural object, and, where called for in a deed as a corner or beginning point, was understood to be permanent evidence of where the boundary was.

And where one of the boundary lines of land conveyed by a deed was described as "beginning at a large pine in B.'s line; thence west 160 poles to two small post oaks," and there was a dispute as to the location of the "two small post oaks," the court in *Echerd v. Johnson*,[111] held that the monument described in the deed was a natural one, and so its location could be aided by parol proof and by reputation.

Also, where the beginning corner of the lot intended to be conveyed was described in the deed as "at a planted stone on Williamsboro street, about 6 feet southeast of a large red oak," and it appeared that there was a stump of a red oak, and not far away a standing red oak, either of which might answer the call in the deed, it was held, again by the North Carolina court, in the case of *Taylor v. Meadows*,[112] that oral evidence of the conduct and acts of the parties to the deed was admissible, in order to ascertain the beginning corner of the lot. The court said that, had it been clear from the description in the deed where the boundary lines were located, then no one of them could be changed by parol evidence.

---

[109] 108 N.C. 507, 13 S.E. 162 (1891).

[110] 142 N.C. 475, 55 S.E. 340 (1906).

[111] 126 N.C. 409, 35 S.E.1036 (1900).

[112] 175 N.C. 373, 95 S.E. 662 (1918).

In the case of *Graybeal v. Powers*,[113] one of the calls in the deed was for "Simeon Graybeal's line." It appeared that Simeon Graybeal owned two tracts, tract number 2 lying west of tract number 1 and distant from it some 30 or 40 poles, so that it was uncertain whether the north and south line bounding tract number 1 on the west, or the north and south line bounding tract number 2 on the east, was the line referred to, and the court held that parol evidence was admissible to locate the true line.

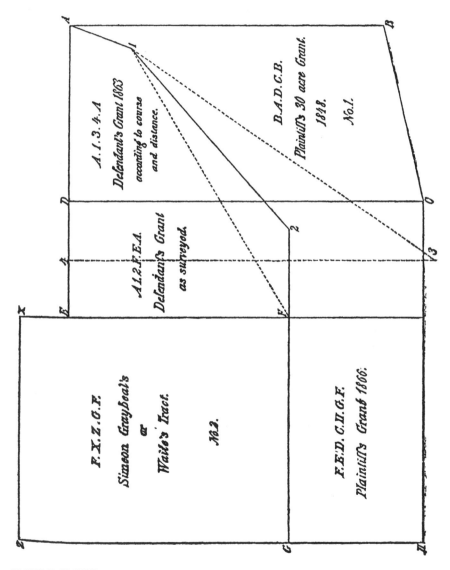

Where each of two deeds describing a boundary line referred to an oak tree at the ford in E. creek above J. V.'s residence, and it appeared that there were two oaks located at that place—one above, and one below, the ford—it was held by the Kentucky court in *Vanover v. Consolidated Coal Co.*,[114] that extrinsic and parol evidence was admissible to explain which of the two oaks was intended as the true boundary limit, the ambiguity being latent.

Where a deed described one of the boundaries of the land conveyed as "Y del camino a las lomas," which translated reads, "And from the road to the hills," and it appeared that, in the community in which the land was located, the words "las lomas" had several different meanings, such as "hills of considerable height" and "hills or bluffs," and it further appeared that there were several hills or elevations to which the words "las lomas" might apply, the court held that parol evidence was admissible to show the meaning of the term "las lomas" as used by the parties to the deed, in order to ascertain to which one of several hills or elevations the deed referred as the boundary line. The New Mexico court stated that the use of the term "las lomas," constituted a latent ambiguity, which was subject to explanation by parol.[115]

And in *Bentley v. Napier*,[116] land was described in a deed as "beginning at a rock near the forks of the Rock House Creek on the east side of said creek; thence up said Rock House with the calls of the patent to the second ford of said creek to a marked line; thence square across said creek each way down to the beginning, containing 125 acres, be the same more or less." The ambiguity in this deed lay in the location of the marked line at the second ford, since it appeared that there were two fords along the creek either of which might have been known as the second ford, and there were marked lines along each of these fords. The Kentucky court held that parol evidence was admissible to show which line was intended by the parties, from the language of the deed, to be the true line referred to in the deed. The court stated that parol evidence was always admissible to show where the objects called for in a deed were located on the ground.

[114] 193 Ky. 616, 237 S.W. 21 (1922).

[115] *Gentile v. Crossan*, 7 N.M. 589, 38 Pac. 247 (1894).

[116] 122 S.W. 180 (Ky, 1909).

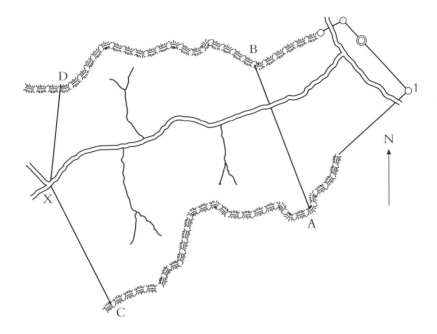

Call for an Artificial Monument. It is common for an instrument, in describing land, to refer to an artificial monument, such as a fence, a post, or a rock placed at a certain point, and so forth, and, in order to identify and locate such a monument, resort must frequently be had to parol and other extraneous evidence. Such evidence is generally admitted.

Thus, where a deed described one of the boundaries of land sold as a log fence south of one Stickney's house, and it appeared that there was at that time a log fence there, erected for temporary purposes, and also the remains of another, it was held by the New Hampshire court in *Clough v. Howman*,[117] that parol evidence was admissible to show which fence was the one intended.

Where some of the corners of land conveyed by deed were designated by posts, the Ohio court held in *Alshire v. Hulse*[118] that parol evidence was admissible to show where the posts were, since the purpose was not to contradict the deed, but to ascertain at what places on the ground the corners called for stood.

And in *Stinchfield v. Gillis*,[119] it appeared that the deed description, in addition to giving the length of each side of the ground designated certain monuments at the

---

[117] 15 N.H. 504 (1844).

[118] Wright (Ohio) 170 (1832).

[119] 107 Cal. 84, 40 Pac. 98 (1895).

corners of the tract conveyed. The California court held that, if these monuments could be determined, they would prevail over the designated measurements and that parol evidence was admissible to locate these monuments, and to show the identity of the land so conveyed with that on which a trespass had taken place.

Again, where a deed to a railroad company of a "right of way to be along the line as surveyed and laid out by H. C. Kellogg, civil engineer," and so forth, it was held in *Thompson v. Southern California Motor Road Co.*[120] that parol evidence was admissible to show that, when the deed was executed, the road had been commenced, and the line of the right of way claimed under the deed, and occupied by the road, had been surveyed and designated by stakes stuck in the ground, so that it could be easily traced, and that subsequently the road was constructed following the exact line of the survey.

Where the beginning corner of land conveyed by a deed called for "a stake set at the extremity of the sand bar of the C. River at the foot of the high river bottom, and corner made for Mrs. C., from which a cluster of willow trees brs. S. 36 W. 4 vrs," and the record call was "thence up the river with the meanders w. 400 vrs," followed by a description of the other boundaries of the land conveyed, and it appeared that, in applying the description to the ground, it was uncertain whether the said bar mentioned in the deed was included in the land conveyed. The Texas court held in *Roberts v. Hart*[121] that the deed contained a latent ambiguity, to which and to identify the land conveyed, parol evidence was admissible. The court said that whether or not the monuments could be found and identified and, if so, whether the lines located by such monuments included the land in controversy, could not be told by reading the grant, but only by the oral testimony as to such facts was always admissible.

But the New York court, in *Seaman v. Hogeboom*[122] held that parol evidence was inadmissible to show the place where were located "a stake and stones," as described in a deed, it having appeared that the monument was no longer in place.

Description as "Adjoining the Land" of Others.

Language "adjoining the land" of others often raises an ambiguity, because it may be uncertain on which side of the land of another the particular tract conveyed is located, or because it may be shown that the land intended to be conveyed, as appears from the remainder of the description, does not in fact "adjoin" the lands specified. Ordinarily, parol evidence has been allowed to be introduced in such cases, but the result in each case must depend on its own particular facts.

In *Wilkins v. Jones*,[123] the North Carolina court held that a description of land in mortgage as "30 acres of land situated in Strong Creek township, adjoining the lands

---

[120] 82 Cal. 497, 23 Pac. 130 (1890).

[121] Tex. Civ. App., 165 S.W. 473 (1914).

[122] 21 Barb (N.Y.) 598 (1855).

[123] 119 N.C. 95, 25 S.E. 789 (1896).

of the late J. W., J. C. J., and R. B.," was not so vague and uncertain that it could not be aided by parol evidence.

And in *Perry v. Scott*,[124] it appeared that a tract of land was described in a sheriff's deed as lying in the county of J., "on the south side of T. river, adjoining the lands of C. and others, containing 360 acres, more or less." It was contended that the word "adjoining" employed in the deed rendered the instrument void for uncertainty, and that, had the words "bounded by" been substituted in its place, the uncertainty would have been removed, and the deed would have been valid. It was held that the description was not too indefinite to preclude the admission of parol evidence to identify the land, and that testimony that at the time of the sale the grantor owned but one distinct tract of land in the county of J., answering the description, was admissible. The North Carolina court, in reaching this conclusion, said: "in the numerous decisions of this court upon the infinite variety of descriptions presented for construction, there can be seen but one clear and unwavering purpose in the minds of the judges, and that is, without contravening the Statute of Frauds, to give effect to the true intention of the parties. In doing this they have found it impossible, as Justice Ruffin once said, to formulate any artificial rules by which in many cases, this intention is to be ascertained, and they have necessarily been compelled to resort, in such instances, to those principles of reasoning which commend themselves to men of plain and ordinary understanding."

A deed describing the land conveyed as "a certain tract of land in N. township, joining the lands of said H. S. and others, said to contain 37 1/2 acres more or less" was held in *Hinton v. Roach*[125] sufficient to admit parol evidence of its identity.

But in *Wilson v. Johnson*,[126] the same court said that a description, "a certain tract of land in B. county, adjoining the lands of J. R. C. and others, containing 50 acres, more or less," was too vague and indefinite to be aided by parol evidence.

Description as "Bounded by" the Land of Others.

When land is described as "bounded by" the lands of certain other persons, an ambiguity may arise from the fact that the ownership of the lands may have changed since the execution of the instrument, or from the fact that the land may not in fact be so bounded, or from various other circumstances that may have existed unknown to or taken for granted by the parties to the instrument. Parol evidence has generally been allowed in such cases.

Where a homestead described exempted land by naming the persons whose lands constituted the respective boundaries of the exempted tract, it was held by the Georgia court in *Byrd v. Olmstead*[127] that parol evidence that the ownership of the adjoining tracts had changed since the setting apart of the homestead was held to be relevant, and it did not vary or contradict the terms of the homestead.

---

[124] 109 N.C. 374, 14 S.E. 294 (1891).

[125] 95 N.C. 106 (1886).

[126] 105 N.C. 211, 10 S.E. 895 (1890).

[127] 150 Ga. 815, 105 S.E. 480 (1920).

And where land intended to be conveyed by a deed was described there as bounded by certain lines and corners of various tracts of land, it was held in *Stevens v. Robb*[128] by the Texas court, that parol evidence was admissible to show the actual location of these lines and corners as they were established and recognized by the owners of such tracts at the time of and prior to the execution of the deed, even though they were not located on the ground at the points called for by the deeds conveying those tracts. The court said that the rule against varying the terms of a written instrument by extraneous evidence was not infringed by extraneous evidence explaining the calls in a deed and identifying the location of the land intended to be conveyed.

In the Virginia case of *Hunter v. Hume*,[129] where a tract of land conveyed by deed was described as situated on the west side of a certain turnpike, bounded on the south by the canal, and on the east by the old military road, and it appeared that the parties to the deed had made a mistake in designating the directions in that "south" should have been "east," and "east" should have been "south," after the true location of the land in dispute had been ascertained, parol evidence was admissible to show the proper location of all the descriptive calls in the deed, and thus give effect to the true intent of the parties.

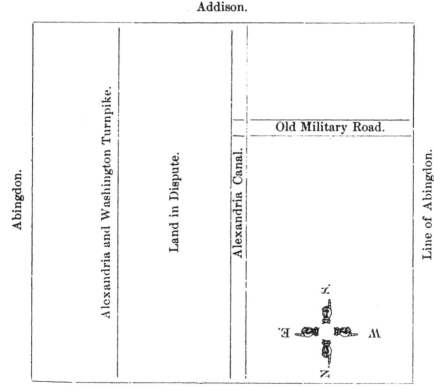

Addison.

Line of Abingdon.

[128] Tex. Civ. App., 229 S.W. 891 (1921).

[129] 88 Va. 24, 13 S.E. 305 (1891).

But where a description was: "Fifty acres of land lying in the county of H. and bounded as follows: By the lands of L., L., and S.," it was held that, since there was nothing in the descriptive clause from which the particular 50 acres of the land mentioned in the deed could be identified, parol evidence was inadmissible to aid the description. The North Carolina court stated in *Blow v. Vaughan*[130] that if there had been testimony tending to show the location of 50 acres of land in H. County, bounded on all sides by the lands of L., L., and S., then such evidence would have been admissible. "The test," said the court, "of the admissibility of evidence dehors the deed is involved in the question whether it tends to so explain some descriptive word or expression contained in it as to show that such phraseology, otherwise of doubtful import, contains in itself, with such explanation, an identification of the land conveyed."

In *Reynolds v. Boston Rubber Co.*,[131] it appeared that a deed from X described the land intended to be conveyed as a "tract of marshland" bounded "northwardly on said B upland, a ditch dividing said marsh from said upland." Later X conveyed the piece of land adjoining the marshland, described in the deed as "a certain piece of upland bounded southerly on marshland . . . a ditch dividing said upland from said marshland." It appeared that the line of demarcation between the marsh and the upland was curved inwardly towards the upland, while the ditch was straight, so that some of the marshland was between the ditch and the line of demarcation that separated the marsh from the upland. The Massachusetts court held that, since the description was uncertain, and not clear, parol evidence was admissible to locate the true boundary line.

Call for "North," "South," "East," or "West" Corner of Land of Another.

As an example, in *Hereford v. Hereford*,[132] it appeared that the defendant claimed under a deed describing the land in dispute as the lot of land known as a part of lot number 5 in square number 20, in that part of the city of Montgomery laid off on the Scott property, "said lot beginning at the north corner of Mary Ann Green's lot, running north, 35 feet, thence, 75 feet west, thence, running south, 35 feet, thence, east, 75 feet, to the corner or place of beginning." The difficulty arose in applying this description to the lot in dispute, when it was discovered that the Mary Ann Green lot had two north corners, a northeast and a northwest corner. The Alabama court held that this uncertainty constituted a latent ambiguity and that, accordingly, parol evidence was admissible to clear the indefinite description, so that a conclusion as to the true meaning of the parties in the use of the words "north corner" might be reached. The evidence sought to be admitted was that, by beginning the northeast corner of the Mary Ann Green lot, the particular calls and courses described in the defendant's deed were clearly met, and that, by beginning at its northwest corner, these calls and courses would not be appropriate, and could not be made to describe any lot ever owned by the plaintiff or the defendant in lot 5, square 20, of the Scott lands.

[130] 105 N.C. 198, 10 S.E. 891 (1890).

[131] 160 Mass. 240, 35 N.E. 677 (1893).

[132] 131 Ala. 573, 32 So. 620 (1902).

And where a deed called for a boundary "beginning at the north corner of R.'s store," and it appeared that the store stood squarely east and west, so that there were two north corners, either of which might equally fit the call, the North Carolina court held in *Lawrence v. Hyman*[133] that in a description containing such a latent ambiguity, parol evidence was admissible to show which of the two corners was meant.

Call for a Road, Street, or Alley as a Boundary.

Where land is described in an instrument as bounded on a certain street, road, or alley, an ambiguity often arises from the fact that the width of the street or road as recorded on a plat or map differs from the width according to its actual use, or from the fact that it is doubtful whether the distance stated in the deed includes the land as far as the center or only to the edge of the road. In such cases, parol evidence is held admissible to determine the true limits of the land intended to be conveyed.

Where land was described in a deed as "the north one half of lot No. 9 and the north one half of lot No. 10 fronting 50 feet on T. Street, and running back on R. Avenue for depth 100 feet;" and it appeared that the lines of the lots and streets as indicated by certain fences which had been built for a long period of years differed from the lines on the original survey when the lots and streets were first laid out, in that the fence lines encroached on the width of the streets according to the survey so that it was uncertain which were the true boundary lines, the Texas court, in *Bell v. Wright*,[134] held that the deed contained a latent ambiguity, to explain which parol evidence was admissible.

And where the deeds under which the plaintiff and the defendant claimed, in describing the boundaries of land intended to be conveyed, designated the following calls: "Continuation of the Odlin Road, so-called," "said Odlin Road," and "Odlin Road extended," and it appeared that a road surveyed and staked out by the engineers from the terminus of the original Odlin Road to a certain boundary line, known as the Herman line, would, if constructed, have answered such calls, but that it was never actually constructed, it was held that there was a latent ambiguity, to explain which parol evidence was admissible.[135] The Maine court said that, where a monument as found on the face of the earth answered a call in the deed in some but not in all particulars, the reference to it in the deed was not to be entirely rejected, but that parol evidence was admissible to show whether the monument partially but erroneously described was the one intended.

---

[133] 79 N.C. 209 (1878).

[134] 94 Tex. 407, 60 S.W. 873 (1901).

[135] *Tyler v. Fickett*, 73 Me. 410 (1882).

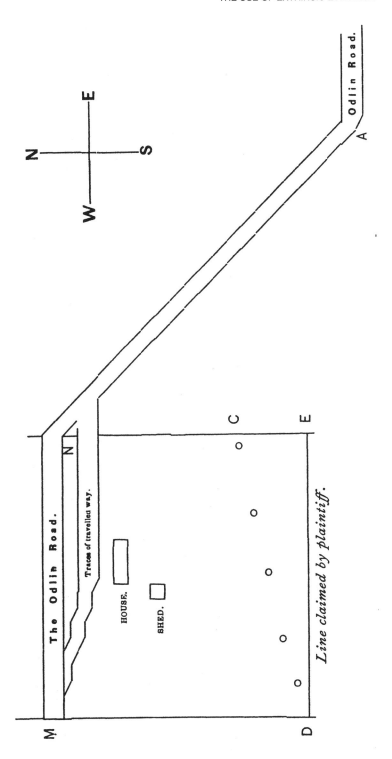

In *Rix v. Smith*,[136] the Michigan court held that, where the description in a deed called for 75 feet of the depth of a lot 132 feet deep, and not for a specified fraction thereof, parol evidence was admissible to show that the street in front of the lot was in fact wider than the plat referred to in the deed indicated, and that the grantee's land extended a corresponding distance farther from the street, since this evidence did not vary the description in the deed, but simply located the boundaries of the land on the ground.

In *Mann v. Dunham*,[137] it was shown that a deed of partition conveyed three lots—the northerly lot bounding on O. Road, the southerly lot on F. Street, the third lot, a parcel lying between these two lots. The dispute arose as to the boundary line between the southerly lot and the third lot. It appeared that, taking the northerly end of F. Street as the boundary of the southerly lot, the distance given in the deed as the length of the two lots in dispute was far in excess of the actual distance, but that, measuring from the center of F. Street, the distances in the deed could be carried out. The Massachusetts court held that the deed contained a latent ambiguity, to explain which extrinsic and parol evidence was admissible.

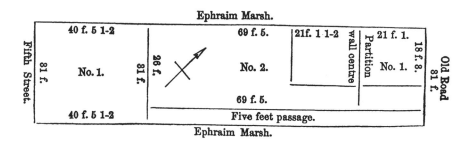

And in *Hunt v. Francis*,[138] it appeared that a lot intended to be conveyed by a deed was described as bounded on one side by "an alley." It appeared that the alley referred to, as opened and used for several years, was 16 feet in width, but that it was recorded as 22 feet. The Indiana court held that parol evidence was admissible to show to which line of the alley the lot conveyed by the deed was intended to extend, the 12-foot alley line or the 16-foot alley line. "In this state of ambiguity," said the court, "we must, it seems, look to the intention of the parties, as deduced from the circumstances which surrounded them."

Call for Railroad as Boundary. Where land is described as bounded by a railroad, and it appears that there is some doubt as to the true boundary of the railroad, parol evidence is held admissible to locate the land intended to be conveyed.

---

[136] 145 Mich. 203, 108 N.W. 691 (1906).

[137] 5 Gray (Mass.) 511 (1855).

[138] 5 Ind. 302 (1854).

Where one of the calls in a deed conveying land was "one hundred and twenty-two (122) feet easterly from the easterly line of" a certain railroad, and it appeared that there was some doubt as to the exact location of the easterly boundary line of the railroad, that is, whether it was along a stone wall parallel to the railroad, or whether it was along a line some feet to the west of the wall, the New York court in *Holden v. Crolly*,[139] held that parol evidence of the mutual understanding and conduct of the parties to the deed as to the actual location of the boundary line at the time of the execution of the deed was admissible. The court said that such evidence did not tend to contradict a written instrument, but purported merely to explain the language therein contained, which was not in itself self-explanatory.

And where, in a deed, one of the boundaries of the land intended to be conveyed was described as running to a certain railroad and running along that road for a number of feet, and it appeared that the railroad owned a strip of land lying next to its road but not within the original location of the road, the Massachusetts court held in *Hoar v. Goulding*[140] that parol evidence was admissible to explain the ambiguity, which was latent, and to show which strip of land owned by the railroad was intended as the true limit.

Call for Fresh Waters as a Boundary.   Where land borders on fresh water such as a river or stream, there exists a legal presumption that the title extends to the center or thread of the stream. And so, where the language of an instrument does not indicate a different intent, parol evidence is inadmissible to rebut this presumption. But where there are words in the writing purporting a different intention parol evidence is admissible.

To illustrate, the question facing the New Hampshire court in *Claremont v. Carlton*,[141] was whether an island located north of the center line of Sugar River was included in a conveyance by a deed describing the lot intended to be sold as bounded on the south by Sugar River. According to New Hampshire law, the owner of the land bordering on fresh water owned to the center or thread of the water. The court held that parol evidence should be admitted to show the exact situation of the island in order to arrive at the intention of the parties to the deed, for it might be shown that the island was of such length and so near the center as to cause the stream on each side of it singly to have been considered and called in common parlance "Sugar River," and not branches or parts of Sugar River, and also that the island might be shown to be so large and of such soil as to be very valuable, and hence that the consideration paid to the grantor in the deed might have extended only to the worth of the lot, independent of the island.

And where a tract of land was described as running "by the side of the mill pond," the Maine court held in *Lowell v. Robinson*[142] that parol evidence was admissible to

---

[139] 153 App. Div. 254, 138 N.Y.Supp. 23 (1912).

[140] 116 Mass. 132 (1874).

[141] 2 N.H. 369, 9 Am. Dec. 88 (1821).

[142] 16 Me. 357, 33 Am. Dec. 671 (1839).

show what kind of a pond was the one referred to—whether a natural pond or a pond artificially raised by a river; since, if the latter was true, a well-established rule of construction carried the land bordering thereon to the thread of the river or stream, while, if the former was true, the rule was that the land extended to the margin of the pond. The pond in question was found to be a river or stream stopped by a milldam.

Again, where a deed, in describing the land intended to be conveyed, designated that a certain pond referred to was a natural one, which had been raised more or less at times by artificial means, it was held that a latent ambiguity was shown to exist, and that parol evidence was admissible to explain and clear it up by proving that a certain line agreed on and understood at the time of executing the deed was the boundary of the pond. The court said that, where a description was employed that had not, by statute, usage, or judicial decision, acquired a fixed legal construction, or a boundary was referred to that was fluctuating and variable, parol evidence must be resorted to in order to ascertain the meaning and construction of the deed. However, where a description had already been given a definite legal construction, parol evidence would not be admissible to control the legal effect of such description. For example, where a description in a deed was that the premises were to bound on the sea, and the legal effect was to give title to low watermark, then such parol evidence would not be admissible to show a different meaning not expressed by the parties in their deed.[143]

Also, in *Wheeler v. Wheeler*,[144] it was shown that in a deed a certain tract of land conveyed thereby was described as bounded on the "west by Pudding Swamp." It appeared that, by a plat made at the time of execution of the deed, the western boundary extended to the edge of the swamp. It was a general rule that, where a stream was given as a boundary, the grant was presumed to extend to the middle of the stream. The court held that this rule was not invariable or inflexible, and that a contrary intention might be shown by other competent evidence appearing either on the face of the deed itself or of the plat made at the time, and by parol proof of the actual location of a different boundary line on the ground, and possession taken and held to the boundary so located.

Call for Salt Waters as a Boundary. An ambiguity frequently arises when land is described as bordering on salt waters, because it is uncertain whether the lands are intended to extend to low watermark or high watermark. In the absence of words in a grant showing a different intention, there appears to be a legal presumption that the lands extend to low watermark, and in such a case parol evidence is admissible. But, when the words of the instrument import a different intention, parol evidence is not admissible to determine the limits of the land.

---

[143] *Waterman v. Johnson*, 13 Pick. (Mass.) 261 (1832).
[144] 111 S.C. 87, 96 S.E. 714 (1918).

Thus, in *Palmer v. Farrell*,[145] it was shown that a deed designated as one of the boundaries of the lands conveyed a line running between certain on the bank of a certain river, and further stated that the lands "were bounded and described according to the survey made thereof by J. H.," and so on. It appeared that, both by the general description in the deed, and the survey referred to therein, the flats or lands between high and low watermarks, which was the land in dispute, were excluded. The court first stated the rule that, when the bank of a navigable stream was called for as a boundary in a deed—that and no more—the law would presume the grantor's intention to have been to carry the line to low watermark, and when the words of a grant were unambiguous, and no fraud or mistake was alleged, the intention of the parties could not be shown to override their obvious meaning, but that, where there was anything in the words of the grant which would indicate probably a different intent, the question, in the absence of fraud or mistake, was one for the construction of the court.

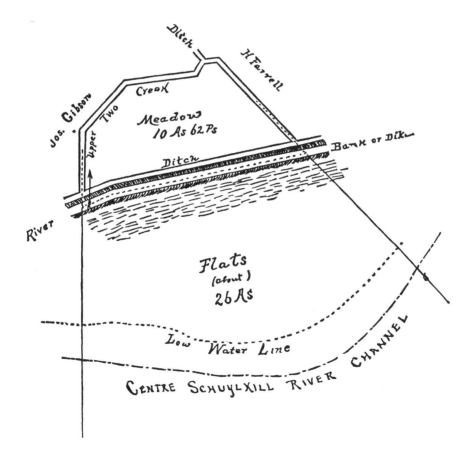

---

[145] 129 Pa. 162, 15 Am. St. Rep. 708, 18 A. 761 (1889).

According to this rule, it followed that only by a legal construction of the deed in question, based on assumed intention of the grantors, could the flat lands be embraced in it. The court held that, under these circumstances, it was competent to admit parol evidence of the extrinsic facts and circumstances attending the transaction, in order to show the true intention of the parties to the deed; for instance, to show that the bank referred to in the deed was an artificial one in the nature of a dike.

In *Storer v. Freeman*,[146] the Massachusetts court said that where, in a deed, a boundary line of the land intended to be conveyed was described as running to a heap of stones, by the shore of Elwell's corner, and it appeared that Elwell's corner was a known monument at low watermark, then a latent ambiguity existed, to explain which parol evidence was admissible. The ambiguity consisted in the fact that the phrase "by the shore" ordinarily meant along the outer edge of the shore, which would be high watermark, the shore consisting of that space on the sea between high and low watermarks; which fact would be inconsistent with the location of the monument at low watermark. The court further said that, in the case of a patent ambiguity, parol evidence was not admissible to explain it.

Ambiguity in Conveyance of Ditch.   There are cases where, in applying a description in a deed conveying a certain irrigation or drainage ditch, an ambiguity arises from the fact that it is uncertain whether the description includes certain lateral or branch ditches of the main ditch. Such are typical cases of latent ambiguity, which may be explained through parol evidence.

Where a deed purported to convey certain ditches known as the Silver Bow Company's ditches conveying water from Silver Bow Creek to the placer mines at Butte City, and more particularly those known as the Humphreys and Allison ditches, the Montana court in *Donell v. Humphreys*[147] held that parol evidence was admissible to show whether a certain ditch known as the Park ditch was included in the ditches conveyed. The evidence sought to be admitted was to show the relation of the Park ditch to the upper and lower ditches, and that the three ditches were known as the Humphreys and Allison ditches. The court said: "Parol contemporaneous evidence is inadmissible to contradict or vary the terms of a valid written instrument. . . . And so the intent of the parties must be gathered from what is written rather than from parol evidence, but the language of the instrument may be construed by the light of surrounding circumstances, and, so far as possible, the court may put itself in the place of the parties, and may interpret the language from this standpoint, but nothing can be added to or taken from the written words. So, extrinsic parol evidence is always admissible to give effect to a written instrument, by

---

[146] 6 Mass. 435, 4 Am. Dec. 155 (1810).
[147] 1 Mont. 518 (1872).

applying it to its proper subject-matter, by proving the circumstances under which it was made, thereby enabling the court to put themselves in the place of the parties with all the information possessed by them, the better to understand the terms employed in the contract, and to arrive at the intention of the parties."

The question arose in *Carman v. Staudaker,*[148] whether parol evidence was admissible to show that a deed conveying a one-third interest in a certain irrigation ditch did not convey any rights in either of two lateral ditches, one on the east side and one on the west side of the irrigation ditch, the deed having made no mention of the lateral ditches. The court, in holding parol evidence admissible, said, "It does not, in our opinion, appear from the terms of the deed of defendant to plaintiff of a one-third interest in the main ditch mentioned therein, that these lateral ditches, or either of them, are necessarily appurtenant to the interest the plaintiff purchased in the main ditch. It does not appear from the terms of the deed in question that the lateral ditches, or either of them, were essential or necessary to the enjoyment of the rights acquired by plaintiff by the purchase of one-third interest in the main ditch. It does not appear from the deed that such lateral ditches were or are even conveniences in the use of the rights acquired by plaintiff by his purchase of an interest in the main ditch. As none of these things appear upon the face of the deed in question, and which we think ought to appear, either expressly or by implication, and as the evidence offered and excluded was not intended to contradict or alter the terms of the deed, we think the court erred in excluding it."

Ambiguity in Description of Division Line in Deed. Where an ambiguity arises in the interpretation of an uncertain boundary line between two deeds, parol evidence is usually admitted in order to determine the true division line.

In the Connecticut case of *Nichols v. Turney,*[149] it appeared that the plaintiff and the defendant, adjoining landowners, both claimed under deeds from a common grantor. In the deed under which the plaintiff claimed, the land conveyed thereby was bounded on "the west, by the due north and south line of the N. tract, deeded to J.T., and her heirs [through whom the defendants claimed]." And in the deed under which the defendant asserted title, the land conveyed was bounded on the "east by a line due north and south." The question in dispute in this case was the location of this north and south line as described in each of the deeds. It appeared from extraneous evidence that there were several lines that might possibly be the "north and south line" described in the deed. The court held that the deeds presented a case of latent ambiguity, to explain which oral evidence was admissible.

---

[148] 20 Mont. 364, 51 P. 738 (1898).
[149] 15 Conn. 101 (1842).

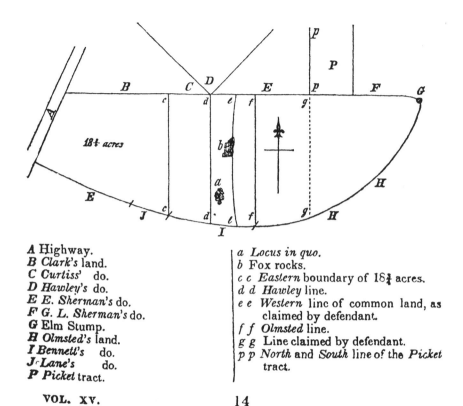

A **Highway**.
B *Clark's* land.
C *Curtiss'* do.
D *Hawley's* do.
E *E. Sherman's* do.
F *G. L. Sherman's* do.
G **Elm Stump**.
H *Olmsted's* land.
I *Bennett's* do.
J *Lane's* do.
P *Picket* tract.

a *Locus in quo*.
b **Fox rocks**.
c c *Eastern* boundary of 18¾ acres.
d d *Hawley* line.
e e *Western* line of common land, as claimed by defendant.
f f *Olmsted* line.
g g **Line claimed by defendant**.
p p *North* and *South* line of the *Picket* tract.

Sloan v. King,[150] where the boundary line between two tracts of land was in dispute, the plaintiff claimed the line as described in his deed—200 varas from the river (a stone mound marking the place) on a course "north 57 east 975 varas to a stake in the east line of survey 64" where a mesquite starts. The defendant claimed the line as described in his deed from the stone-mound corner 200 varas from the river; north, 40 degrees east, to the northeast corner of survey 64. It was held by the Texas court that parol evidence was admissible to explain the ambiguity and to show the true location of the boundary line.

Where it appeared that there was doubt as to the location of the western boundary of lot "number 11" as described in a deed conveying it, and also uncertainty as to the eastern boundary of the adjoining lot, "number 12," as described in a deed conveying that lot, it was held by the Kansas court in *Parks v. Baker*[151] that a latent ambiguity existed in each of the two deeds, and accordingly parol evidence was admissible to explain this ambiguity. The deed conveying lot number 12 described

---

[150] 77 S.W. 48, 33 Tex Civ. App. 537 (1903).
[151] 81 Kan. 351, 105 P. 439 (1909).

the boundary line as commencing at a point directly north of the west end of the house on lot 11, and the evidence showed that this point would not be in the north line of lot number 11. This variance constituted the ambiguity in the deed conveying lot number 12. The same ambiguity as to this eastern boundary of lot number 12 existed in the deed conveying lot number 11, which described the land intended to be conveyed as "lot No. eleven (11) . . . and all that part of lot No. twelve (12), which was not deeded to Charlotte Baker" (one of the defendants). The court said that statements of the grantors made while negotiating the sale respecting the location of the boundaries and corners were competent as tending to identify the property they intended to convey and to aid in the uncertain description in the deeds.

And in *Dorkins v. Montgomery*,[152] it appeared that the land intended to be conveyed to the defendant was described in the deed as "the northeast quarter (N E 1/4) of the northwest quarter (N W 1/4) of section seventy-one (71) . . . containing forty (40) acres, more or less, with all the buildings and improvements thereon." The plaintiff's deed described the land intended to be conveyed to him as "all and singular the following described property with the improvements thereon . . . and designated as being the southeast quarter of the northwest quarter of section seventy-one." The dispute in this case arose over the determination of the boundary line between the two tracts conveyed, the defendant claiming land south of the southeast quarter section line up to a telegraph pole line running east and west, which piece of land included the buildings mentioned in his deed, while the plaintiff claimed this same tract of land, alleging that his northern boundary line ran up to the southeast quarter section line. There was no mention of buildings in the plaintiff's deed. The Louisiana court held that the deeds under which the defendant and plaintiff claimed were ambiguous and that parol evidence was accordingly admissible to explain and remove the ambiguities.

Discrepancy between Description and Deed Referred to Therein. In *Port Wentworth Terminal Co. v. Equitable Trust Co.*,[153] the facts showed that a mortgage, which encumbered certain property "described in and conveyed by" a certain deed, gave the same description by metes and bounds as was given in the deed. It appeared that the deed excepted from its provisions a certain tract of land known as the N. Lead, which the grantor at that time did not own. At the time of the execution of the mortgage, the grantee in the deed owned the N. Lead, but had leased it to a third person. The mortgage was made subject to this lease. The court held that the description in the mortgage was ambiguous and that parol evidence of the attendant facts and circumstances was admissible in order to ascertain the intention of the parties.

---

[152] 2 La. App. 292 (1924).
[153] 18 F.2d 379 (1927: C.C.A. 5th).

Discrepancy between Description and Map or Plat Referred to Therein.

A frequent occurrence is the discrepancy between a description of land in a deed or mortgage and the plat or map referred to therein. In such cases, the courts seem to have held uniformly that parol evidence is admissible to explain the variance, and to identify the land intended to be conveyed.

The case of *Detroit, G.H. & M.R. Co. v. Howland*,[154] illustrates this very well. It appeared in this case that the plaintiff railroad company had bought a strip of land described as "Part of lots numbered two hundred forty-nine (249) and two hundred fifty (250) of the Ferry Farm addition to the city of Pontiac, described as follows: 'A strip of land 20 feet wide off the southwesterly side of said lots 249 and 250, and bounded on the southwesterly side by said second parties' right of way.'" The survey of the subdivision, not accurately drawn, made the addition overlap on the plaintiff railroad's right of way several feet. The map or survey, being referred to in the deed, was held thereby incorporated in the deed as fully as though delineated in the instrument. The ambiguity was not perceived until an attempt was made to lay the description of the tract purchased on the survey of the subdivision. It then became apparent that there were two conflicting descriptions—one, a strip of land 20 feet wide off the southwesterly side of the lots, including, however, 8.85 feet of the land the plaintiff railroad owned; and the other, a strip of land 20 feet wide, whose southwesterly boundary was the plaintiff railroad's right of way. The court held that the ambiguity was a latent one, and, since the conflicting descriptions could be reconciled, the one that best comported with the manifest intention of the parties would be adopted, and that parol testimony was admissible to explain the intention of the parties and clear up the ambiguity.

The same conclusion was reached in a similar Michigan case. Thus, where a deed described a tract of land by lot numbers referring to the town plat, and also by boundaries, designating as the southerly boundary a "public road," and it appeared that, according to a plat existing at the time of the execution of the deed, the southerly boundary of the lots did not reach the "public road," the court in *Purkiss v. Benson*[155] held that there was an ambiguity not on the face of the deed but that was brought to light by extraneous evidence, and that parol evidence of the acts and conversations of the parties to the deed at the time of its execution, showing their intention that the land was to run to the road, was admissible as a part of the *res gestae*. The court said that the whole *res gestae* furnished the best light on all ambiguities and doubts arising out of any act that was a part of them, as was the drawing of the deed.

And, in *O'Farrel v. Harney*,[156] the California court examined a case where it appeared that a deed described the land intended to be conveyed thereby as "lot No. 30 in block 13, said lot being 26 feet 8 inches on A. street," and also referred to the official map of the town in which the lot was located for a further description. It was further shown that a stake had been placed at the northwest corner of lot No. 30, at the time when the survey was made from which the map was drawn. According to the map, the eastern line of the lot was several feet east of the eastern line accord-

---

[154] 246 Mich. 318, 224 N.W. 366 (1929).

[155] 28 Mich 538 (1874).

[156] 51 Cal. 125 (1875).

ing to the distance, 26 feet, 8 inches, from the stake, as fixed by the survey. The court held that parol evidence showing the location of the stake at the time the survey was made was admissible, since the map was intended as a representation of the actual survey, and the evidence only proved the position of the lines as run—locating the calls mentioned in the map.

In *Lawver v. Anderson*,[157] it was shown that the land conveyed by a deed was described therein as follows: "A quadrangular piece of ground marked on the plan of D.'s heirs as lots numbers 21 and 22, bounded by S. street on the one side and by alleys on the other three sides." The court first held that the map became an essential part of the deed, with the same force and effect as if copied therein. It then appeared from the map that lot numbers 21 and 22 constituted a triangular piece of ground, not a quadrangular; and further that they were bounded on one side by S. street and on the other two sides by alleys. This discrepancy existing between the map and the descriptive words in the deed, the court held that parol evidence was admissible to aid the jury in determining the subject matter of the deed. The court quoted with approval from *Kountz v. O'Hara Street R. Co.*,[158] wherein it was held that, where the subject matter of a grant was insufficiently described in a deed, parol evidence might be admitted to show precisely what was intended to be conveyed.

In *Doe ex dem. Morton v. Jackson*,[159] it appeared that the land conveyed by a deed was described therein as "a certain piece or parcel of land situate on the Y. river, in Y. county, above the lots and commons of the town of M., as designated on the map thereof, and being the remainder of the tract of land on which said town was laid out, exclusive of lots, commons, streets, etc., being designated more particularly on the map of said town, as swamp land, and a tract not numbered," etc. It was further shown that in fact there was no land designated on the map referred to in the deed as swamp land. The court held that parol evidence was admissible not to explain, limit, or vary the deed, but to point out the subject-matter on which it was to operate.

In *Ferris v. Coover*,[160] the language in a grant was as follows: "The land granted to him consists of 11 square leagues (*sitios de ganada mayor*), comprehended in the extent designated in the plat which accompanies the expediente, without including the land inundated by the impulse and currents of the rivers, its boundaries being: On the north, the Three Peaks, and latitude 39° 41' 45" north; on the east, the margins of Feather River; on the south, latitude 38° 49' 32" north; on the west, the river Sacramento."

The map referred to in the deed was for the purpose of identification, and was regarded by the court as incorporated into the deed. It appeared that there was a difference in the location of the boundaries as given in the grant and as found on the map. The court held that parol evidence was admissible to explain the boundaries, and to fix the location of the land conveyed. The court said: "Reference will be had to the circumstances under which the grant was made and the intention of the parties, and parol evidence is admissible, in such case, for that purpose. That portion will be rejected, and that construction adopted which will give effect to the intentions of the parties.

---

[157] 77 Pa. Super. Ct. 208 (1921).

[158] 48 Pa. Super Ct. 132 (1911).

[159] 1 Smedes 8 M. (Miss.) 494, 40 Am. Dec. 107 (1843).

[160] 10 Cal. 589 (1858).

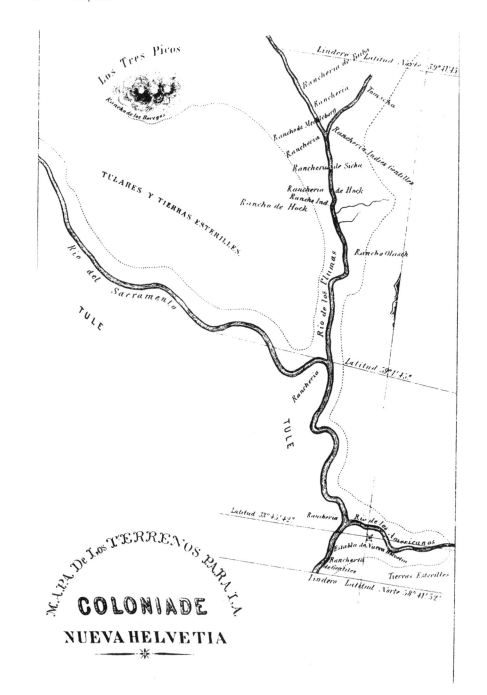

Los Tres Picos

Rancho de los Borregos

Lindero Buba Latitud Norte 39°41'45

Rancheria de Buba

Rancheria

Rancheria

Tonscha

Rancho de Meddleburg

Rancheria

Rancheria Indios Gentiles

Rancheria de Sicha

Rancheria de Hock

TULARES Y TIERRAS ESTERILLES

Rancheria de Hock
Ranche Ind

Rancho de Hock

Rancho Olasch

Rio del Sacramento

Rio de los Plumas

TULE

Rancheria

Latitud 39°1'45"

TULE

Latitud 38°45'42"   Rancheria   Rio de los Americanos

Establa de Nueva Helvetia

Rancheria de Gegiltes

Tierras Esterilles

Lindero Latitud Norte 38°41'52"

MAPA DE LOS TERRENOS PARA LA

COLONIADE

NUEVA HELVETIA

Discrepancy between Monuments and Courses and Distances. This type of ambiguity naturally arises only on an application of the description to the surface of the earth. When the conflict between the monuments and the courses and distances is thus made apparent, parol evidence is admissible to explain the variance.

As an example, where the land intended to be conveyed by a deed was described both by courses and distances, and by well-defined monuments, and it appeared that, in applying the description to the surface of the earth, the lines by courses and distances were longer than the lines by monuments, it was held in the Pennsylvania case of *Pringle v. Rogers*[161] that the deed presented a case of latent ambiguity, and that parol evidence was admissible to show the true intention of the parties as expressed in the language of the deed. The court also held that the description by monuments must control and that no evidence could have the effect of changing this well-established rule.

And, in the California case of *Colton v. Seavey*,[162] it was held that parol evidence was admissible to fix the location of the monuments described in a deed and thus fix the boundary lines of the tract conveyed. It appeared that the first and second lines of the tract. conveyed were described as follows: "Beginning at a stake near the old corral of Jose' Jesus Lopez, running thence easterly to the head of the canada, thence westerly, including the canadas, to a stake, so that a line running from thence to the Dos Pedros will pass about 200 yards from the present new corral of the said Jose' Jesus Lopez," and so forth. The parol evidence admitted was to show that this second line, instead of being in a westerly course, was about northeast, as shown by the natural landmarks referred to, and that, if this line was run a west course, it would not include the premises in controversy.

Again, where the boundaries were described as beginning "at stake 154 vrs. S. 60° W. from the northeast corner of a 20-acre tract owned by D. B. from which a live oak marked 'S' 12 in. dia. brs. N. 13 1/2° E. 9-4/10 vr.," and it appeared that there was no oak tree at the distance called for from the northeast corner of the D. B. land, as that corner was located by a surveyor, it was held in the Texas case of *Dilworth v. Buchanan*[163] that parol evidence was admissible to locate and to determine which was the true call intended by the parties to the deed—the call for distance or the call for the oak tree.

And in a Pennsylvania case,[164] where one of the calls for a boundary in a deed was "thence along line of J.B., south 1/4° west, 29 perches and 10 feet, to corner of J. B.'s land," and it appeared that the line "south 1/4° west, 29 perches and 10 feet"

---

[161] 193 Pa. 94, 44 A. 275 (1899).

[162] 22 Cal. 497 (1863).

[163] 249 S.W. 835 (1923).

[164] *Kostenbader v. Peters*, 80 Pa. 438 (1876).

did not reach "J. B.'s corner," parol evidence was held admissible to show that the words "J. B.'s corner" were left in the deed by mistake, and that the true monuments were marked on the ground at the time of the execution of the deed. It is to be noted that in this case the court apparently allowed a preference to the description by the monument "J. B.'s corner."

And where one of the boundary lines described in a deed was "thence N. 70° E. 6 perches to a post," and it was sought to have admitted parol evidence that the true boundary was an old fence, as established by the parties to the deed at the time of its execution, which ran "S. 70° E. 6 perches to a post," and that the surveyor by mistake inserted in the deed "N. 70° E. 6 perches to a post," the Pennsylvania court in *Rook v. Greenewald*[165] held that such evidence was admissible, and it was a question for the jury to determine, from such evidence, the true boundary line. The court considered this ambiguity, arising from collateral matter, to be latent.

Failure to Specify Town, Section, Range, County, or Similar Geographical Information. The courts do not appear to be uniform in their decisions on the question whether parol evidence is admissible to aid the description of land in a deed or mortgage, where there is no mention therein of one or more of the following: Town, township, range, section, county, or state. In the majority of cases, however, the rule appears to be that, where there is a description in the instrument sufficient to point out with some definiteness the land intended to be conveyed, the fact there is a failure to designate the town or township or range, and so forth, does not vitiate the instrument, and parol evidence is admissible to identify the land. But there are a few cases where the courts, clinging to the old distinction, between latent and patent ambiguities, have held that a failure of the writing to mention the town or county or state, or the like, constitutes a patent ambiguity, to explain which parol evidence is inadmissible. It is also to be noticed that the more of these designations that are lacking, the more ambiguous the instrument becomes, and the less likely parol evidence can be allowed to be admitted.

Failure to Specify County. In the Georgia case of *Chauncey v. Brown*,[166] it appeared that the deed in question, which had been executed in the county of Twiggs, in describing the land, referred to the county in which it was located as "said county" and the "county aforesaid." It further appeared that the county of Montgomery was referred to in the deed, and later it recited that on a given day L., as administrator, was authorized to sell the tract of land belonging to the estate of the deceased lying in the "county aforesaid," and, further, that thereafter the land was duly advertised in conformity to law, and put up and exposed for sale to the highest bidder at the courthouse "in Jacksonville, in said county," and that, the plaintiff's

---

[165] 22 Pa. Super. Ct. 641 (1903).
[166] 99 Ga. 766, 26 S.E. 763 (1896).

interstate being the highest bidder, the same was knocked off to him, and that accordingly the administrator conveyed to him the said tract of land No. 309, in the thirteenth district of said county. Jacksonville was not in Twiggs county or in Montgomery county but was in Telfair county. The plaintiff's interstate bought from one L., administrator, a tract of land corresponding in number and district to the one of which the premises sued for were a part. It was in doubt, however, whether the land lay in the county of Montgomery or the county of Twiggs or the county of Telfair, and in this consisted the ambiguity. The court held that parol evidence was admissible to explain the ambiguity.

Failure to Specify County and State. Where a deed described lands by section and township numbers but failed to name the county and state in which the lands were situated, and it appeared that the state and county boundaries were not considered in the making of governmental surveys of public lands, it was held by the Mississippi court in *Ladnier v. Ladnier*[167] that the ambiguity was latent, to explain which parol evidence was admissible. The court said that on the face of the deed there was no apparent ambiguity; that the ambiguity lurked in the thing conveyed, and was brought to light by extrinsic facts disclosed in an effort to apply the description; and that the uncertainty thus created might be removed, ordinarily by evidence of the same character The court, in reaching its conclusion, applied the rule that, where the parol evidence is offered for the purpose of adding a material term to an instrument, or where the court, having looked to the circumstances of the parties, the subject-matter of the instrument, and all proper collateral facts, remains uncertain as to what the meaning of the written words is, a patent ambiguity arises, to explain which, parol evidence cannot aid.

In *Butler v. Davis*[168] the Nebraska court held that a deed was not void for uncertainty that failed to name state and county and that parol evidence was admissible to identify the land intended to have been conveyed, the ambiguity being a latent one.

And in the Alabama case of *Chambers v. Ringstaff,*[169] it was held that, where a deed described lands by government numbers but failed to designate in what county and state, land district, or government survey they were located, oral evidence was admissible to show that the party making the deed was living in a certain county in Alabama, on lands answering to the description.

Where a tract of land was described in a deed by sectional subdivision, township, and range, but the deed failed to mention the county, state, and land district, the Alabama court in *Webb v. Elyton Land Co.*[170] held that parol evidence that the grantor owned certain land corresponding to the description in the deed, that he and his grantee lived near it, and that he never owned any other land, was admissible.

---

[167] 75 Miss. 777, 23 So. 430 (1898).

[168] 5 Neb. 521 (1877).

[169] 69 Ala. 140 (1881).

[170] 105 Ala. 471, 18 So. 178 (1895).

Failure to Mention Town, County, and State. Where a description of land in a mortgage represented the land as being situated in a certain section, township, and range, and contained other designations by distance and monuments so as to earmark the land mortgaged, but failed to mention the town, county, and state, it was held by the Michigan court in *Slater v. Breese*[171] that extrinsic evidence was admissible to locate and identify the tract mortgaged. The court said that it was always competent to fix and identify by extrinsic evidence the natural monuments and other badges of identity, and connect the description in the deed with the material subject matter dealt with by it.

Failure to Mention Township, Range, County, and State. In *Wilson v. Calhoun*,[172] a mortgage that described land as bounded on the north, east, and west by the land of named individuals, and on the south by a named river, but that did not mention the county or state in which the land was situated, was held to be so definite that parol evidence would be admitted to supply the defects in the description, under the rule, as stated by the Tennessee court, that when the description in the writing is so definite and exclusive as to afford means of positive identification, its location may be designated by intrinsic proof.

In Arkansas, in *Fuller v. Fellows*,[173] where a deed described lands by subdivision of certain sections only, without giving the township, range, county, or state, it was held that the description contained a patent ambiguity, and that parol evidence was not admissible to supply what was omitted in the description. Said the court: "There were no ambiguous terms in this deed, nothing to interpret; the terms were all plain, and the description perfect as far as it went. But it was patent that a further description was necessary. And whilst parol is admissible to explain what is meant by the use of terms, as, for instance, if the grant be of a tract called 'Blackacre,' evidence is admissible to show what tract was intended to be conveyed by that description, but where there is a lack of description, a neglect to describe, there is nothing which can be explained or made plain by parol evidence; such evidence, if admissible, would add to the description, but not explain that given in the deed."

Failure to State Meridian. In *Dougherty v. Purdy*,[174] it appeared that a description in a deed was as follows: "An undivided half of the northwest quarter of section 1, in township 1 north, in range 2 west, in the state of Illinois." It also appeared that, by looking outside the deed, examining the acts of Congress and the public surveys of the state, there was an ambiguity in this description, because it did not refer to any meridian. The court held that this was a latent ambiguity and that parol evidence was admissible to explain the description and to identify the land

---

[171] 36 Mich. 77 (1877).

[172] 157 Tenn. 667, 11 S.W.2d 906 (1928).

[173] 30 Ark. 657 (1875).

[174] 18 Ill. 206 (1856).

intended to be conveyed. The court said that, when an ambiguity was duly made to appear by the introduction of proof outside the deed, it was a latent ambiguity, and might be explained in the same way that it was shown.

Failure to Specify Township.   Where there is a mere failure to state in the description of land in an instrument, the township in which the land lies, parol evidence is held admissible to locate the land described.

In *Ozark Land & Lumber Co. v. Franks*,[175] wherein it appeared that a deed described the land intended to be conveyed as the S.E. 1/4 of section 26, range 5, and named the grantor therein as the patentee of that land but failed to designate the township. The Missouri court, in holding that a record of this deed did not impart notice to third persons, said that as between the parties the deed was not so indefi-nite as to prevent the admission of parol evidence to identify the land described and intended to be conveyed, since it appeared that the grantor was the patentee of only one section 26, in range 5.

Failure to Specify Township and Range.   In the majority of cases where the instrument failed to point out the township or range, the courts have held admissible parol evidence to locate the land. A few jurisdictions, however, have considered the ambiguity as patent, to explain which parol evidence was inadmissible.

Thus, where a description in a deed, after pointing out the county in which the land lay, continued as follows: "One acre and a half in the northwest corner of sec-tion five (5), together with the brewery," and so forth, and failed to state the town-ship and range of the county, and it appeared that there were several sections 5 in the county, the Illinois court held in *Bybee v. Hagemar*,[176] that parol evidence was admissible to explain the ambiguity, which was a latent one.

And where the description of land in a deed failed to state the township and range wherein the land intended to be conveyed was located, it was held that extrinsic and parol evidence was admissible to identify the property; that the description might be aided by extrinsic proof not inconsistent with the description contained in the deed itself.[177]

Failure to Mention Section, Township, or Range.   In the Arkansas case of *Dorr v. School Dist.*,[178] it appeared that a deed described the land conveyed as "a certain parcel of land lying and being in the county of Independence, commencing at a black gum tree standing near the road and graveyard, near the residence of

---

[175] 156 Mo. 673, 57 S.W. 540 (1900).

[176] 66 Ill. 519 (1873).

[177] *Richardson v. Sketchley*, 150 Iowa 393, 130 N.W. 407 (1911).

[178] 40 Ark. 237 (1882).

B. E., running north 40 poles to a stone, thence east 12 poles to a stone, thence south 40 poles, thence west 12 poles to the place of beginning, containing 3 acres, including said schoolhouse and graveyard." The contention was that the deed was void for uncertainty, in that the description failed to mention the section, township, and range, and because it had never been recorded. The court said: "Here the ambiguity is latent. Is the description so defective that it is impossible by the aid of parol evidence to locate the land? It is in a certain county, and in a certain school district, which has definite boundaries, is parcel of the tract upon which stood the residence of Benjamin I. Edwards; contains 3 acres and is described by metes and bounds, and by visible monuments, to wit, the graveyard, the schoolhouse, the highway, corner stakes, and an initial tree from which to start. And defendant had gone into possession. A competent surveyor could have found the land without much difficulty."

Some of the Calls Missing.   There are a few cases where an ambiguity in the description in an instrument consists of the fact that either one or more of the calls for course, distance, or monument was missing. In most of the cases, the ambiguity was apparent on the face of the instrument, but in several it appeared not to have arisen until an application of the description to the surface of the earth was attempted. Where there is a sufficient description remaining to make possible an identification of the land intended to be conveyed, the courts have had no hesitancy about admitting parol evidence to supply the missing calls and to identify the land. In so doing, the courts do not seem to be inserting new material matter into the deed, but merely to be interpreting the intention of the parties to the writing as it is vaguely expressed therein, nor does there seem to be any altering or varying of the contents of the writing.

As an illustration, in the Texas case of *Fortenberry v. Cruse*,[179] a deed describing the land conveyed, as a tract of land "situated in T. county, said state [Texas], on the waters of B. creek: Beginning at the S.W. corner of W.P. land; thence east 1,220 varas to the L. league; thence south 1,050 varas to the beginning, containing 226 acres, more or less," was held not to be so vague and uncertain as to preclude the admission of parol and extrinsic evidence to identify the land conveyed. The court said that the calls in the deed corrected themselves, since by taking the two lines given, recurring to the beginning point and running south a distance equal in length to the second line, and closing these second and third lines by a straight line, a parallelogram containing exactly the 226 acres called for in the deed would be constructed; and that parol evidence was admissible to show that this tract of land was the land conveyed by the deed. Recognition was made of the correctness of the rule that, where the defect in the description was patent on the face of the instrument, and was such as did not convey any land described, the instrument was void, and extrinsic proof could not be resorted to in order to make certain that which appeared uncertain from the instrument; but

---

[179] 199 S.W. 523 (Tex. Civ. App., 1917).

it was held that there was no patent ambiguity in the description in the deed under consideration.

Also, where, in a deed describing the land intended to be conveyed thereby, the second call was omitted, and it appeared that, by reversing the calls in the deed, the missing call could readily be supplied, the Texas court held in *Montgomery v. Carlton*[180] that the land was capable of identification and that extrinsic evidence was admissible in aid thereof.

Where, in attempting to apply a description of land as set out in a deed, it appeared that the closing call had been inadvertently omitted, it was held in *Snow v. Gallup*[181] by the Texas court that parol evidence was admissible to supply this necessary call and then to identify the land intended to be conveyed—the ambiguity being latent.

Calls Failing to Provide a Closed Figure.

Where the calls as specified in an instrument fail to close, parol evidence is usually allowed to explain the ambiguity and to identify the land intended to be conveyed.

In the Illinois case of *Stevens v. Wait*,[182] a description of land in a deed contained the south, east, and north lines, and no uncertainty existed as to their location, but that the statement describing the west line as running from the west end of the south line "thence north 60° 30' west, sixty-nine and one-quarter poles to a stake," was meaningless, since such a line would not close with the other lines, but would run in an opposite direction, enclosing no land. The court held that parol evidence was admissible to identify the land intended to be conveyed, and to show that by striking out the words "60° 30' west" in the description of the fourth line, the description of the land was certain and definite, and answered the call, making the call in all four of the lines harmonious.

And so where, on the face of a deed, the description purported to be a strict survey, but it appeared that, when plotted, it would not close by 5 chains, and that, by forcing the closing line to the beginning, it would contain thirty-eight hundredths acres, instead of twenty-six and fifty-five hundredths acres, as described, the New Jersey court in *Fuller v. Carr*[183] held that parol evidence was admissible on the question of location of the land conveyed.

And where land was described in a deed as "[running] to a remarkable rock; from thence, first, north, 63 degrees east, 110 rods; then northeast 122 rods; then, turning and running northwest, 43 rods; from thence southwest to the northeast corner of the aforesaid W. B.'s lot," and it appeared that there was error, either in the distance, 110 rods and 122 rods, or in the direction of W. B.'s northeast corner from the end of the 43 rods, since the calls did not close, it was held by the Maine court in

---

[180] 56 Tex. 431 (1882).

[181] 123 S.W. 222, 57 Tex. Civ. App. 572 (1907).

[182] 112 Ill. 544 (1884).

[183] 33 N.J.L. 157 (1868).

Greeley v. Weaver[184] that parol evidence was admissible to locate the northeast corner of the W. B. lot, and thus identify the land conveyed.

Ambiguity in Locating an Easement.
Where a deed conveys an easement over certain land but fails to locate exactly the line over which the easement extends, parol evidence is admissible to locate it.

Thus, where a deed conveying a certain tract of land also conveyed a right of way over another tract of land, but did not limit or define the way, it was held in *Kinney v. Hooker*[185] by the Vermont court that parol evidence of an agreement by the parties to the deed, as to the place where the way was to run, was admissible. The court said that such evidence did not vary or contradict the deed, provided the way was located within the boundaries of the land over which the right was granted.

And where a deed conveying certain land also granted an easement but failed definitely to locate the line over which such easement should run it was held by the Indiana court in *Shedd v. American Maize Products Co.*[186] that parol evidence was admissible to point out the location of the line. The court said that an easement was an interest in land, and a deed by which it was granted, like other written contracts, became the repository of the entire agreement, and, in the absence of fraud or mistake, could not be varied by parol evidence; but that, notwithstanding this rule, the actual consideration might be shown by parol evidence, even though it differed from that expressed in the instrument.

Technical and Ambiguous Expressions Such as "Forty," "To," "At," "Et cetera," and so forth.
Where the language in an instrument has a technical meaning, or is capable of being interpreted in several ways, or is translated from a foreign language, and so on, parol evidence is usually admitted to explain the uncertainty.

For example, where a deed described certain property as "the southeast forty of the northeast quarter," and so forth, it was held by the Illinois court in *Evans v. Gerry*[187] that parol evidence was admissible to show that about the date of the deed (1847) it was a common expression to say "a forty" or "an eighty," and so on, to indicate 40 acres or 80 acres, and that accordingly the description, "a southeast forty," would readily mean the southeast 40 acres.

The phrase "in front of" sometimes is an ambiguous expression, as was shown in the Massachusetts case of *Graves v. Broughton*,[188] wherein it appeared that a deed of partition purported to divide a house and the land surrounding it into two parts, referred to as the "lower half" and the "upper half" of the premises. The deed, after designating the rooms of the house that formed the "lower half," described the land

---

[184] 13 Atl. 575 (1888).

[185] 65 Vt. 333, 36 Am. St. Rep. 864, 26 A. 690 (1892).

[186] 60 Ind. App. 146, 108 N.E. 610 (1915).

[187] 174 Ill. 595, 51 N.E. 615 (1898).

[188] 185 Mass. 174, 69 N.E. 1083 (1904).

forming a part of that half as "one half of the small piece of land in front of the house, viz., that part which is below the bank or break in said land," and, after stating that the remainder of the house formed part of the "upper half," described the land intended to be conveyed as a part of that half, as "upper half of said piece of land in front of the house." The court held that the language used was in such general terms that resort must be had to parol evidence to ascertain what was meant by the words used, especially the words "in front of the house." And so, according to the court, evidence of the actual use and occupation of the land by different owners was admissible to determine the boundaries and the extent of the ownership of each tenant.

The Texas case of *Linney v. Wood*[189] illustrates the uncertainty that arises from a translation of a foreign language. There, it was shown that the beginning corner of a certain tract of land conveyed by a deed was translated from the Spanish at one time as: "With 13,305 Mexican varas, running a line from the aforesaid creek [Arkansas creek] at the point called 'El Alamo,' about one and a half miles above, where a landmark was set, at the place where the late Martin De Leon's rancho stood." Another translation made a few years later was: "With 13,305 Mexican varas, drawing a line from the indicated creek [Arkansas creek] at the point called 'El Alamo,' about a mile and a half more above where a landmark was fixed, and at which place the old rancho of the deceased Don Martin De Leon was found." It appeared that there were different interpretations of these translations. The defendant contended that the language used meant that the landmark was set at the rancho of De Leon, a mile and a half above the point called "El Alamo." The plaintiff contended that the meaning was that the beginning corner was at a point on the A. creek called "El Alamo" and that this point was 1 1/2 miles above where a landmark was set at the rancho of De Leon. It also appeared that the term "El Alamo" had several different meanings or interpretations; thus, a latent ambiguity developed. The court held that parol evidence of the subsequent acts and declarations of the parties to the deed, showing the practical construction put by them on the words of the description, was admissible for the purpose of ascertaining their intention, and then to aid in the identification of the land intended to be conveyed.

But in an early Virginia case,[190] where a deed described the land intended to be conveyed thereby as 230 acres of land in C. county, on the south side of P. swamp, "bounded by the lines of E., the said G., T., and the above said P. swamp, etc.," it was held that parol evidence was not admissible to show what was meant by "et cetera," which constituted a patent ambiguity.

Ambiguous Description in a Deed Executed at a Judicial Sale.

Because the courts have usually applied a stricter rule of construction in the case of a deed executed at a judicial sale, such as tax deeds, than in the case of a deed between individuals, it is considered necessary to treat such cases together. There appears to be no general rule as to the admissibility of parol evidence in this type of case, each state seemingly applying its own peculiar rules of construction.

---

[189] 66 Tex. 22, 17 S.W. 244 (1886).
[190] *Gatewood v. Burrus*, 3 Call (Va.) 194 (1802).

Where land was described in a tax assessment and tax deed as "2/3 of square 39 in Fisher tract," it was held that the description was too indefinite and uncertain to admit of parol evidence to explain the ambiguity, which was patent. According to the court, the ambiguity might mean "an undivided two-thirds interest, held by the owner, by way of a tenancy in common; or (2) it might mean an entirety of two thirds in area of the whole square;" and which of the two it did mean, it was impossible to say. This case, *Dane v. Glennon*,[191] was decided by the Alabama court under a statute defining the certainty with which lands conveyed by tax deeds must be described, providing: "In case of land surveyed, or laid out as a town, city, or village . . . it shall be described by the designation of the number thereof. If it be a part of a lot or block, it may be described by its boundaries, or some other way by which it may be known."

In Arkansas, a tax deed containing the description, "E. part N. 1/2 S.E. 1/4 sec. 27, town, 2 range 12 W., containing 754 acres," was held void, there being no circumstance mentioned in the description, as of ownership, to assist in the identification of the land, although it was agreed that the land claimed under the deed was in the form of a trapezoid. Said the court: "In this case we find nothing in the description itself or in the circumstances to indicate who owned the land sold for taxes; and, as ownership was not disclosed as a means of identification, it may be questioned whether the description before us could in any case be adjudged sufficient."[192]

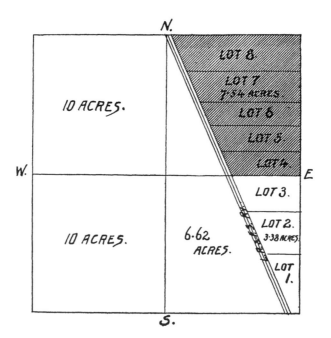

Figure 8.9 Plat of SE. ¼ of SE. ¼ of sec. 27, T. 2 N., R. 12 W.
The land in controversy is represented by the black in the diagram.

[191] 72 Ala. 160 (1882).

[192] *Schattler v. Cassinelli*, 56 Ark. 172, 19 S.W. 746 (1892).

Also in Arkansas, in *Tatum v. Croom*[193] it appeared that a sheriff's deed described the land as "the N.E. part of S.E. 1/4 of S.E. 1/4 of section 16, township 6 N., range 30 west, 20-36/100 acres, also N.E. part of S.W. 1/4 of S.E, 1/4 of said section, township, and range, containing 2-95/100 acres." The court held that parol evidence was admissible for the purpose of enabling the court to understand the terms used in the description, but that parol evidence was not admissible to show what the sheriff intended to sell. The description in this case was held too indefinite and uncertain to allow the passage of title.

In the Colorado case of *Sullivan v. Collins*,[194] where it appeared that land was described in a tax deed as "lot 5, block 144, East Denver, Arapahoe County, Colorado," and in a city plot as "lot 5, block 144, Clement's Addition to the city of Denver," it was held that parol evidence was admissible to show that the two descriptions applied to the same property, and that the property was as well known by one description as by the other.

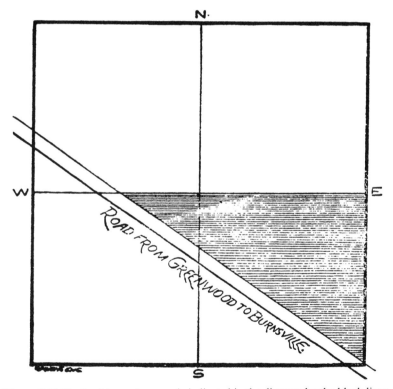

Figure 8.10 The land in controversy is indicated in the diagram by the black lines.

[193] 30 S.W. 885, 60 Ark. 487 (1895).
[194] 20 Colo. 528, 39 P. 334 (1895).

Where land claimed by virtue of a tax deed that, according to the official plat in the general land office was described as the N. 1/2 of the Elisha Lott claim, certificate 55, section 28, township 6, range 17, and the assessment and tax deed under which it was claimed described the said land as the S. 1/2 of section 6, township 6, range 17 W., and there appeared on such official plat of the township in which said lands were situated, land answering the latter description, it was held that parol testimony was not admissible to show that the land described as the N. 1/2 of the Elisha Lott claim was intended to by assessed and sold under the description of S. 1/2, sec. 6, T. 6, R. 17.[195]

However, where a tax deed purported to convey certain lots by specified number in "Arndt's Addition," and it appeared that lots corresponding to those numbers were to be found in the plot of "Arndt's Second Addition," but none in "Arndt's Addition," it was held in *Curtis v. Brown County*[196] by the Wisconsin court that, since a tax deed was the sole evidence of the land conveyed, parol evidence could not be received to aid or correct a description which was false or mistaken, and the deed was declared void for uncertainty. It is to be noticed that the Wisconsin statute allowing a more liberal construction of tax deeds was passed in 1866, after the execution of the deed in question.

WILLS

Two possible situations arise in the interpretation of wills: where there is no ambiguity, and where the property devised is inaccurate or there is a latent ambiguity.

Unambiguous Descriptions. Where the description of the land devised is definite and unambiguous, and applies with reasonable certainty to specific property, extrinsic evidence is not admissible to show an intention on the part of the testator to devise other property or to include property not covered by the description.[197]

In the Pennsylvania case of *Thompson v. Kaufman*,[198] the property described in the will was situate "in Moss Street, No. 336," was enclosed by a fence and adjacent buildings, and was occupied and used by tenants in this manner up to the time of the testator's death. The description in the will was held to be unambiguous, so that the testator's declarations were incompetent to show his intention to include other land in the devise.

---

[195] *W.H.Weston Lumber Co. v. Strahan*, 128 Miss. 54, 90 So. 452 (1922).

[196] 22 Wis. 167 (1867).

[197] *Bradley v. Rees*, 55 Am.Rep. 422, 113 Ill. 327 (1885).

[198] 9 Pa. Super. Ct. 305 (1899), affirming 6 Pa. Dist. R. 522 (1897).

Likewise, in the New Jersey case of *Hand v. Hoffman*,[199] a devise of "all that part of cedar swamp to the eastward of the aforesaid run and branch below said sawmill" was held not to be ambiguous, and extrinsic evidence was held not admissible to show that the testator owned several tracts of cedar swamp, and intended to devise only one of them.

Also, as stated by the Georgia court in *Napier v. Little*,[200] "It is competent to adjust by parol testimony the description of the land to the boundary, or boundaries, but it is not competent to disprove by parol testimony the boundary, or boundaries, fixed by the will itself."

Ambiguous Descriptions. Where the description of the real property devised is inaccurate, or there is a latent ambiguity with respect thereto, extrinsic evidence is allowable to resolve the ambiguity and identify the property designated.

In the Georgia case of *Flannery v. Hightower*,[201] the court said: "If the intention of the devise be to convey all of the property of the testator, such general description will suffice, and extrinsic evidence is admissible to show such property as was in the testator and such as was necessarily included in the general terms employed by him in devising it. At last, the question of description is one of degree only, and if the conveyance be of an entire estate, parol evidence is admissible to ascertain the geographical extent and limit of the property covered thereby."

Location of Land in General: County, State, City, Parish, Township, or Range. Where a will described land only by section number, township, and range and did not specify the county where it was located, extrinsic evidence as to the county of the testator's residence and the location of his real estate was held to be admissible to identify the land devised.[202]

Similarly, in *Black v. Richards*,[203] extrinsic evidence was held admissible to identify the land intended to be devised by a description giving the location within the section, the section number, township and range number, and acreage, but not designating the state or county in which the land was located, nor whether the township in question was north or south of some base line from which congressional townships were numbered.

In *Pemberton v. Perrin*,[204] the land devised was described as 40 acres in a certain section and township, in range 9. Evidence was held admissible to show that the testatrix had no land in range 9, but owned 40 acres in range 10, otherwise conforming

---

[199] 8 N.J.L. 71 (1824).

[200] 137 Ga. 242, 73 S.E. 3, 3 L.R.A. (N.S.) 91, Ann. Cas. 1918A (1911).

[201] 97 Ga. 592, 25 S.E. 371 (1895).

[202] *Myher v. Myher*, 224 Mo. 631, 123 S.W. 806 (1909).

[203] 95 Ind. 184 (1884).

[204] 94 Neb. 718, 144 N.W. 164, Ann. Cas. 1915B (1913).

to the description in the will, and that it was this land, which was the only land she owned, that was intended to be devised.

Where no tract of land could be located by the description given in a will, because there was no section of the number stated in the township specified in the will, because the township was erroneously designated, but, by way of further description, the will recited that the testator had contracted to sell the tract in question, there was held to be a sufficient description to allow parol evidence to further identify the property.[205]

Additionally, where the land devised was described with reference to a stream, as lying "on the south side of Beaver Dam branch," extrinsic evidence was admissible to show the true location of the branch and, therefore, the land.[206]

In *Parsons v. Fitchett*,[207] the Virginia court stated that extrinsic evidence of the surrounding circumstances was admissible to aid in locating the land devised under the description, "the 20 acres of land adjoining."

And where a will described land as "a small farm in Wayne county, Iowa, near Missouri line," the Iowa court held in *Christy v. Badger*[208] that extrinsic evidence was not admissible to show that the testator intended to devise land lying in Lucas county, about 6 miles from his land in Wayne county.

Section Number. Extrinsic evidence is admissible to show that the testator did not own the land described as in the section named in the will but did own land in another section, which he intended to describe and devise, where, after rejecting the false part of the description, that is, the section number, sufficient description remains to identify the property, in light of the extrinsic circumstances.[209]

And where land was described in a will as being bounded on one side by a certain named railroad, and on another side by the land of named persons, and lying in section 21 in a certain town and range, containing 8 acres, and it appeared that there was no land in section 21 answering to that description or bounded as there described, extrinsic evidence was held admissible to show that fact, and that the testator never owned any land in section 21, but did own land in section 22, the description of which, except for the section number, corresponded with the description given in the will, by the Wisconsin court in *Hanley v. Kraftczyk*,[210] treating this as a latent ambiguity.

Location within a Section, Range, or Tract. Parol evidence was held to be admissible in *Eagle v. Oldham*,[211] to determine the location within the section of the

---

[205] *Christman v. Magee*, 108 Miss. 550, 67 So. 49 (1914).

[206] *Walston v. White*, 5 Md. 297 (1853).

[207] 148 Va. 322, 138 S.E. 491 (1927).

[208] 72 Iowa 581, 34 N.W. 427 (1887).

[209] *McMahan v. Hubbard*, 217 Mo. 624, 118 S.W. 481 (1909).

[210] 119 Wis. 352, 96 N.W. 820 (1903).

[211] 116 Ark. 565, 174 S.W. 1176 (1915).

land intended to be devised by the description of "the southwest part of section 31,150 acres, in township 1, range 8 west."

And in *McIlhattan's Will*[212] extrinsic evidence, including declarations by the testator, was held by the Wisconsin court to be admissible to show the location of 3 acres of land excepted from a devise of a tract of 80 acres, where the 3 acres were not definitely located or described.

However, in *Pickering v. Pickering*,[213] there was a devise of 5 acres in the northwest corner of the testator's field, to be laid off as nearly square as was convenient, and opposite to 5 acres devised to another person, the New Hampshire court held the ambiguity to be patent, so it was not allowable to show by parol evidence the acts and declarations of the testator as to his intention concerning the location of the land devised, such as the fact that about the time of the making of the will the testator measured off a tract of land in a corner of the field, and set up stakes, and declared it to be for the devisee in question.

Where the question involved was the location of a chapel in a cemetery, for the erection of which the testatrix set apart a fund, by a provision directing its erection "just beyond the present space," evidence was held admissible by the Kentucky court in *Carroll v. Cave Hill Cemetery Co.*[214] to describe the topography of the cemetery.

Number of Lot, Block, or Tract. In *Patch v. White*,[215] the majority of the court, in a five-to-four decision, adopted the view that there was a latent ambiguity, explainable by extrinsic evidence, as well where the misdescription consists of the designation of a thing not in existence, or, if in existence, not belonging to the testator and not intended to be designated, as where the description applies to more than one piece of property. It was held that in this case, where the will purported to devise "lot No. 6 in square 403," and the testator did not own lot No. 6 in square 403, although such a lot did exist, but did own lot No. 3 in square 406, parol evidence was admissible to show these facts so as to enable the court to determine, in the light of the other provisions of the will and of such extrinsic circumstances, whether or not lot No. 3 in square 406 was the lot intended.

And where a will devised "36 acres, more or less, in lot 37 in the second division, in Barnstead, being same I purchased of John Peavey," extrinsic evidence was held admissible to show that there was no such lot 37 in the second division of the town named, but that the testator owned a tract answering the description in lot 97, the court pointing out that by disregarding the false part of the description, namely, the lot number, enough remained to identify the land intended to be devised.[216]

[212] 198 Wis. 518, 224 N.W. 713 (1929).

[213] 50 N.H. 349 (1870).

[214] 172 Ky. 204, 189 S.W. 186 (1916).

[215] 117 U.S. 210, 29 L. Ed. 860, 6 S. Ct. 617 (1886).

[216] *Winkley v. Kaime*, 32 N.H. 268 (1855).

In the Indiana case of *Cruse v. Cunningham*,[217] extrinsic evidence was held admissible to show that a devise of "part of the donation lot No. 158 in township No. 3 north, of range No. 8 west, containing 200 acres," was intended to cover the northwest half of donation lot No. 58.

Street Name or Number. Where real estate is described in a will by its street number, extrinsic evidence is admissible to determine what land was included in the property at that number, and what monuments had been established upon the ground by the testator in his lifetime.[218]

And, extrinsic evidence was allowed in order to show what property was "known as" No. 250 Fifth avenue and what lands and improvements were included therein.[219]

Ambiguity, Indefiniteness, or Error in Boundaries, Landmarks, Corners, Courses, or Distances. Where a will contained a latent ambiguity as to the line of division between lots devised, the Kentucky court, in *Williams v. Williams*,[220] held that extrinsic evidence was admissible.

Also, where a will described a tract of land as "beginning in Joseph Wall's line at the corner between him and Chambers," and there were two corners in Wall's line answering to the description in the will, a latent ambiguity arose, which may be resolved by the introduction of extrinsic evidence showing the testator's intention.[221]

And where the will called for a cross fence as a boundary, and it turned out that there were two cross-fences, there was held to be a latent ambiguity, rendering admissible extrinsic evidence to determine which of the fences was intended by the testator.[222]

In the Illinois case of *Smith v. Dennison*,[223] a testator devised to different children tracts of land in adjoining quarter sections, and there was uncertainty as to the location of the boundary line separating the two quarter sections, evidence that the testator, long before his death or the making of his will, had agreed with an adjoining owner as to the location of such boundary, and had recognized and adopted the line agreed upon, was held to be admissible, not as tending to show that the boundary thus agreed on was the true quarter, but because of its tendency to show what land was intended to be devised to the respective grantees.

[217] 79 Ind. 402 (1881).

[218] *Myers v. Myers*, 16 Pa. Super. Ct. 511, 18 Lanc. L. Rev. 195 (1901).

[219] *Clark v. Goodridge*, 51 Misc. 140, 100 N.Y.S. 824 (1906).

[220] 182 Ky. 738, 207 S.W. 468 (1919).

[221] *Cubberly v. Cubberly*, 12 N.J.L. 308 (1830).

[222] *McCall v. Gillespie*, 51 N.C. 533 (1859).

[223] 112 Ill. 367 (1884).

In Connecticut, in the case of *Nichols v. Lewis*,[224] extrinsic evidence was held to be admissible to locate the eastern boundary of a tract of land embraced in a larger tract owned by the testator, bounded "east on the harbor at the foot of the bank," since "parol evidence is always admissible to explain a latent ambiguity, and to locate land and monuments referred to as boundaries."

And in the Oregon case of *Kerr v. Duvall*,[225] where a boundary line was designated as "by and along the foot of said hill," the court admitted testimony of a surveyor to the effect that the foot of the hill did not mean the place where the land began to slope up but the beginning of an abrupt rise.

However, extrinsic evidence was not admissible to establish the boundary line intended to be described by the testator where there was no ambiguity.[226]

Ambiguity, Indefiniteness, or Error in Area, Quantity, or Number of Parcels. Where the subject of a devise is described by quantity, it may be identified by extrinsic evidence.[227]

In *O'Donnell v. O'Donnell*,[228] extrinsic evidence was held not to be admissible to show that a devise of a certain number of "acres" of a named tract of land, the testator meant Irish acres and not statute acres.

Ambiguity, Indefiniteness, or Error in Source of Title or Time or Manner of Acquisition. Parol evidence was held admissible to identify land described in the will as "the tract of land I bought of James and Charles McCartney, lying in Greene County," in the Tennessee case of *McCorry v. King*.[229]

Where a will devised "all that certain farm or tract of land" located at a certain place, "it being the same farm or tract of land which was devised to me by the last will and testament of my late father," and the question arose as to whether this devise included a small field which was not devised to the testatrix by her father but was conveyed to her later, extrinsic evidence was held to be admissible showing that the field was at all times after its acquisition used in connection with and as a part of the farm, was annexed thereto, and treated by the testatrix as one of the fields constituting the farm.[230]

---

[224] 15 Conn. 137 (1842).

[225] 62 Or. 470, 125 P. 830 (1912).

[226] *Best v. Hammond*, 55 Pa. 409 (1867); *Cadmus v. Vreeland*, 28 N.J.Eq. 356 (1877).

[227] *Graves v. Rose*, 246 Ill. 76, 92 N.E. 601, 30 L.R.A. (N.S.) 303 (1910); *Stevens v. Felman*, 338 Ill. 391, 170 N.E. 243 (1930).

[228] Ir. L.R.13 Eq. 226 (1882).

[229] 3 Humph (Tenn.) 267, 39 Am. Dec. 165 (1842).

[230] *Knight v. Roe*, 5 Boyce (Del.) 570, 96 A. 32 (1915).

Description by Generic or Customary Name or Terms. Where land is devised as "Black acre," parol evidence is competent to show what land comprises Black acre.[231]

Similarly, where the will devised all the testator's "back land," extrinsic evidence of the custom of the testator, his family, and the neighborhood, of referring to certain land as "back land," was held admissible to identify the land devised.[232]

"My farm," "my land," "my lot," "plantation," "home farm," "home place," "homestead," "land where I now reside," or similar designation. Where a will describes land devised merely as "my farm," extrinsic evidence is admissible to identify the land intended to be devised.[233]

Where a will devised "my two lots of ground lying on the east and west sides of Leadenhall street, 'Ridgely's addition' to Baltimore town," extrinsic evidence was held admissible to show the sense in which the word "lots" was used, that is, whether it was used to designate lots of the identical size and location as shown on the plat of Ridgely's addition, or whether it embraced a large parcel of ground and was not intended to refer to the small subdivision designated on that plat.[234]

"Home farm" or "home place." Extrinsic evidence has been held to be admissible to determine what land was included in a devise of the testator's "home farm."[235]

"The land [farm, etc.] where I now reside." Extrinsic evidence has also been held to be admissible to identify the property covered by a devise of realty described in the will as the land, farm, or house where the testator resides.[236]

Thus, where there was a devise of "the farm on which I now live consisting of about 130 acres," extrinsic evidence was held admissible to identify the land devised, and to show that a lot not immediately adjacent to the main tract was formerly a part thereof and was used as a part of the farm.[237]

"Homestead." Extrinsic evidence has been held admissible to determine what land is embraced in a devise of the testator's "homestead."[238]

---

[231] *Cleverly v. Cleverly*, 124 Mass. 314 (1878).

[232] *Ryerss v. Wheeler*, 22 Wend. (N.Y.)148 (1839).

[233] *Bell v. Couch*, 132 N.C. 346. 43 S.E. 911 (1903).

[234] *Warner v. Miltenberger*, 21 Md. 264, 83 Am. Dec. 573 (1864).

[235] *P'Simer v. Steele*, 32 Ky. L. Rep. 647, 106 S.W. 851 (1908).

[236] *Aldrich v. Gaskill*, 10 Cush. (Mass., 1852) 155; *Thomson v. Thomson*, 21 S.W. 1085, 115 Mo. 56 (1893); *Horton v. Lee*, 5 S.E. 404, 99 N.C. 227 (1888).

[237] *Aldrich v. Gaskill*, 10 Cush. (Mass.) 155 (1852).

[238] *Morrall v. Morrall*, 86 N.E. 578, 236 Ill. 640 (1908); *Hopkins v. Grimes*, 14 Iowa 73 (1862).

Extrinsic evidence is also admissible to identify land included in a devise of "the old homestead, house, and a garden in which I now live and occupy."[239]

Tract Named for a Former or Present Owner or Occupant. Where land is described with reference to its present occupant or the former owner, extrinsic evidence has been allowed to identify it.[240]

For example, where there was a devise of "the dwelling house and stable which my said brother now occupies" in a certain town, parol evidence of the acts and declarations of the testator, and his conveyances by description, were admissible to identify the property devised.[241]

And in the case of *Coleman v. Eberly*,[242] the Pennsylvania court held that extrinsic evidence was admissible to show the extent of "the McKinstry farm occupied and farmed by William Brown."

Parol evidence was also held to be admissible in the South Carolina case of *Jones v. Quattlebaum*,[243] in order to identify the tract of land described in the will as "all my tract of land, a part of which is known as the Henry Raiford place, containing 100 acres more or less."

Description Applicable to More Than One Piece of Property. It is well settled that where the description in a will applies indifferently to two or more pieces of property, a latent ambiguity exists, to resolve which extrinsic evidence is admissible to show which was meant by the testator.[244]

Where a will gave as a boundary of the land devised, "the barnyard," and it appeared from extrinsic circumstances that there were two barnyards that could answer the description in the will, it was held to be a latent ambiguity, therefore allowing extrinsic evidence, which included declarations of the testator made at the time of the execution of the will, to determine which barnyard the testator intended.[245]

Where There Is No Description; or There Are Blanks. Parol evidence is not admissible where there is no description of the thing devised.[246]

---

[239] *Robertson v. Schermerhorn*, 47 Hun. 637, 14 N.Y.S.R. 309 (1888).

[240] *Douglas v. Blackford*, 7 Md. 8 (1854); *Chace v. Lamphere*, 51 Hun. 524, 21 N.Y.S.R. 676, 4 N.Y.S. 228 (1889).

[241] *Cleverly v. Cleverly*, 124 Mass. 314 (1878).

[242] 76 Pa. 197 (1874).

[243] 31 S.C. 605, 9 S.E. 982 (1889).

[244] *Patch v. White*, 117 U.S. 210, 29 L.Ed. 860, 6 S.Ct. 617 (1886); *Hoffner v. Custer*, 86 N.E. 737, 237 Ill. 64 (1908); *Kinney v. Kinney*, 34 Mich. 250 (1876); *French v. French*, 14 W.Va. 458 (1877).

[245] *Morgan v. Burrows*, 45 Wis. 211, 30 Am. Rep. 717 (1878).

[246] *Warner v. Brinton*, (C.C.) Fed. Cas. No. 17,179 (1835).

The description of property cannot be entirely supplied by extrinsic evidence, where there is nothing in the language of the will itself to show what property is intended.[247]

Translation or Explanation of Foreign Language, Code, Cipher, Symbol, or Abbreviation. Evidence to decipher or translate a will written illegibly, or in code, symbols, or foreign language, does not, strictly speaking, relate to the construction of the language or the meaning of the words as used in the will, but rather to the meaning of the words or characters in the abstract.

It was stated by Vice Chancellor Wigram: "Words cannot be ambiguous because they are unintelligible to a man who cannot read; nor can they be ambiguous merely because the court which is called upon to explain them may be ignorant of a particular fact, art, or science which is familiar to the person who used the words, and a knowledge of which is therefore necessary to a right understanding of the words he has used."[248]

Therefore, extrinsic evidence is admissible to translate a will written in a foreign language.[249]

Extrinsic evidence is also admissible to explain technical phrases, terms of art, and scientific words.[250]

Local Usage or Reputation with Respect to Words, Phrases, or Names. Where there is an ambiguity or inaccuracy in the name or designation of a devisee or legatee, extrinsic evidence is admissible to show that a certain claimant is generally known and called in the community by the name given in the will.[251]

In *Ayres v. Weed*,[252] the Connecticut court stated that if a legacy is given, or a devise is made to a person by his nickname, parol evidence is admissible to show that he is generally known in the community by that name.

Deeds, Plats, or Other Instruments. The testator's deeds and plats have been held to be admissible to determine the land intended by an ambiguous devise.[253]

---

[247] *McDonald v. Shaw*, 92 Ark.15, 121 S.W. 935, 28 L.R.A. (N.S.) 657; *Hawman v. Thomas*, 44 Md. 30 (1876).

[248] Wigram, *Extrinsic Evidence in Aid of the Interpretation of Wills*, 2d Am. ed. p. 253.

[249] *Hart v. Marks*, 4 Bradf. (N.Y.) 161 (1856).

[250] *Spencer v. Higgins*, 22 Conn. 521 (1853).

[251] *Ayres v. Weed*, 16 291 (1844); *Equitable Trust Co. v. Banning*, 149 A. 432, 17 Del.Ch. 95 (1930); *Hinckley v. Thatcher*, 1 N.E. 840, 52 Am. Rep. 719, 139 Mass. 477 (1885); *Tallman v. Tallman*, 3 Misc. 465, 23 N.Y.S. 734 (1893).

[252] 16 Conn. 291 (1844).

[253] *Eagle v. Oldham*, 116 Ark. 565, 174 S.W. 1176 (1915).

For example, a deed conveying certain real property to the testator was held admissible to supplement the language of the will by providing a more particular description of the property devised.[254]

And former conveyances, by description, have been held admissible in evidence to determine the property included in the devise of a "dwelling house and stable."[255]

Instruments, Objects, or Circumstances Referred to in Will. It is well settled that an extrinsic instrument, book, object, or matter specifically referred to in a will, and duly identified, is admissible to aid in the interpretation of the will.[256]

Where land devised is described by reference to a deed conveying it to the testator, such deed is admissible in evidence to identify the land devised.[257]

The New Hampshire court in *Pickering v. Pickering*,[258] stated that where a testator marks out a tract of land and sets up monuments at the corners, and devises land by description referring to such monuments, parol evidence is admissible to identify the monuments.

Establishment of Will Lost before Testator's Death. The general rule is that where a will is lost before the death of the testator it continues to be legally effective if the loss was unknown to the testator and may be established in probate proceedings.[259]

The Maine court stated in the case of *Rich v. Gilkey*,[260] "if a will is made and adhered to by a testator till his death, and he desires it to exist, or supposes it to, then it does legally exist till his death, unrevoked, though prior thereto it has been lost or mislaid, or accidentally or fraudulently despoiled. The writing or script may be gone, but the will remains."

The general rule was declared by the Kentucky court in *Chisholm v. Ben*:[261] "No written will having been produced for probate, and none having been in existence at the death of Chisholm, it was incumbent on the party offering to prove by parol the substance of his will, to establish: (1) The fact that he had made a valid will; (2) the contents or substance of that will, or of such portion as might be recorded as his will; and (3) that the will, though not in existence at his death, had not been revoked by him."

---

[254] *Re Wolf*, 128 Cal. App. 305, 17 P.2d 1052 (1932).

[255] *Cleverly v. Cleverly*, 124 Mass. 314 (1878).

[256] *Warner v. Brinton*, Fed. Cas. No. 17,179 (C.C., 1835); *Temple v. Bradley*, 87 A. 394, 119 Md. 602, *Pickering v. Pickering*, 50 N.H. 349 (1870); *Harris v. Harris*, 72 A. 912, 82 Vt. 199 (1909).

[257] *Williams v. Bailey*, 178 N.C. 630, 101 S.E. 105 (1919).

[258] 50 N.H. 349 (1870).

[259] *Gaines v. Hennen*, 24 How. 553, 16 L.Ed. 770 (1861); *Hodge v. Joy*, 92 So. 171, 207 Ala. 198 (1921); *Harris v. Camp*, 76 S.E. 40, 138 Ga. 752 (1912); *Rich v. Gilkey*, 73 Me. 595 (1881); *Davis v. Sigourney*, 8 Met. 487 (Mass., 1844); *Scoggins v. Turner*, 3 S.E. 719, 98 N.C. 135 (1887).

[260] 73 Me. 595 (1881).

[261] 7 B. Mon. (Ky.) 408 (1847).

In the above case of *Rich v. Gilkey*, the court further stated, "it may be that the will was destroyed by the testator in a fit of insanity, or that it was lost, or accidentally or fraudulently destroyed. Such accidental or fraudulent destruction will not deprive parties of their rights under its provisions, if they can produce the evidence necessary to establish the will."

Proof of Contents.   The contents of a will lost before the testator's death may be established by secondary evidence.[262]

Again, in the foregoing case of *Rich v. Gilkey*, the Maine court stated that the contents of a will may be proved by a copy, testimony of the subscribing witnesses, or by any other competent and clear evidence thereof.

In the Alabama case of *Jaques v. Horton*,[263] it was held that the contents of a last will might be proved by parol evidence or by a true copy thereof, after diligent search had failed to reveal the original will.

FOR FURTHER REFERENCE

12 ALR 1179	*What included in devise of "house," "dwelling house," or the like.*
34 ALR 1304	*Establishment of will lost before testator's death.*
100 ALR 1465	*Admissibility of parol evidence as to meaning of cryptic words, abbreviations, signs, symbols, or figures appearing in written contracts or other writings.*
102 ALR 287	*Comment note—Rule that latent ambiguities may be explained by parol evidence but that patent ambiguities may not.*
152 ALR 938	*What passes under devise of "farm" or "plantation."*
38 ALR2d 840	*What passes under, and is included in, devise of "home" or "home place."*
61 ALR2d 1390	*Admissibility of parol evidence with respect to reservations or exceptions upon conveyance of real property.*
12 ALR4th 795	*Which of conflicting descriptions in deeds or mortgages of fractional quantity of interest intended to be conveyed prevails.*

[262] *Jaques v. Horton*, 76 A. 238 (1884); *Re Johnson*, 40 Conn. 587 (1874); *Rich v. Gilkey*, 73 Me. 595 (1882); *James v. Parker*, 102 A. 760, 131 Md. 466 (1917); *Lane v. Hill*, 73 Am.St.Rep. 591, 44 A. 393, 68 N.H. 275 (1895); *Harris v. Harris*, 26 N.Y. 433 (1863).

[263] 76 Ala. 238 (1884).

REFERENCES

American Jurisprudence 2d. Volume 23. *Deeds*. Rochester: Lawyers Cooperative Publishing Company. 1956.

American Jurisprudence 2d. Volume 30. *Evidence*. Rochester: Lawyers Cooperative Publishing Company. 1956.

Corpus Juris Secundum. Volume 26. *Deeds*. Brooklyn: American Law Book Co. 1938.

68 ALR 4 (1930) *Admissibility of parol evidence to explain ambiguity in description of land in deed or mortgage.* The Lawyers Co-operative Publishing Co., Rochester & Bancroft–Whitney Co., San Francisco.

94 ALR 26 (1935) *Admissibility of extrinsic evidence to aid interpretation of will.* The Lawyers Co-operative Publishing Co., Rochester & Bancroft–Whitney Co., San Francisco.

CHAPTER 9

MAPS, PLATS, PLANS, AND CHARTS

It is a common and proper practice in courts of justice
to receive models, maps, and diagrams, or sketches,
drawn on paper, or traced with chalk on a blackboard,
for the purpose of giving a representation of objects
and places which cannot otherwise be as conveniently
shown or described by the witnesses to the jury.

—Webster Gas Constr.
Co. v. Danner
C.A.9th Cal, 97 F. 882 (1899)

Terminology. There are a number of different terms used to denote drawings and portrayals of information and data in map form.

Cadastral map. Strictly, a map for the purpose of making a cadastre. A *cadastre* is an official register used to apportion taxes. Hence, it is a map showing the value and relationship of land for taxing purposes (Brown, et al).

Chalk. A drawing or sketch produced in court to illustrate and supplement verbal testimony. Today, formal paper exhibits and models are used, but at one time sketches were made on a blackboard, hence the term "chalk."

Chart. Generally used to mean a special-purpose map for navigation purposes, either nautical or aeronautical, along with maps of the heavens.

Map. A representation of the earth's surface or of some portion of it, showing the relative position of the parts represented, usually on a flat surface (*Black's Law Dictionary*).

Plan. A map, chart, or design; being a delineation or projection on a plane surface of the ground lines of a house, farm, street, city, and so forth, reduced in absolute length, but preserving their relative positions and proportion. *Jenney v. Des Moines*, 72 N.W. 550, 103 Iowa 347; *Wetherill v. Pennsylvania R. Co.*, 45 A. 658, 195 Pa. 156.

Plat. In land surveying, a map, or representation on paper, of a tract of land subdivided into blocks, lots, and so on, usually drawn to a scale. In the survey of the public lands, the term "plat" refers to the drawing that represents the particular area included in a survey, such as a township, private land claim, or mineral claim (Shalowitz).

Plat, or Plot. A map, or representation on paper, of a piece of land subdivided into lots, with streets, alleys, and so on, usually drawn to a scale. *McDaniel v. Mace*, 47 Iowa 510; *Burke v. McCowen*, 47 P. 367, 115 Cal. 481.

Plate. The term generally applied to a shop drawing or design in mechanical drafting; equivalent to the term sheet.

Plot. The act of placing survey data on a plat or map. Often used as *plot plan* or *plot* meaning the plat itself. Preferred usage is to use the word plot only as the act of placing data on a plat or map (Brown, et al).

Plot Plan. Correctly used, a *plot plan* denotes the product produced in a property inspection for lending purposes. Also know as a mortgage plan or, incorrectly, "mortgage survey." The latter term implies that a *survey* was made when usually only an inspection of the property was done. *Plot plan* is also sometimes used to denote *survey plat*. The interchanging of these terms has led to much confusion as to what a drawing actually represents.

Sketch. A simple drawing, done rapidly and without much detail, describing the major parts or points of a land description. These drawings may be freehand, not to scale, and should be construed as being for illustrative purposes only. In document research, sketches are used to illustrate a land parcel or how several parcels are related to one another.

Ancient Maps. The history of cartography is a fascinating study, yet it seems that only a small percentage of today's population appreciates the scope and scale of the subject. This is surprising, since we are all dependent on maps in our everyday lives. We read and hear of current events in faraway places, we drive or fly from place to place, we give and take directions almost on a daily basis, and we are intrigued by places like Atlantis, the solar system, Treasure Island, Middle Earth and the Northwest Passage.

Ancient plats are valuable from the standpoint of depicting conditions at a point in time and in giving names contemporaneous with other, earlier instruments. Names, particularly those of roads and streets, are subject to change, and it is the conditions and names at the time of the instrument that must be considered.[1]

Ancient maps, like many other ancient writings, have traditionally been considered by the courts as those which are more than thirty years old. Presently, the ancient document rule under the Federal Rules of Evidence, is twenty years. See Rules 803(16) and 901.

Evidence. Having a history of the maps pertaining to an area is as important as having a history of the title. Therefore, every attempt should be made to view all documents relating to the land in question, not just deeds and other instruments of title.

Many older maps are not recorded and part of the public record at the court house or other typical land record repository. Nontypical record sources should be consulted, such as archives, libraries, and historical societies. *The earlier the map or survey plan, the less likely it is to be part of the public record.*

Only in relatively recent times have reprographic facilities such as photocopiers, blueprint machines, and plotters been in use. Early maps, particularly those resulting from simple surveys were generally given to the client and are usually now lost or unavailable.

Figure 9.1 Field notes indicating that the plan was "left with Coll⁹ Jennifs."

[1] In construing a deed, court has duty to place itself as nearly as possible in situation of parties at the time instrument was made. *Sanborn v. Keroack,* 171 A.2d 25, 103 N.H. 297 (1961).

Occasionally, however, these old maps were preserved with other valuable personal records and can be found. Every attempt should be made to preserve and protect these valuable records in a safe and secure place where they will be available for inspection and use by researchers. Some state surveying societies have taken an assertive leadership role in this task.

Since these maps are usually found in obscure places, the researcher needs to have some of the qualities of a detective. An understanding of family records is important, for the old story of the key document "being found in so and so's trunk in the attic" is a true and familiar one. Sometimes a clue may be found in a probate record, so it is important to check this source of evidence when examining the title documents.

Another key to locating early records is knowing who may have practiced surveying or law in the area at a particular time. Finding survey information such as bearings or measured distances in a land description is a clue that someone made measurements and, therefore, may have made notes or a map. An historical study of surveyors in the subject area and the location of their records, if still available, is often critical in solving a land problem. One of the greatest services a professional land surveying society can provide is locating and protecting surveying records after a surveyor has ceased to practice.

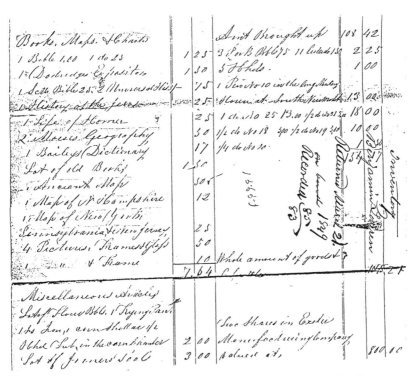

Figure 9.2 Section of an 1849 probate inventory. Listed are "Books, Maps & Charts," 2 geography books, an "Ancient Map," and maps of New Hampshire, New York, Pennsylvania, and New York.

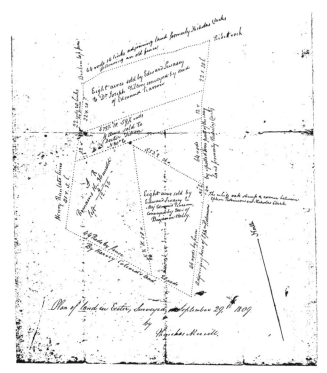

Figure 9.3 Survey plan found at a local historical society. Note such things as corner monuments and calls for fences, which do not appear in any of the succeeding deed descriptions, many of which were drawn from this plan.

Interpreting Maps. Since maps are a special category of land description, they deserve separate treatment. In the practical construction and drawing of maps, there are generally accepted standards. Standards are rules, and they should be understood in order to adequately interpret the symbols and intent of the map. These standard rules and symbols are not always followed, so a legend should appear on the document to explain the meaning of the symbols used. Most current, published standards for surveying practice require that a legend be shown on all maps, plats, or plans. Further, these standards generally address the issue of map accuracy. In interpreting maps, it is extremely important to understand the accuracy of the document as it relates to the map user's perspective and to the actual *intent* of the map. For example, the accuracy and intent of a survey map depicting subdivided lots to be sold is completely different from the accuracy and intent of a community land use map.

Legend. Symbols are used to depict features on different types of maps, and there should be a legend on every map. A legend is a description, explanation, table of symbols, and other information that is printed on a map or chart for a better understanding and interpretation of it. An example of typical map legend symbols is shown in Table 9.1.

Usually a circle symbol [O] indicates a round object such as a pipe, or a pin. Variations such as ● or O may indicate whether it was already existing and found or was set during the course of the survey. A square symbol [□] generally indicates a square object, such as a stone or concrete bound. A square could also indicate a wooden post, or a stake. On forestry maps and town line surveys, a post or bound is commonly shown as illustrated in Figure 9.4.

A triangle symbol △ often designates a traverse point, but also may indicate a triangulation station or similar object.

Due to the fact that standard symbols are not always used, it is important not to assume as to what one is. Either consult the legend, if there is one, or make some other independent verification. There are literally hundreds of available books and manuals that show and explain standard mapping symbols. Government maps and highway maps have used standard symbols for some time. It is most often the private surveyors and other independent cartographers who are responsible for the wide variation in the use of map symbols. This is all part of the detective work that may be necessary to fully interpret the map.

Survey maps should also indicate whether a piece of evidence was "found" or "set." The other possible category is "searched for, not found" when something is called for in a title document and is unable to be found. It is not advisable to say "searched for, found to be destroyed," or "missing." Sometimes, the marker does, in fact, exist but it is not found for various reasons. For instance, initially it may have been set incorrectly and really does exist a considerable distance away, making it very difficult to find. "Not found" is a fact; "destroyed" is a conclusion but in some cases may be a fact.

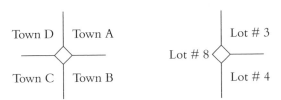

Figure 9.4 Property corner symbols.

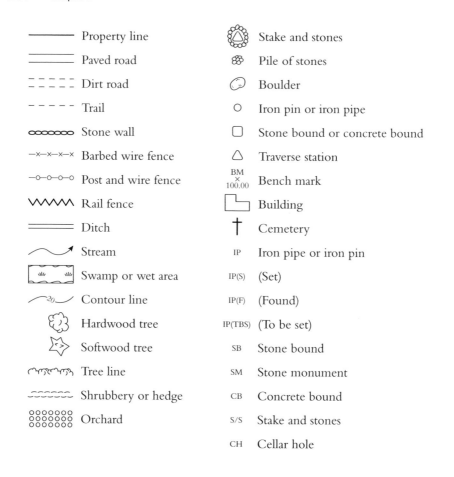

————	Property line	Stake and stones
═══════	Paved road	Pile of stones
- - - - -	Dirt road	Boulder
– – – – –	Trail	Iron pin or iron pipe
∞∞∞∞∞	Stone wall	Stone bound or concrete bound
–x–x–x–x	Barbed wire fence	Traverse station
–o–o–o–o	Post and wire fence	BM × 100.00 Bench mark
WWWW	Rail fence	Building
════════	Ditch	Cemetery
	Stream	IP Iron pipe or iron pin
	Swamp or wet area	IP(S) (Set)
	Contour line	IP(F) (Found)
	Hardwood tree	IP(TBS) (To be set)
	Softwood tree	SB Stone bound
	Tree line	SM Stone monument
	Shrubbery or hedge	CB Concrete bound
	Orchard	S/S Stake and stones
		CH Cellar hole

Table 9.1   Some Common Symbols Found On Property Maps

Orientation. Maps are oriented toward some direction, preferably NORTH. The direction NORTH has many different meanings in cartography. The specific type of north orientation should be noted on the map, whether "true," magnetic, grid, assumed, or otherwise. Ideally, if the north is "true," the basis of the determination is stated; if magnetic, the date of the observation and which line(s) was(were) used is stated; and if from another source, such as a railroad plan, a highway plan, or a tie to geodetic control, that fact is stated.

Older plans usually did not state the basis of the bearing. Customarily, however, private surveys were done with magnetic bearings, therefore the orientation of the map was magnetic north.[2]

---

[2] The courses in a deed are to be run according to the magnetic meridian, unless something appears to show that a different mode is intended. *Wells v. Jackson Iron Mfg. Co.,* 47 N.H. 235, 90 Am. Dec. 575 (1866).

Sometimes the design of the north arrow signifies the type of orientation. A full head on the arrow is used if the basis is true north, while a half-head appears on an arrow designating magnetic north. The "flag" on the head of the arrow is drawn on the appropriate side to indicate the declination.

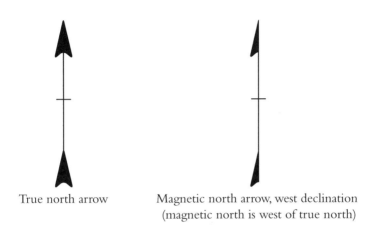

True north arrow

Magnetic north arrow, west declination
(magnetic north is west of true north)

How to Determine Orientation. If the orientation of a map is not shown, it can be found in a number of ways.

- *Comparison with other maps.* Usually, topographic maps can be used for comparison, particularly if the same features appear on each map, such as a road, railroad, stream, or other body of water.
- *By computation.* If any directions are shown on the map, the north arrow can be derived. For example:

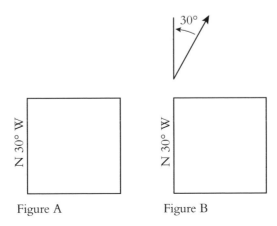

Figure A   Figure B

A parcel described or shown like Figure A, with only one direction mentioned, can be used to derive the north orientation, as shown in Figure B.

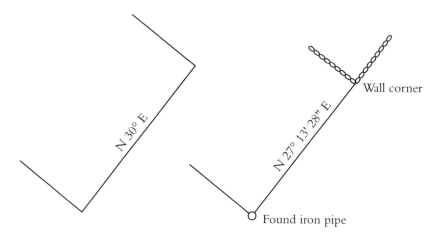

A. Sketch of Bearing
   from Deed Description

B. Part of Survey Plat
   Showing Description

No two surveys of the same line are likely to exactly agree, except through some strange set of circumstances.[3] First, without further study, it may not be known if one bearing is based on true north and the other on magnetic north. Second, the two observations were taken at different times, sometimes many years apart, so the change in magnetic declination will result in a difference between the two. Third, two different compasses under two different sets of conditions will produce two different bearings. Fourth, the early description was probably based on a survey or set of bearings that relied on direct observation, while the recent definition of direction would probably be the result of a closed traverse adjusted by some accepted rule or set of rules to remove inherent error, causing it to "close" mathematically. As a result of doing this, even two traverses or two independent surveys of the same plot of ground would not agree exactly because they would each have their own set of errors. If the surveyors used different rules of adjustment, each would generate different results, and even if they chose the same set of rules, their results would be slightly different. However, since only a relatively small amount of error is allowable in a survey, results are going to be reasonably close.

---

[3] It is a matter of common knowledge that surveys made by different surveyors seldom, if ever, completely agree and that, more than likely, the greater number of surveys the greater number of differences. *Erickson v. Turnquist,* 77 N.W.2d 740, 247 Minn. 529 (1956).

The property line does not shift, only its frame of reference, or its definition, changes. If the same points are in agreement, and if the directions and distances are reasonably close to each other, the two lines are most likely one and the same.

One way to make a valid comparison between two descriptions is to compare interior angles or angles between the bearings. If two descriptions of the same property are on different reference systems or on different bearing systems, the angles between bearings should be the same since the physical location of the lines themselves do not change. See comparisons in the following diagrams.

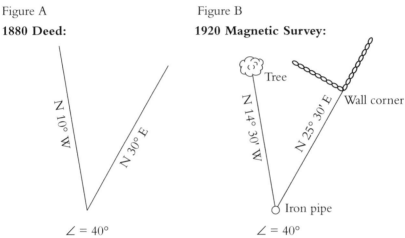

Figure A

1880 Deed:

N 10° W

N 30° E

∠ = 40°

Figure B

1920 Magnetic Survey:

Tree

N 14° 30' W

N 25° 30' E

Wall corner

Iron pipe

∠ = 40°

Figure C

True bearings in 1920 with 10° E declination:

Tree

N 04° 30' W

N 35° 30' E

Wall corner

Iron pipe

∠ = 40°

Figure D

Recent, very precise, adjusted survey:

Tree

N 16° 43' 10" W

N 23° 10' 40" E

Wall corner

Iron pipe

∠ = 39° 53' 50"

≈ 40°

The same problem arises in comparing two descriptions, an early one and a later one, if they are based on two sets of observations (see Figures A–D above). There is always a difference in magnetic declination, so two sets of observations are bound to differ. In addition, when descriptions are based on estimates, how the viewer saw the property has a profound effect on the resulting description. For convenience, people often view roads as running either north-south or east-west, and property lines as running at right angles to or parallel with the roads, streams, or other features. In reality this is unlikely, so differences are commonly encountered when later, more accurate, definitions are attempted.

Determining Scale. When the scale of a map is not known, it can be determined in many ways, if it is drawn to scale. Some maps are mere sketches with no intention on the part of the draftsperson to have a common scale throughout, and should be so labeled. For those maps that are drawn to scale, the following methods can be used to determine and derive the actual scale of the map.

1. **Comparison with real conditions on the ground.** Select two well-defined points on the map and measure the distance between them in appropriate units. Then compare the distance measured on the map with the same distance measured on the ground.
2. **Comparison with a map of known scale.** Measurements can be made on two maps, and then compared. Using the map of known scale, the scale can be calculated for the other map.
3. **Comparison with an aerial photograph of known scale.** This procedure is the same as in (2).
4. **Comparison with some other known base.** Surveys and deed measurements can be used as a known scale to solve for an unknown scale.

Obviously, some methods are more accurate than others, but if only an approximate scale is needed, it can be readily and easily determined.

Examples:

1. Two well-defined points such as iron pins on the map and on the ground:

Map distance = 1.2"
Ground distance = 1,200'
Map scale: 1.2" = 1,200'
1" = 1,000'
that is, 1 inch on the map represents 1,000 feet on the ground

2. Comparison with map of known scale:

Map Distance	Map Distance
unknown scale	known scale: topographic quadrangle sheet, 1:24,000
1.2"	1.6"

The formula to use is measurement/unknown scale = measurement/known scale.

1:24,000 is known as a representative fraction, commonly abbreviated RF. What this means is that 1 *unit* on the map is equivalent to 24,000 *units* on the ground. There are no measurement units attached to an RF; it is merely a ratio and if units are desired, both numbers carry the same units. That is,

1" on the map = 24,000" on the ground, or, converting, the scale would be 1" = 2,000', determined by dividing 24,000 by 12 to convert the inches to feet.

Older quadrangle sheets were at a scale of 1:62,500, or, 1" = 5,208' (approximately, but not quite, an inch to the mile).

1:24,000 is the RF

1" = 2,000' is the scale

In the example,

$$\frac{\text{unknown map distance}}{\text{unknown scale}} = \frac{\text{known map distance}}{\text{known scale}}$$

or put another way,

$$\frac{\text{unknown map distance}}{\text{known map distance}} = \frac{\text{unknown scale}}{\text{known scale}}$$

$$\frac{1.2"}{x} = \frac{1.6"}{2,000} \quad \text{comparative items must be in the same units}$$

$$1.6x = 2,400$$
$$x = 1,500$$

The scale of the unknown map is 1" = 1,500'
The RF of the unknown map is 1 : (1,500 × 12) or
1:18,000

When discussing map scale in RF terms the smaller the number, the larger the scale. That is, 1:18,000 is a larger scale than 1:24,000, since 18,000 is a smaller number than 24000. When discussing map scale in verbal or fractional terms the smaller the number (denominator) the larger the scale. For example, 1/100 or 1" = 100' is a larger scale than 1/200 or 1" = 200'. Smaller scale maps cover larger areas, but with less (smaller) detail.

As general rule, the longer the line used as a basis for measurement, the more accurate the result is likely to be. However, practical measures are certainly a factor. For example, if you have a choice between a 3-inch line or a 7-inch line on a map for comparison purposes, the 7-inch line will probably be more reliable.

Scale 1 to-	Inches per mile	Miles per inch	Feet per inch	Gunter's chains per inch	Engineer's chains per inch	Acres per square inch	Square miles per square inch
500	126.72	0.0079	41.67	0.631	0.417	0.0399	0.00006
600	105.60	0.0095	50	0.758	0.500	0.057	0.00009
800	79.20	0.013	66.67	1.010	0.667	0.102	0.00016
1000	63.36	0.0158	83.33	1.263	0.833	0.159	0.00025
1200	52.80	0.0189	100	1.515	1	0.230	0.0004
1500	42.24	0.0237	125	1.894	1.250	0.359	0.0006
2000	31.68	0.0316	166.67	2.525	1.667	0.638	0.0010
2400	26.40	0.0379	200	3.030	2	0.9183	0.0014
2500	25.34	0.0395	208.33	3.157	2.083	0.997	0.0016
3000	21.12	0.047	250	3.788	2.500	1.435	0.0022
3500	18.10	0.0552	291.67	4.419	2.917	1.953	0.0031
3600	17.60	0.057	300	4.545	3	2.067	0.0032
3960	16	0.0625	330	5	3.300	2.500	0.0039
4000	15.84	0.063	333.33	5.050	3.333	2.551	0.0040
4800	13.20	0.0758	400	6.061	4	3.674	0.0057
5000	12.67	0.0789	416	6.313	4.167	3.986	0.0062
5280	12.00	0.0833	440	6.667	4.400	4.445	0.0069
6000	10.56	0.095	500	7.576	5	5.741	0.0090
6500	9.748	0.103	541.67	8.207	5.417	6.736	0.0105
7000	9.051	0.110	583.33	8.838	5.833	7.813	0.0110
7200	8.800	0.114	600	9.091	6	8.266	0.0129
7920	8	0.125	660	10	6.600	10.002	0.0156
8000	7.920	0.126	666.67	10.100	6.667	10.205	0.0159
8400	7.543	0.133	700	10.605	7	11.251	0.0176
9000	7.041	0.142	750	11.363	7.500	12.916	0.0202
9600	6.600	0.152	800	12.121	8	14.696	0.0230
10,000	6.346	0.158	833.33	12.626	8.333	15.946	0.0249
10,800	5.877	0.170	900	13.635	9	18.599	0.0291

Table 9.2   Map Scale and Vertical Aerial Photo Relations

12,000	5.280	0.189	1,000	15.152	10	22.957	0.0359
12,672	5	0.200	1,056	16.000	10.560	23.892	0.0400
13,200	4.800	0.208	1,100	16.666	11	27.784	0.0434
14,400	4.400	0.227	1,200	18.181	12	33.065	0.0516
15,000	4.224	0.237	1,250	18.938	12.500	35.878	0.0560
15,600	4.062	0.246	1,300	19.695	13	38.806	0.0606
15,840	4	0.250	1,320	20	13.200	40.01	0.0625
16,000	3.960	0.253	1,333.33	20.200	13.333	40.82	0.0638
16,800	3.771	0.265	1,400	21.210	14	45.01	0.0703
18,000	3.520	0.284	1,500	22.725	15	51.66	0.0807
19,000	3.335	0.300	1583.33	24.000	15.833	57.55	0.0899
19,200	3.300	0.303	1,600	24.240	16	58.78	0.0918
20,000	3.168	0.316	1,666.67	25.253	16.667	63.77	0.0996
20,400	3.106	0.322	1,700	25.755	17	66.36	0.1037
21,120	3	0.333	1,760	26.664	17.600	71.13	0.1111
21,600	2.933	0.341	1,800	27.270	18	74.40	0.1162
22,200	2.854	0.350	1,850	28.030	18.500	78.57	0.1228
22,800	2.779	0.360	1,900	28.785	19	82.89	0.1295
24,000	2.640	0.379	2,000	30.303	20	91.83	0.1435
25,000	2.534	0.395	2,083.33	31.563	20.833	99.64	0.1557
30,000	2.110	0.473	2,500	37.879	25	143.51	0.2242
31,680	2	0.500	2,640	40	26.400	160.04	0.2500
36,000	1.760	0.568	3,000	45.455	30	206.61	0.3228
38,400	1.650	0.606	3,200	48.485	32	235.08	0.3673
40,000	1.580	0.631	3,333.33	50.505	33.333	255.13	0.3985
45,000	1.408	0.710	3,750	56.818	37.500	322.90	0.5044
48,000	1.320	0.758	4,000	60.606	40	367.39	0.5739
50,000	1.267	0.789	4,166.67	63.131	41.667	398.56	0.6228
60,000	1.050	0.947	5,000	75.758	50	398.65	0.8967
62,500	1.014	0.986	5,208.33	78.914	52.083	574.05	0.9730
63,360	1	1	5,280	80	52.800	640.15	1
70,000	0.905	1.105	6,333.33	88.384	63.333	781.35	1.4386
80,000	0.792	1.263	6,666.67	101.010	66.667	1,021	1.5942
90,000	0.704	1.420	7,500	113.636	75	1,291	2.0173
96,000	0.660	1.515	8,000	121.212	80	1,469	2.2957
100,000	0.633	1.578	8,333.33	126.263	83.33	1,595	2.4909
120,000	0.528	1.894	10,000	151.515	100.00	2295.68	3.5870
125,000	0.506	1.973	10,416.67	157.828	104.17	2,491	3.8922
126,720	0.500	2	10,560	160	105.60	2,561	4
144,000	0.440	2.273	12,000	181.818	120.00	3305.79	5.1653
200,000	0.316	3.157	16,666.67	252.525	166.67	6,378	9.9639
250,000	0.253	3.946	20,833.33	315.657	208.33	9,966	15.5686
253,440	0.250	4	21,120	320	211.20	10,242	16
316,800	0.200	5	26,400				
380,160	0.167	6	31,680	480	316.80	23,045	36
500,000	0.127	7.891	41,666.67	631.313	416.67	39,856	62.7244
760,320	0.083	12	63,360	960	633.60	92,181	144
1,000,000	0.063	15.783	83,333.33	1262.626	833.33	159,459	249.0976
$\dfrac{63,360}{scale}$	$\dfrac{scale}{63,360}$		$\dfrac{scale}{12}$	$\dfrac{scale}{66 \times 12}$	$\dfrac{scale}{100 \times 12}$	$\dfrac{scale^2}{43560 \times 144}$	$\dfrac{(ft/in)^2}{(5280)^2}$

Table 9.2   (Continued)

Elevations and Depths. Maps can also show relief, which is in a vertical direction. Maps showing elevations are known as *contour maps* or *topographic maps*, while those indicating depths are either *depth maps*, *charts*, or more properly, *bathymetric maps*.

Contours, which are lines of equal elevation, or vertical distance above a defined plane known as a *datum*, are often shown on a map or plat. The distance between successive contours is the *contour interval*. These are commonly used for planning any type of development and also can be found on lot surveys when buildings and/or septic facilities are planned.

The universally accepted datum upon which elevations are based is mean sea level. However, for individual lots or small local areas such as a building site, an arbitrary datum may be selected and defined. Sometimes a nearby, semipermanent point is chosen and designated as an assumed elevation, for example, "assumed elevation 100.00 feet." Elevations and contour lines are then designated as some number of feet, tenths and hundredths above or below that defined point or datum plane.

Accuracy and Precision. The terms "accuracy" and "precision" are often confused and are used interchangeably, when they should not be, since they mean different things. Precision has to do with refinement of measurement, while accuracy denotes nearness to the truth.

Precision of survey is expressed as a ratio (the Standard Fraction of Precision), 1 part in _____ parts. For example, a survey having a relative precision of 1 part in 10,000 merely says that the measurements expressed are expected to have an error of no more than 1 unit per 10,000 units measured. Obviously, then, a survey with a precision of 1 in 10,000 is a more precise (relatively speaking) survey than one which is 1 in 5,000, for in the former the error could be 1 unit per 10,000 units whereas the latter could contain as much as 2 units per 10,000 feet.

Figure 9.5 Example of a section of topographic map showing both topography (contour lines) and bathymetry (depth curves).

Bear in mind, however, that a 1 in 10,000 survey is not necessarily a *better* survey than one which is 1 in 5,000, it is merely a more *precise* survey. The question is not whether a survey is good or bad, but whether or not it is *adequate*. Depending on the conditions, circumstances, and purpose of a survey, one which is low in precision may suffice perfectly.

The other consideration has to do with *accuracy*. A 1 in 10,000 survey made of the wrong lot, or using the wrong corners or boundaries, may be very precise, but inaccurate, and probably worthless. That is one reason the courts have emphasized monumentation and the determination of correct boundaries rather than whether a survey was 1 part in 5000 versus 1 part in 10,000.

Plat Certification. Notations on a survey plat frequently address the precision of a survey. This is sometimes misleading since surveyors and computational processes do not always derive the figure in the same way. The expression itself is often misleading as well. "I certify that this survey has an accuracy of not less than 1 part in 10,000" is stating one thing, but means something entirely different. First, it is an attempt to express the *precision* of the work, it has nothing to do with the *accuracy* of the work. Second, the expression of relative precision is a statement of *relative* error in the measured traverse, not necessarily the survey itself, or the true error. And third, the statement is not a positive one, that is, it tells us how bad the survey work might be as opposed to how good we might expect it to be.

One part in 20,000 is a lesser amount and a smaller relative error than 1 part in 10,000. Therefore, if the statement were to read "an accuracy of *less than* 1 in 10,000," we would then know that it was at least that good, that we could expect no more than a foot of error per 10,000 feet of measured distance, or traverse.

Relative Error. The expression—1 in 10,000—is misleading in another way. A survey with a Standard Fraction of Precision of 1 in 10,000 may be acceptable for many purposes. All or most of the (allowable) error, however, could be in one place, that is, in one line or at one corner. In addition, errors are compensating as well as cumulative. The expression of relative precision tells us the arithmetical total error, not the true total error. For example, suppose that there is an error of plus 1 foot in one line and an error of minus 1 foot in another. The arithmetical error is zero; however, the measurements of those two lines are not free of error.

Another consideration is, that measurements of a small lot with a precision of 1 inch in 10,000 may not contain the same amount of error as the measurements of a large lot with the same relative precision. Consider that the small lot has a perimeter of 1,000 feet, with an uncertainty of ± 0.1 foot. The large lot with a perimeter of 20,000 feet would have a probable uncertainty of ± 2.0 feet. It is also to be noted that a precision of 1 in 10,000 is difficult to attain in small tracts, such as 40 x 100 feet. Large tracts, such as one 2,600 feet x 1,600 feet, can be measured with that precision much easier. This is especially true when the measurements are made with an EDMI (electronic distance-measuring instrument) as opposed to a steel tape.

Accuracy vs. Precision. Consider measuring a square lot with a tape that is one foot short due to a manufacturer's mistake. Even though it is graduated correctly, and is marked "0" on one end and "100" on the other, it is missing a foot somewhere in the middle and is in reality only 99 feet long.

While the measurements may be made very carefully resulting in a very small discrepancy, those measurements are *precise* but are not *accurate*. Not only are all the sides in error by about one foot, but the lot is also reported to be larger than it really is. Using the incorrect measurement the lot computes to contain 10,201 square feet, while in fact it only contains 10,000 square feet.

TOPOGRAPHIC MAPS

Topographic maps, used by nearly everyone, have a variety of professional and recreational applications. The use of these maps can be expanded by taking advantage of the different scales available as well as in comparing different editions.

Scales. The two most common topographic map scales are 1:62,500 and 1:24,000. Since 24,000 is a smaller number than 62,500, this is a larger scale map and it encompasses only 1/4 of the area of the 62,500 scale map. The former is known as the 15-minute series, while the latter is known as the 7 1/2-minute series, both based on being defined by minutes of latitude and longitude.

The older, earlier topographic maps published by the United States Geological Survey (USGS) were generally shown at the 1:62,500 scale. Over time, the U.S. has been remapped at the larger 1:24,000 scale. It takes four maps at the 1:24,000 scale to cover the same area shown by one 1:62,500 map. It is obvious that the larger-scale maps can show much greater detail and higher resolution than the smaller scale 1:62,500 maps.

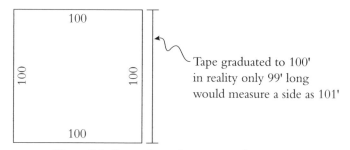

Figure 9.6 Comparison of *accuracy* and *precision*.

Editions. For many years there have been two editions of maps, for a few years there have been three, since the institution of the 1:24,000 scale. As there are often many years between editions, here is a very helpful tool to view changes that have taken place over time. It may astound the reader as to features existing on one map but not found on another.

Chapter 4 contains a discussion comparing two editions of topographic maps, 1957 and 1919, with changes that occurred between the two years. Roads, railroads, and similar features come and go, rivers and streams change course or become dammed creating new ponds, and some old ponds are enlarged, while others disappear. Local names, as they appear on maps, change, with new ones added and others deleted. Any or all of these could serve to explain description calls or demonstrate existing conditions that are mapped at a given time.

Series	Scale	1 inch represents	Standard quadrangle size (latitude–longitude)	Quadrangle area (square miles)	Standard paper size width-length (inches)
7½ minute	1:24,000	2,000 feet	7½ × 7½ min.	49 to 70	[1]22 × 27
Puerto Rico 7½-minute	1:20,000	about 1,667 feet	7½ × 7½ min.	71	29½ × 32½
15-minute	1:62,500	nearly 1 mile	15 × 15 min.	197 to 282	[1]17 × 21
Alaska 1:63,360	1:63,360	1 mile	15 × 20 to 36 min.	207 to 281	[2]18 × 21
U.S. 1:250,000	1:250,000	nearly 4 miles	[3]1° × 2°	4,580 to 8,669	[4]34 × 22
U.S. 1:1,000,000	1:1,000,000	nearly 16 miles	[3]4° × 6°	73,734 to 102,759	27 × 27

[1]South of latitude 31° sheets are 1 inch wider.
[2]South of latitude 62° sheets are 17 × 21 inches.
[3]Maps of Alaska and Hawaii vary from these standards.
[4]North of latitude 42° sheets are 3 inches narrower. South of latitude 34° sheets are 2 inches wider. Alaska sheets are 30 × 23 inches.

Table 9.3   National Topographic Maps

Symbols. Topographic maps, like other maps previously described, have a legend showing standard map symbols. Since the most common topographic maps are published by the USGS, the symbols used thereon are more standardized than those found in maps produced by a wide variety of private vendors. The chart shown as Table 9.4 illustrates commonly used topographic map symbols.

Hard surface, heavy-duty road		Boundary: national	
Hard surface, medium-duty road		State	
Improved, light-duty road		County, parish, municipio	
Unimproved dirt road		Civil township, precinct, town, barrio	
Trail		Incorporated city, village, town, hamlet	
Railroad: single track		Reservation, national or state	
Railroad: multiple track		Small park, cemetery, airport, etc.	
Bridge		Land grant	
Drawbridge		Township or range line, U.S. land survey	
Tunnel		Section line, U.S. land survey	
Footbridge		Township line, not U.S. land survey	
Overpass—underpass		Section line, not U.S. land survey	
Power transmission line		Fence line or field line	
Telephone line, landmark line (labeled as to type)		Section corner: found—indicated	
Dam with lock		Boundary monument: land grant—other	
Canal with lock			
Large dam			
Small dam: masonry—earth		Index contour	Intermediate contour
Buildings (dwelling, place of employment, etc.)		Supplementary cont.	Depression contours
School—church—cemetery		Cut—fill	Levee
Buildings (barn, warehouse, etc.)		Mine dump	Large wash
Tanks; oil, water, etc. (labeled only if water)		Dune area	Tailings pond
Wells other than water (labeled as to type)		Sand area	Distorted surface
U.S. mineral or location monument—prospect		Tailings	Gravel beach
Quarry—gravel pit		Glacier	Intermittent streams
Shaft—tunnel entrance		Perennial streams	Aqueduct tunnel
Campsite—picnic area		Water well—spring	Falls
Located or landmark object—windmill		Rapids	Intermittent lake
Exposed wreck		Channel	Small wash
Rock or coral reef		Sounding—depth curve	Marsh (swamp)
Foreshore flat		Dry lake	Inundation area
Rock: bare or awash		Woodland	Mangrove
Horizontal control station		Submerged marsh	Scrub
Vertical control station		Orchard	Wooded marsh
Road fork—section corner with elevation		Vineyard	Bldg. omission area
Checked spot elevation			
Unchecked spot elevation			

Table 9.4   Standard Topographic Symbols

PANORAMIC MAPS

Commonly known as "bird's-eye views," panoramic maps depict an area from an oblique angle. These maps usually pictured a town or city as an individual would see it at a particular point in time.

Panoramic maps were popular during the Victorian age of the United States, and even today they are commonly found preserved in archives and libraries. Turn-of-the-century town histories (at which time there were many of them published) frequently contain maps of this type.

Most of these depictions were prepared with great effort. Equipped with a plan of the city or town, the artist would literally walk the streets and sketch detail such as trees, buildings, and other features. Buildings were drawn with the correct number of doors and windows. If ships appeared in a body of water, or if the map showed other similar details, it was likely that the scene was as the artist saw it at the time of drawing the map.

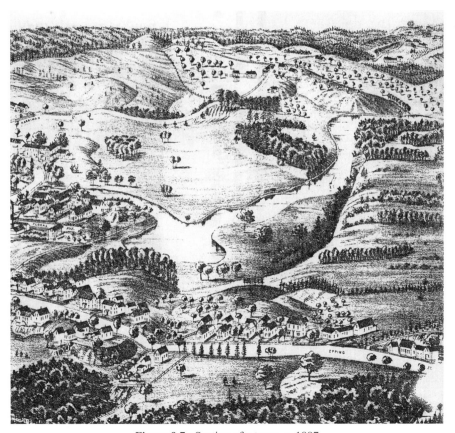

Figure 9.7 Section of a town, c. 1887.

FIRE INSURANCE MAPS

During the late nineteenth century and early twentieth century, many cities were mapped for fire insurance purposes. These maps show city streets, buildings, and other improvements. They document conditions a hundred years ago, give or take a few years, and when produced at about ten-year intervals, reflect changes around that time. Comparison of an early set of conditions with today's information will demonstrate subtle, as well as gross, changes during the past one hundred or more years.

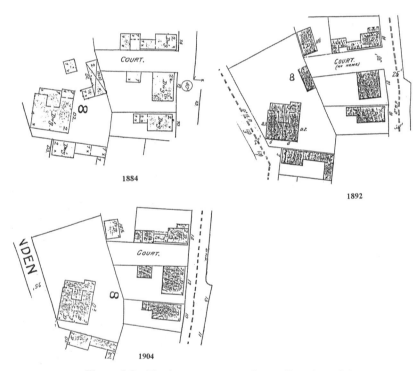

Figure 9.8 Fire insurance maps of a small section of city depicting changes over a twenty-year period. Sanborn™ Maps are copyright protected by the Sanborn Co., EDR Sanborn, Inc. Further reproductions are expressly prohibited without written permission from EDR Sanborn, Inc.

SKETCHES

A sketch can be of great importance in the decision-making process, in interpreting documents, or in determining what parties had in mind when they owned land, or transferred it from one to another. An important precaution to consider, however, is that when someone makes a sketch, usually it is done from their individual perspective of how they view the premises. Consequently, many sketches owe their foundation to the geometrical shapes of squares and rectangles, with occasional triangles. The old saying about a layperson's view of land and property descriptions is often as much truth as it is humor: that (1) all lines that do not visibly intersect are parallel, (2) those that do intersect, intersect at right angles, (3) all roads run north-south or east-west, and, (4) all lines running back from a road do so at right angles to the road. These not being real situations, one must be cautious that someone else may have seen it that way, or did it that way for their individual convenience.

Frequently, sketches are used to compile deed and related descriptions, but they rarely accompany the document or become a matter of public record. To locate these rare documents, local inquiry should be made of the owners at the time or their relatives. While most compilation sketches have been lost or destroyed, occasionally one is found, so it is always worth a reasonable investigation.

For example, the following notation appeared in a foreclosure document: "The above two parcels of land being formerly known as Levi Brown's Farm according to a survey and plan made February 21, 1835 by John Kimball and in the possession of the said George E. Frost." Discussion with the client revealed that George Frost had died but that he had a nephew who lived nearby. A call to the nephew resulted in locating the drawing, and a copy was made for submission as evidence to the court in a right-of-way case.

It is not unusual to find a sketch in a probate file accompanying either a will or an inventory. Surveys and plans are usually the result of a probate division or set-off such as dower, but sometimes rough drawings or sketches are included to illustrate the property. The examples in Figures 9.9 and 9.10 illustrate the situation.

This division of property was made by the administrator of an estate to divide the land into smaller parcels for subsequent sale. The deeds describing these later sales contain metes and bounds descriptions of the lots, including corner monumentation. Individually, neither the sketch nor the description tells the whole story, so it is necessary to have both. The obvious value of the sketch is to show how all the lots relate to one another without having to individually research and compile them. This would be a very difficult task if the record was incomplete or inaccurate, as it was in this case because some of the deeds were not recorded.

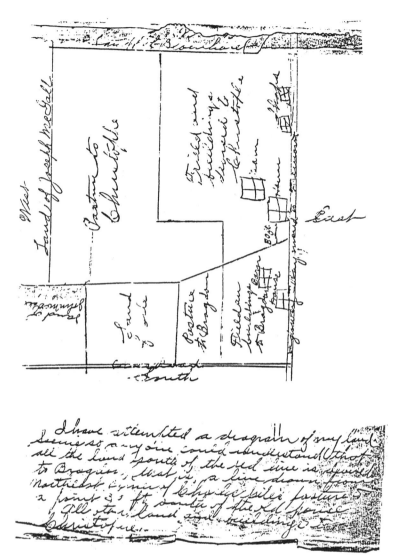

Figure 9.9 This sketch was drawn by the maker of a will to illustrate his farm and how it was to be divided.

CREATION AND USE OF BASE MAPS

A base map can be a very important research tool in a particular area when working with more than one parcel. Ordinarily, some form of existing map containing prominent features such as roads and bodies of water can be located at an appropriate size or scale. Once an acceptable map is found, it can be reduced or enlarged to a useable size and/or scale.

Figure 9.10   Sketch of the division and
sale of parcels from an estate.

Base maps can be constructed using a number of possible approaches. Some of
the more common source documents are described as follows.

Assessors Maps. Using assessors maps, also known as property maps or tax
maps, can be a quick and easy way to create an overview of an area. They are usually
drafted at an acceptable scale and show roads, water bodies, and occasionally, other
prominent features. These maps are generally produced from air photos, and since
property lines are often defined as planimetric features such as stonewalls, streams,
or roads, and since these features often can be seen on the airphotos, it follows that
some of the property lines on the assessors maps will be representative of lines in

the record property description, or "deed lines." Caution must be exercised here, however, as some lines on such maps represent mere fence lines, while others have been derived, based on frontage or acreage calls in deed descriptions.

Aerial Photographs. Again, roads, water bodies, and other features appear on aerial photographs and base maps may be produced merely by tracing photos. The problem that arises immediately is that unless the photos have been rectified to remove scale differences and distortion, adjacent photos and resulting maps will not be scale accurate. However, these may suffice for some purposes, where scale accuracy is not a big issue.

Topographic Maps. USGS quadrangle maps can be used to depict large areas. Enlarging them can provide a reasonably accurate map from which to trace and plot large parcels.

Combining Survey Plans. Existing survey plans may be fitted together to form a base map. All or only relevant portions of the plans can be traced to produce a base, which can be reduced or enlarged to a desirable size or scale.

Combinations. Any or all of the previously mentioned sources as well as any other valuable resources can be combined to produce a desirable result. With modern reproduction technology, the ability to assemble and display different source documents at varying sizes and scales is both an efficient and effective process.

ADMISSIBILITY OF RECORDS

A witness testifying to physical conditions or events may do so diagrammatically as well as verbally. One may represent points and objects by marks or shapes on a drawing, and distances, directions, and movements by lines thereon. This may be done during the witness's testimony, or through a map, plat, or other drawing made by the witness or found during a records search.

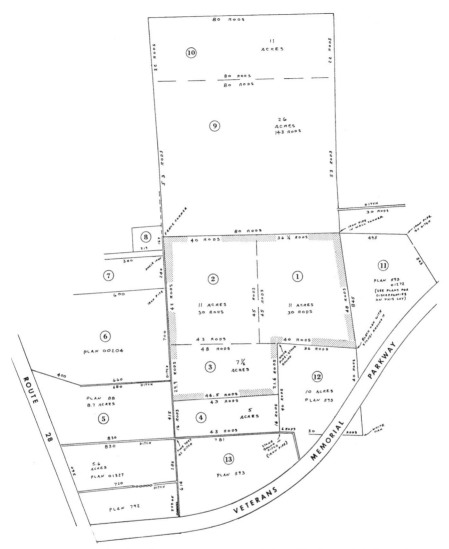

Figure 9.11 Overlay of an assessors map consisting of land records information obtained from deed descriptions and survey plans. The shaded parcel is the subject tract and is related to all abutting boundaries and corners. The circled numbers relate to chains of title that document the information shown on the base map.

It is common, and not improper, for a witness to make a diagram or map for the court in order to give them a better understanding of the place of which the witness speaks, or to use one he has made for the purpose, or one he knows to be correct; however, generally ex parte maps are not of themselves admissible and can only be used in the circumstances and in a supplementary and ancillary way—that is, to go with and explain the testimony of the particular witness.[4]

Assistance to the jury is emphasized in the following: "A map, plan or picture, whether made by the hand of man or photography, if verified as a true representation of the subject about which testimony is offered, is admissible in evidence to assist the jury in understanding the case. They are frequently formally admitted in evidence, and in so far as they are shown to be correct, are proper for the consideration of the jury, not as independent testimony, but in connection with other evidence, to enable the jury to understand and apply such evidence.[5]

The two classes of cases in which the foregoing principle applies, though often without a clear line of distinction, are (1) those in which the witness draws as he testifies, or uses a drawing as he testifies, to illustrate what he says, and (2) those in which a witness testifies to a drawing or diagram as embodying his knowledge or memory in detail, and which is, upon his mere verification of it as correct, admissible in evidence. In the latter class are survey maps and plats, architect's plans of a building, and similar diagrammatic evidence when used to express the physical details of property or situations involved.

An unofficial plat of property involved in a case is admissible as identifying and locating the property "upon the same theory that a pencil sketch, made by a witness while on the stand, to show the location of property with reference to some known object or permanent monument, would be admissible."[6]

The general rule permitting counsel on either direct or cross-examination to introduce the testimony of a witness as to physical conditions, locations, or directions by having him make a drawing to describe them diagrammatically is so thoroughly established in practice as to be unquestionable.[7] Witnesses often make drawings while testifying to illustrate the relative location of objects, and this is always permissible. It is simply the testimony of the witness in graphic form.[8]

---

[4] *Hoge v. Ohio River R. Co.,* 14 S.E. 152, 35 W. Va. 562 (1891), followed in *Poling v. Ohio River R. Co.,* 18 S.E. 782, 38 W. Va. 645 (1893), 24 LRA 215.

[5] *Adams v. State,* 28 Fla 511 (1891). Maps or diagrams, shown to be correct representations of physical objects about which testimony is given, can be exhibited before the jury, and witnesses will be permitted to use them in explaining their evidence.

[6] *Drew v. Butte,* 44 Mont 124, 119 P. 279 (1911).

[7] See 9 ALR2d 1048, § 3. *Evidence: use and admissibility of maps, plats, and other drawings to illustrate or express testimony.*

[8] *Clinchfield Coal Corp. v. Hayter,* 108 SE 854, 130 Va. 711 (1921).

The rule permitting drawings to be made by witnesses while testifying, to explain or illustrate their descriptions, is extended in practice to include drawings, or markings upon drawings, by counsel under the direction of the witness, or in response to his answers and confirmed by him as correct.[9] A measure of leading is clearly customary in practice, as where counsel makes a sketch on a blackboard and elicits from the witness the statement that it is a fair representation of the place in question. The degree of leading is, of course, subject to the discretion of the presiding judge.[10]

A map, plat, or other drawing made by a witness prior to coming into court, but testified by him to be a true representation, may be used by him in his testimony or admitted as embodying his testimony, upon the same general principle as that which would support his making the drawing before the court.[11] It was not an objection, though it is probably subject to some discretion of the trial judge, in the Texas case of *Redman v. Cooper*[12] that the witness was not skilled in making drawings of this type. An owner with no qualifications as a surveyor or engineer could support an issue as to what land was in his adverse possession by measuring his fences, making a diagram of them, and producing and testifying to it in court. Also, in a condemnation proceeding by a railroad company, the owner of the property was permitted to introduce in evidence, as part of his testimony, a map or diagram of his property made by him, showing the location of his various improvements—his house, barn, orchard, meadow, cultivated land, and so forth—and the location of the railroad across the land relative to these improvements. He was also permitted to testify by referring to this drawing.[13]

Survey Maps and Plats. Formal survey maps are the cause of much misunderstanding in land disputes. A plat produced by an expert surveyor and supported by his testimony that it is correct, especially if it purports to show boundaries in favor of the party who called him, challenges objection because of its seeming legal effect. Unless some statutory authorization, official recognition, admission by the opposing party or his predecessor, or documentary reference to it gives it such effect, it is admissible only as any other diagrammatic medium may be, to illustrate the testimony of the surveyor who made it. This is so whether it is offered in its entirety upon his mere testimony that it correctly represents a survey by him, or offered after he has testified to each point, measurement, or other detail represented on it.[14]

---

[9] *State v. Emory,* 246 S.W. 950 (Mo., 1922).

[10] Ibid.

[11] *Napier v. Matheson,* 68 S.E. 673, 86 S.C. 428 (1910).

[12] 160 SW.2d 318 (Tex Civ App, 1942).

[13] *Chicago, K. 8 Western R. Co. v. Dill,* 21 P. 778, 41 Kan 736 (1889).

[14] See 9 ALR2d 1055, § 8. *Evidence: use and admissibility of maps, plats, and other drawings to illustrate or express testimony.*

"A plan taken ex parte can never be used but as chalk, unless by consent." *Bearce v. Jackson,* 4 Mass. 408 (1808).

A plat is a mere illustration or graphical description of what the surveyor found on the ground. "A map is a picture of a survey; field notes a description thereof. The survey is the substance, and consists of the actual acts of the surveyor."[15]

A surveyor cannot be required to repeat from memory the calls and distances for a particular tract of land. That being so, he may summarize them in a survey map or plat instead of reciting them, from his memory or his notes.[16]

If the surveyor testifies that a plat correctly represents the location of the objects marked thereon and the measurements made by him of distances, it is rendered admissible as part of his testimony.[17] But it is merely a description and illustration of his work on the land.[18] It is admissible only as an expression or explanation of his testimony of physical facts.[19]

A surveyor may use his map merely to illustrate his verbal testimony as to locations of monuments, and as to distances and directions from or between them, referring to points and lines on the map as indicating them, or he may use it as indicating boundaries, as determined by him, if other surveys in evidence were followed. In such a case the entire map may be admitted as illustrating his testimony in detail.[20]

It is the practice of surveyors, when engaged to survey a parcel of land, to get the calls and monuments which determine the title. It is proper for a surveyor to do this, and the fact that he has done it does not make his survey inadmissible as representing legal conclusions.[21] The practice of examining the title leads many surveyors to regard their function as including the settling of boundaries from the legal standpoint. The frequent objections by counsel to the admission of their plats are based on the same false aspect of them as having legal effect, especially when they are marked with the names of owners placed on the parcels as delineated thereon. An unofficial plat, if admitted on the principle of illustrating testimony, is not evidence itself of title, or even the expression of expert opinion as to title, but only evidence of the physical objects and points on the land and the relations between them in distances and directions.[22]

---

[15] *Outlaw v. Gulf Oil Corp.*, 137 S.W.2d 787 (Tex.Civ.App., 1940), revd 136 Tex 281, 150 S.W.2d 777 (1941).

[16] *Hackney v. Louisville 8 N.R. Co.*, 1 KY LR 357 (1880).

[17] *Seidschlag v. Antioch*, 69 NE 949, 207 Ill 280, (1904).

[18] *Wilson v. McCoy*, 117 S.E. 473, 93 W. Va. 667 (1923).

[19] *Sudduth v. Central of Georgia R. Co.*, 77 So. 350, 201 Ala. 56 (1917); *Goldsborough v. Pidduck*, 54 NW 431, 87 Iowa 599 (1893).

[20] *Pickering Light & Water Co. v. Savage*, 69 P. 846, 6 Cal. Unrep 985 (1902), *McMichael v. Eastern Hydraulic Press Brick Co.*, 78 A. 144, 80 N.J.L. 398 (1910), *Griffith v. Rife*, 12 S.W. 168, 72 Tex. 185 (1888).

[21] *Henrietta Egleston Memorial Hospital v. Groover*, 43 SE2d 246, 202 Ga 327 (1947).

[22] *Rose v. Davis*, 11 Cal. 133 (1858), *Roberts v. Atlanta Cemetery Assn.*, 91 S.E. 675, 146 Ga. 490 (1917), *Wilson v. Stoner*, 9 Serg. & R. 39, 11 Am.Dec. 664 (Pa., 1822), Cf. *Cartwright v. Cartright*, 74 S.E. 655, 70 W.Va. 507 (1912), Ann.Cas. 1914A 578.

Maps and Plats from Public Offices or in Public Use. The mere fact that a map or plat is kept in a public office, and is generally used there for reference, is not enough to warrant its admission in evidence without proof of its correctness in the matter for which it is to be used.[23]

Use of Different Scales. If a plat or diagram is made up of different views of the same subject, and these views are drawn to such different scales that they are likely, when shown together, to create an erroneous impression in the minds of the jury, it is within the discretion of the trial judge to exclude the whole from use or admission.[24]

Notations in General. Objections to a map, plat, or other drawing on the ground that it includes prejudicial notations not properly admissible upon the testimony of the draftsman or other witness called to verify it, or that may be mistaken by the jury for evidence in themselves, are not governed by any general rule. In particular cases, the question is dealt with by the trial judge as within his discretion, reviewable or not according to the practice in the jurisdiction concerned. The distinction between diagrammatic representations that are offered in evidence as themselves embodying the testimony of a witness, such as survey plats, may be treated differently from those that are used as mere chalks. The marking of measurements and directions on a surveyor's plat, with the names of owners of the different parcels and such designation of monuments and other objects as may be necessary to an understanding of the plat, is usual and permissible.[25]

Ancient Document Rule. Some evidence may be admissible under the "ancient document rule," or the "ancient document" exception to the hearsay rule. The rule admits documents, by reason of their age, without proof or authentication by the testimony of the person who executed the same.

---

[23] *Chirac v. Reinecke,* 2 Pet. 613, 7 L.Ed. 538 (U.S., 1829), *Wright v. Louisville & Nashville R. Co.,* 82 So. 132, 203 Ala. 118 (1919), *Kearce v. Maloy,* 142 S.E. 271, 166 Ga. 89 (1928), *White v. Eden,* 91 S.E. 601, 173 N.C. (1917), *Franey v. Miller,* 11 Pa. 434 (1849), *Harris v. Commonwealth,* 20 Gratt 833 (Va., 1871).

[24] *Hale v. Rich,* 48 Vt. 217 (1876), Cf. *Marcy v. Parker,* 62 A. 19, 78 Vt. 73 (1876).

It is the duty of the trial judge to reject any plan that would tend clearly to mislead the jury on a material point in issue. *Hale v. Rich,* 48 Vt 217 (1876).

[25] *Kentucky Virginia Stages v. Tackett's Admr.,* 171 S.W.2d 4, 294 Ky. 189 (1943).

Where a map or diagram of the place of an event bears marks and figures placed on it during the trial of an action against others arising out of the same occurrence, the trial judge may properly exclude it. *Martin v. Leatham,* 71 P.2d 336, 22 Cal.App.2d 442 (1937).

A judge may permit the objectionable part of a plat to be erased or cut off, if the value of the remainder is not seriously impaired and the result is not otherwise prejudicial. *Atlantic Coast Line R. Co., v. Dawes,* 88 S.E. 286, 103 S.C. 507 (1915).

Maps, surveys and related documents purporting to be, in most jurisdictions, twenty years[26] or more old are said to prove themselves without the ordinary requirements as to proof of execution or handwriting if relevant to the inquiry, when produced from proper custody, on their face free from suspicion, and authorized or recognized as official documents.[27]

In *Wynne v. Tyrwhitt*,[28] the court held that the rule that a document coming from proper custody and being over thirty years old was not "confined to deeds or wills, but extends to letters and other written documents coming from the proper custody. It is founded on the antiquity of the instrument, and the great difficulty, nay impossibility, of proving the handwriting of the party after such a lapse of time."

A map, undated, bearing on its face every appearance of age and shown by the evidence to have, in fact, long antedated defendant's purchase of lots abutting on streets, and which lots had been conveyed over forty years ago, was held admissible in the Alabama case of *Hamilton v. Warrior*,[29] as an ancient document, even though the map had never been placed on the public record. While it was not shown that the map in question was made by anyone authorized to make the same, it was shown that the corporate limits of the town of Warrior had been plotted and lots sold according to the map referred to in the evidence, and admitted, as the "old survey of the Town of Warrior."

In holding certain maps and field books admissible as ancient documents, the court stated in *Hart v. Gage*[30] that "these books, and the maps made from them, as they ripen by time, and monuments perish, may, like the *Doomsday Book* be the best, if not the only evidence of many ancient surveys." And in *Lowell v. Boston*,[31] where, in determining the nature of the city's title to the Boston Common, the court considered "an exhaustive history of the Common from the earliest colonial times to the present, with ancient maps, plans, town records, deeds, layout of streets, frequent references to historical works, old photographs, and various other documents," in passing upon the truth of facts found in other ancient documents.

---

[26] The old rule, based on the *Federal Rules of Evidence* "ancient document" exception, was thirty years, which was adopted by most states. The *Federal Rules* requirement is now twenty years, and has been adopted by many states. For some years, New York has had the twenty-year rule and other states have made certain exceptions under particular circumstances.

[27] *Steele v. Fowler,* 41 N.E.2d 678, 111 Ind.App. 364 (1942), *Enfield v. Ellington,* 34 A. 818, 67 Conn. 459 (1896).

In *Hostetter v. Commonwealth,* 80 A.2d 719, 367 Pa. 603 (1951), the court pointed out: "When a right or title is of ancient origin or where the transaction under investigation is so remote as to be incapable of direct proof by living witnesses or by the ordinary documentary evidence, the law, of necessity, relaxes the rules of evidence and requires less evidence to substantiate the fact in controversy. For example, ancient maps, records, surveys, ancient town plots, historical books which have been generally treated as authentic, reports made by disinterested parties apparently conversant with the facts and now dead, have been held admissible as furnishing evidence of remote transactions. . . . Maps, surveys, monuments, pedigree and even reputation evidence have been held to be admissible to establish boundaries."

[28] 4 Barn 8 Ald 376, 106 Eng Reprint 975 (1821).

[29] 112 So 136, 215 Ala 670, (1927).

[30] 6 Vt 170 (1834).

[31] 79 NE2d 713, 322 Mass. 709, app dismd 335 US 849, 93 L Ed 398, 69 S Ct 84 (1948).

Where the original of an ancient, officially made or recognized map is no longer in existence, or has become so defaced as to be unintelligible, a copy or tracing thereof, properly authenticated, may be admissible in evidence.[32]

Concerning public surveys, an ancient survey made by competent authority, recorded or accepted as a public document, and produced from proper custody, is admissible in evidence to prove the location of boundary lines.[33] In the Vermont case of *Aldrich v. Griffith*,[34] the boundary between two towns was in dispute, and a field book of one of the towns, in the town clerk's custody, was held admissible in evidence as an ancient document. The book was shown to have come into the possession of the town clerk a "large number of years ago," and deeds of land in the town, executed more than forty-five years prior to the trial, were referred to in it. "The genuineness of a document of this kind," said the court, "on its face purporting to be sufficiently ancient, is shown prima facie by proof that it comes from proper custody."

FOR FURTHER REFERENCE

DeVorsey, Louis, Jr. "The Use of Historical Maps in Litigation." Technical Papers of The American Congress on Surveying and Mapping. 1981 ASP-ACSM Convention, Washington, D.C. 1981.

Ehrlich, S. G. "Preparing and Using Maps." *2 Am Jur Trials:* 669. Rochester: Lawyers Cooperative Publishing Co.

Gillen, Larry, Ed. *Photographs and Maps Go to Court.* Falls Church: American Society of Photogrammetry and Remote Sensing. 1986.

Greenhood, David. *Mapping.* Chicago: The University of Chicago Press. 1964.

Makower, Joel, Ed. *The Map Catalog.* New York: Vintage Books. 1986.

Monmonier, Mark. *How to Lie with Maps.* Chicago: University of Chicago Press. 1991.

Richason, Benjamin F. *Atlas of Cultural Features.* Northbrook: Hubbard Press. 1972.

---

[32] *St. Louis Public Schools v. Risley,* 10 Wall 91, 19 L.Ed. 850 (U.S., 1869), *Ayers v. Watson,* 137 U.S. 584, 34 L.Ed. 803, 11 S.Ct. 201 (1891), *Burns v. United States,* 160 F. 631 (C.A.2d, NY, 1908), *Brown v. Metcalf,* 102 N.E. 413, 215 Mass. 289 (1913), *Gibson v. Poor,* 53 Am.Dec. 216, 21 N.H. 440 (1850), *Schatz v. Guthrie,* 132 N.Y.S.2d 665 (1954), *Birmingham v. Anderson,* 40 Pa. 506 (1861).

[33] *Talbot v. Lewis,* 6 Car & P. 603, 172 Eng. Reprint 1383 (1834), *Morris v. Harmer's Heirs,* 7 Pet. 554, 8 L.Ed. 781 (U.S., 1833), *Burgin v. Simon,* 65 So. 128, 135 La. 213 (1914), *Lexington v. Hoskins,* 50 So. 561, 96 Miss. 163 I1909), *Aldrich v. Griffith,* 29 A. 376, 66 Vt. 390 (1893).

In *Susi v. Davis,* 133 Me 354, 177 A 610 (1935), see 97 ALR 1222, the court said: "A survey shall govern the plan. The plan is a picture, the survey the substance. The plan may be all wrong, but that does not matter if the actual survey can be shown."

[34] 29 A 376, 66 Vt 390 (1893).

Shalowitz, Aaron L. *Shore and Sea Boundaries.* Volumes 1 and 2. Publication 10-1. U.S. Department of Commerce. Washington: U.S.G.P.O. 1962.

Thompson, Morris M. *Maps for America.* Washington: U.S.G.P.O. 1979.

29 ALR 630.   *Dispensing with proof of proper custody as condition of admission of ancient document.*

108 ALR 1415.   *Use of photograph, plan, map, cast, model, etc., as evidence as affected by marking or legends thereon.*

9 ALR2d 899.   *Authentication or verification of photograph as basis for introduction in evidence.*

9 ALR2d 1044.   *Evidence: use and admissibility of maps, plats, and other drawings to illustrate or express testimony.*

46 ALR2d 1318.   *Admissibility in evidence of ancient maps and the like.*

57 ALR2d 1351.   *Admissibility in evidence of aerial photographs.*

REFERENCES

American Society of Civil Engineers. *Definitions of Surveying Mapping, and Related Terms.* Manual of Engineering Practice No. 34. New York: ASCE. 1954.

Brown, Curtis M., Walter G. Robillard, and Donald A. Wilson. *Brown's Boundary Control and Legal Principles,* fourth edition. New York: John Wiley & Sons, Inc. 1995.

CHAPTER 10

PICTURES

> One picture is worth more than ten
> thousand words.
>
> —Chinese proverb

Pictures tell a story. They show visually what a map or plan shows graphically. Pictures have perspective and dimension, while maps are in one plane, and other types of description are mere words. None can replace the other; they can supplement one another and, occasionally, one may substitute for another.

Pictures exist in several forms, and photographs generally come to mind when the word is mentioned. However, in the world of documents and land records, sketches, drawings, paintings, and anything else that might illustrate, tell a story, or supplement some other type of record should be included as pictures. Carvings and inscriptions may also be included.

Photography. The origin of photography has been traced to 1839 and Louis J. M. Daguerre. In the early 1840s, William H. Fox Talbot improved on the techniques and originated the process in use today.[1]

The first aerial photograph was taken in Paris in December of 1858. The oldest existing airphoto is of a section of Boston, Massachusetts, and was taken October 13, 1860.[2] With this resource available for the last century and a half, the interpreter of land records should be prepared to investigate its existence and availability at least to the early part of the century.

[1] Avery, T. Eugene. *Interpretation of Aerial Photographs.* 1968.
[2] Streb, Jack M. *Photography from the Air—Then and Now.* 1967.

What Pictures May Show

- *Existence of objects.* The fact that a particular thing, or feature, existed at a point in time.
- *Conditions.* Documentation of the appearance of an area at a point in time.
- *Boundary evidence.* Location of features such as buildings, fences, roads, bodies of water, hedgerows, timber type changes, lines between wood and open areas, and other opportunity features which existed at a point in time.

The operative words in the phrases above are *"at a point in time."* A picture is a snapshot at one instant, and one instant only. Drastic changes could occur momentarily after a picture is made, and those changes would never be known unless another snapshot is made which shows them.

Comparison of Photos. By comparing photographs taken at different times, it is possible to document changes that have occurred. This is particularly true in areas where government airphotos have been taken at regular ten-year intervals. In those areas where crops are monitored, yearly photography is available.

For example, some of the information that can be obtained is:

- Bodies of water

    Shoreline erosion
    Changes in location of bed or channel of river or stream
    Silting
    Artificial fill
    Size of pond due to erection or removal of dam
    Flooding
    Evidence of accretion, avulsion, erosion, or reliction

- Roads

    Existence or nonexistence
    Change in location

- Timber

    Harvesting
    Trespass

• Soil

    Excavation
    Trespass

• Building

    Existence
    Erection, removal, or alteration

• Landscape

    Planting
    Clearing
    Cultivation
    Succession of field to forest
    Erosion

• Improvements

    Sheds, pools, and so on
    Parking
    Evidence of easements

Photo Interpretation. Photo interpretation is the art and science of the identification of objects on photographs and the determination of their meaning or significance. It can be done with either *aerial* or *terrestrial photographs*. By using photographs, an area's visual image can be seen, showing features that might not otherwise be available or visible. Sometimes, the successful interpretations or explanations of other land records are made possible through the use of photographs as supplementary evidence.

Aerial Photographs. The interpretation of aerial photographs first became important during World War II when they were used extensively to locate enemy targets and installations. Since then, many nonmilitary applications have been developed and great technical advances have been made in both photo interpretation and in the quality of photography.

The basic format for aerial photographs is the contact print, a 9" x 9" paper image at the photo negative scale. Enlargements, made from the contact print scale, are routinely made and readily available to the land records researcher. Enlargements can be functions of the size and/or scale desired. A common size for land records use is 24" x 24" at a scale of 1" = 660'. This typical size is easily handled and stored and the image is clear. If photos are enlarged too much, photographic quality is lost because the image becomes fuzzy or grainy.

Most federal agencies and many state agencies have had aerial photography taken over many years resulting in a vast resource for researchers. There is also a considerable amount of photography available through local governments and private vendors.

Types of Photography. Several types of photography could be encountered during a search for available pictures.

Panchromatic. The most common and widely used photography is panchromatic, using a black-and-white negative material having approximately the same range of sensitivity as the human eye. Images are varying shades of gray.

Infrared. Infrared black and white has a different sensitivity and when exposed through various filters results in strong contrast between images. The difference between hardwood and softwood tree species can be seen by comparing light and dark, while water appears black.

Conventional color. Color film can be advantageous in that contrast is often different and items of color present themselves in an obvious way. Being more expensive than traditional black-and-white film, it is used mainly for special projects such as vegetation analyses.

Infrared color. Also known as camouflage detection film, the developed photography shows false color. The contrast between living and inanimate objects is often very apparent in this type of picture. It is used mainly for disease detection on vegetation, and the differences between hardwoods and softwoods, or between living and dead vegetation, is easily observed.

Measurements. Measurements can be made on airphotos, the same as they are made on maps. The science of making measurements using aerial photography is called *photogrammetry*, the subject of several textbooks. For purposes of land record research and interpretation, measuring on airphotos can at times be quite useful. However, a word of caution is warranted. A raw, uncontrolled aerial photograph is actually a distorted image in spite of the fact that it shows all visible features. The inherent distortion is due to changes in topography, camera lens, aircraft variables, and other factors. Therefore, measuring directly on these "distorted" images can produce "distorted" results.

There are techniques available that can be used to eliminate the distortion inherent in airphotos. One process, known as rectification,[3] results in a product called an *ortho-photograph*, a distortion-free image upon which accurate measurements can be made.

In all following sections of this chapter dealing with the determination of partic-ular measurement parameters, the reader is cautioned that unless the aerial photo-graphs are rectified or scale corrected in some manner that all measurements will be distorted and biased to some degree. For many purposes, however, the approximate measurements are satisfactory, but for some they are not, and much more sophisti-cated photogrammetric techniques should be used in those instances.

Generally, the easiest way to make measurements on a photo is to do it in con-junction with a map of the same area. Using a map, the photo scale and its orienta-tion can be determined quickly. Without a map, bases of measurement must otherwise be obtained for comparative purposes.

As an illustration, measurements will be made on the woodlot illustrated in Figure 10.1. Basic measurements can be made directly on the photograph or from an overlay tracing of the photograph. Often, a tracing with neat, crisp ink lines is eas-ier to work with than the original photo print. This is especially true when a plot of a description is made for comparison with a photograph.

Determination of Photo Scale. Sometimes, the scale of an aerial photograph is printed at the top or on the back of the photo. It is usually expressed as a representative fraction (reference Chapter 9). However, if the photo scale is not shown, it can be calculated easily if the values of two variables are known; the focal length of the camera lens and the flying height of the aircraft. Alternatively, if any two of the variables in the formula shown below are known, then the third can be quickly calculated.

---

[3] A rectified photograph is one where the picture has gone through a special process to adjust for scale differences. This done, it can be used like a map since the scale is consistent throughout. Adjacent pho-tos can also be fitted together. Advantages are obvious, the disadvantage is that it requires another process and is, therefore, more expensive.

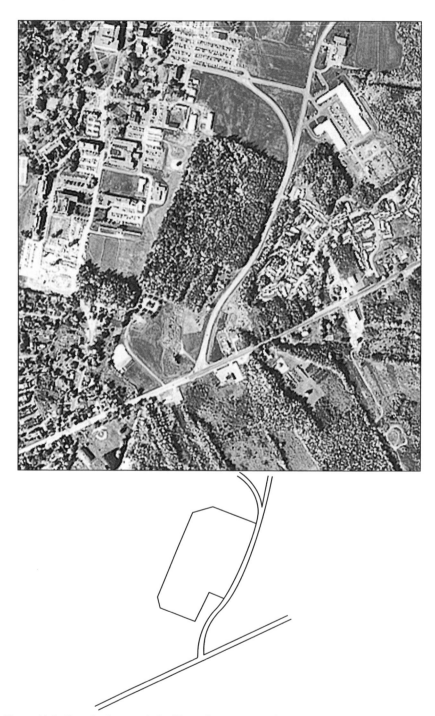

Figure 10.1 Sample photograph for illustration purposes of a developed area and accompanying diagram of the area of interest. Photography by James W. Sewall Company, Old Town, Maine.

Method 1:

Photo Scale = Focal Length of Camera Lens/Flying Height

If Focal Length = 152.57 mm. (calibrated)

= 6.01 inches

And Altitude = 8300 feet (by altimeter)

5° Correction for temperature

True Altitude = 8070 feet

Mean elevation of project = 150 feet

Flying height = 7920 feet

Then Photo Scale = 6.01" / 7920'

= 0.50' / 7920'

= 1 / 15840

or 1" = 1320'

An alternative method of calculating photo scale is to use the following formula:

Method 2:

Photo Scale = Photo Distance/Ground Distance

If distance between two control points measured on photo = 0.58"

And ground distance between same two points = 756.6'

Photo scale = 0.58" / 756.6' × 12

= 0.58 / 9187.20

= 1 / 5840

or 1" = 1320'

An additional alternative using another variable to determine photo scale is:

Method 3:

Photo Scale = Photo Distance/Map Distance

If photo distance between two points = 0.609'

And USGS' map distance between same two points = 0.154', then

Photo Scale = 0.609 / 0.154 = 3.95

Convert 3.95' on USGS scale to RF:

62,500/3.95 = 1:15,823

1:15,840

Determination of Direction. To find the direction of a line, the photograph must be oriented to a known direction. Then, using a topographic map of the area or a direct field observation the direction of a base line on the photo can be found. Once the base line direction is established, other lines can be referenced to it, and the angles between lines can be utilized to determine the other line directions. For example, on the photograph in Figure 10.2, the road angling at the bottom was found to run North 45° East. Based on that determination, the direction of the west line of the woodlot was determined to be North 03° East.

Determination of Distance. Once the photo scale has been determined, distance measurements can be made. If a photo-scale protractor is available at the appropriate scale, distances can be measured directly (see Figure 10.3).

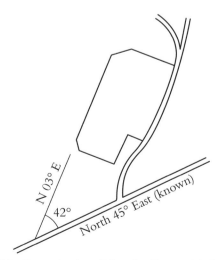

Figure 10.2 Determination of direction from aerial photo overlay.

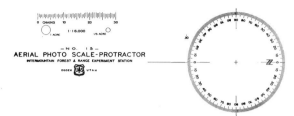

Figure 10.3 Photo-scale protractor at 1:16,000, which approximates 1:15,840. Not to scale.

If the photograph scale is not an even number, measurements can be made in some units and the appropriate conversion made. Critical measurements should be made to the nearest decimal unit, since millimeters on a photo represent feet on the ground. Therefore, the smaller the scale of the photo, the more error there is likely to be due to inaccurate photo measurements.

Length of sideline = 246 mm

= 0.246 meters

$0.246 \times 15840 = 3{,}896.64$ meters on the ground

$3{,}896.64$ m./$3.281 = 1{,}188$ feet

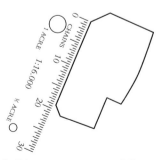

Figure 10.4 Direct measurement of distance. Length of west line of woodlot is 19 chains, or 1,254 feet.

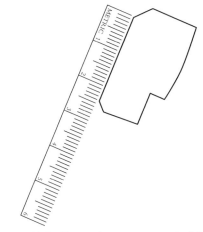

Figure 10.5 Conversion measurement of distance.

Determination of Area. When the scale of a photo is known, area conversion factors can be utilized. On the example photograph, the scale is 1:15,840.

1 acre = 43,560 square feet

= 10 square chains

= 0.0015625 square miles

= 4,046.87 square meters

= 0.404687 hectares

= 0.004047 square kilometers

Area is determined in several ways. A planimeter can be used to find the area of any plotted figure. Directions and distances can be used to compute area, and there are several relatively quick procedures that give an area sufficiently accurate for many purposes.

Method 1: Use of Measured Dimensions

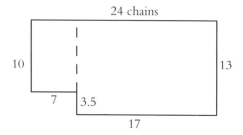

$$10 \times 7 \quad = \quad 70 \text{ square chains}$$

$$17 \times 13 \quad = \quad 221 \text{ square chains}$$

$$\text{Sum} \quad = \quad 291 \text{ square chains}$$

$$\text{Area} \quad = \quad 29.10 \text{ acres}$$

Method 2: Use of a Dot Grid

A dot grid is a transparent overlay with dots at regular intervals, each grid being labeled by the number of dots per square inch. Knowing the scale of a photo, and counting the number of dots covering a specific outline, the area of a tract can be determined. Common overlays are 16 dots per square inch, known as a 16-dot grid, 64 dots per square inch, known as a 64-dot grid, and 256 dots per square inch, known as a 256-dot grid. The recommended intensity, or the number of dots per square inch, depends on the scale of the photo, the size of the area involved, and the desired precision. For tracts of a square mile or less, the desirable dot intensity is one that will result in a conversion factor of 1/4 acre to 1 acre per dot. Grids of 64 dots per square inch are commonly used for photos around 1:20,000. With this, 1 square inch is equivalent to 63.77 acres, each dot representing 0.996 acres.

In using a dot grid, it should be overlaid on the area of study at random to eliminate any bias. While it is easier to align parcel boundary lines with lines of dots on the grid, this procedure is not recommended. Any dot which falls exactly on a boundary should be treated as 1/2 of a dot in the count. The grid can be placed directly on the photo or on a tracing of the subject area.

Scale of photo   =   1:15,840

1" = 1,320'

1 sq. in. = 1,742,400 sq. ft.

= 40 acres

16 dots = 40 acres

13 dots = 32.5 acres

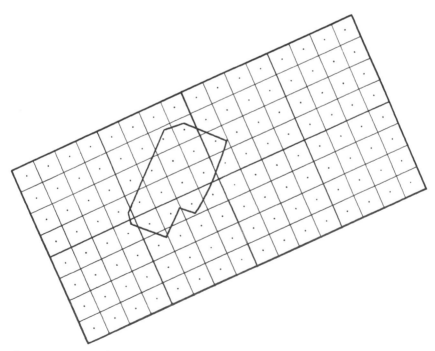

Figure 10.6 Area of interest overlaid with a 16-dot grid. The difference in acreage is accounted for by the fact that while using a 16-dot grid, acreage results are not as accurate as when using one with a greater number of dots per square inch. However, it should give an approximation sufficient for some purposes, especially when dealing with large areas.

1 sq. in. = 40 acres

256 dots = 40 acres

186 dots = 29.0 acres

Photointerpretation Aids. As previously shown, a photo-scale protractor is a valuable tool for making measurements, and dot grids are useful in area determination. In addition, the following tools can be useful measuring devices for particular purposes:

• Standard protractor
• Scales in desired units
• Parallax wedge for height determination
• Slope-percent scale

For use with photography at a scale of 1" = 660', a template is very useful. This scale, on a 24" × 24" format, is readily available from ASCS, and lends itself especially well to boundary interpretation. This commonly used scale makes the template shown in Figure 10.8 very beneficial.

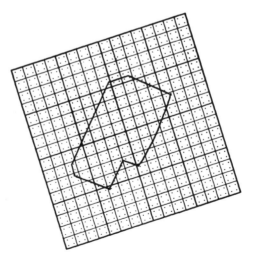

Figure 10.7 Area of interest overlaid with a 256-dot grid. Results are more consistent with actual measurements since a grid with a greater number of dots tends to result in more accurate determinations of area.

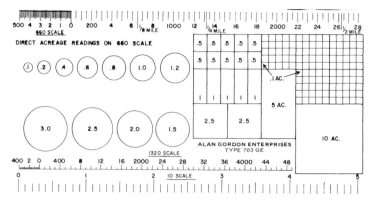

Figure 10.8   Template designed for use with 1" = 660' photography.
Not to scale. Courtesy Alan Gordon Enterprises.

Object Identification. Object count and identification on photographs is sometimes valuable. The identification of roads, buildings, cemeteries, streams, and other items of boundary and title importance can be made and is a function of the following:

- Size and shape of the object
- Scale and resolution of the photography
- Spacial arrangement of the objects
- Tonal contrast between objects and associated background
- Type of film
- Use of a stereo pair versus single prints

Better success is generally obtained using photos in conjunction with topographic maps.

Tree Identification. Individual trees can be identified on airphotos, especially if they contrast with other species in the same area and if the photo scale is large enough. Obviously, the larger the photo scale, the easier it is to use. Photos smaller than 1:15,840 make it very difficult to identify individual trees.

There are two common ways to identify tree species; crown shape from a vertical perspective and tree shadow/silhouette from an oblique perspective.

Knowing the geographic range of a tree species is important to the researcher. This knowledge can help determine the probability of the existence or nonexistence of a species in a particular area. The usual habitat of a species is also helpful information in eliminating choices. Reference material at the end of this chapter includes keys to the identification of tree species, along with diagrams of crowns and silhouettes for comparative purposes.

Figure 10.9   Diagram of a white pine (*Pinus strobus* L.) shown as the shadow/silhouette appears from the side and the crown from above.

Locating Land Descriptions on Aerial Photographs. Land patterns frequently retain their original configuration for a long time as a result of man's dependence on, and respect for, boundaries. Fence lines tend to influence vegetation growth and land use, and the topographic character of the land dictated initial lot layout in many towns. Current photography often presents sufficient evidence to generally reconstruct land use patterns, lot and boundary lines, and related features.

The original rectangular lotting of towns can sometimes be shown by using a combination of measurements and interpretation. Comparison of airphotos with assessors maps,[4] deed descriptions, and surveys will often result in the placement of individual parcels on the photo. Especially effective is the construction of an area map from descriptions and plat such that it is referenced to existing roads or other topographic features. The sketch is either drawn to scale or reduced/enlarged on a photocopy machine to the appropriate scale, then overlaid on the photograph.

A useful technique, when the location of parcel is not known, is to plot its description from a deed or related document, or from a survey, at the same scale as a photograph and make an overlay of the plot. Slide the parcel around the photograph(s) until it corresponds with visible features. Of course, some verification must later be made to verify its correct location. This technique works especially well if the parcel in question is plotted together with its neighboring parcels, particularly if they connect to some prominent feature such as a road or body of water.

Sources of Aerial Photography. Agencies that are regular users of aerial photography include:

---

[4] Assessors maps are generally compiled using a combination of airphotos, deed descriptions, survey plats, and private sketches.

Figure 10.10 1972 photograph showing land lines and its comparative lot plan. The integrity of many of the original boundary lines is preserved by tree lines as a result of fences and can easily be seen in the photo. Fences themselves cannot ordinarily be seen, but potential evidence of fences and other types of land use (cutting lines, field/vegetation lines, etc.) are often visible. *Photography by James W. Sewall Company, Old Town, Maine.* Lot plan from *Background and History of the University of Maine Forest* by Dwight B. Demerritt. University of Maine Bulletin 696, 1972.

Federal agencies, such as:

Agricultural Stabilization & Conservation Service (ASCS)

Bureau of Land Management (BLM)

Soil Conservation Service (SCS)

United States Forest Service (USFS)

United States Geological Survey (USGS)

State agencies, such as:

State department of transportation (DOT)

Planning offices

Forestry and agricultural offices

Private sources:

Photographic firms

Forestry companies

Surveying firms

Mapping companies

Occasionally, old photographs are disposed of when new photography is taken, but more often, older photography is preserved. The National Archives has custody of some of the earliest government photography dating back to the 1930s and early 1940s. Many times it will be the earlier photography that is of most value to the user and reader of land records, so it is important to know what is available and where it may be obtained.

Other photography factors that affect usability and utility include the time of year the pictures were taken, the negative scale, the time of day, weather conditions, and the reprographic quality. Aerial photography can be one of the most reliable tools for the researcher of land records.

Terrestrial Photography. Photography taken of ground conditions is a lot more difficult to obtain and often comes in the form of family pictures and owners' random shots. Nearly everyone takes pictures, for any number of different reasons. Usable photographs of sites abound, the main problem being the lack of documentation. It is generally not known when a photograph was taken, and frequently it is not known by whom it was taken. Depending on the need or the use for the photograph, this documentation may not be crucial. If the photograph can be identified, and related to present conditions, it is probably useful.

A situation occurred where a survey was being performed to determine the dividing line between two city lots. The deeds were not very descriptive, but the parties remembered that at one time there was a stone post on the dividing line, but it had since been removed and no one could remember its exact location. A search through family albums produced a photograph of a family gathering taken in the front yard some years before. The stone post was visible in the background and after determining an approximate scale of the photograph using the building dimensions, both full size and from the photo, the surveyor determined the former location of the post within a small margin of error.

Photographs can document the existence of evidence and objects, their locations, and sometimes their relative or approximate sizes. If the photograph can be dated, the conditions at that time may also be useful.

Picture Postcards. Since postcards have become collector's items in recent years, there are possibly many pictures of sites preserved in private collections or for sale at antique shops, book fairs, and flea markets. Images on postcards may come from photographs, lithographs, or drawings and paintings. Frequently they document conditions at a given point in time.

The science of making postcards is known as *deltiography*. Postcards first came into existence around 1870, with the first picture postcard being produced about 1890. The period of 1895 to 1914 saw the increase and popularization of the use of postcards. Lithography had been around since 1796 and the photo wet plate process since 1851, so the necessary processes were in place at this time and ready for use. With the invention of roll film in the 1880s by George Eastman, commercial photography took over and traveling photographers captured scenes at an astounding rate. The invention of flash bulbs in 1929 and color film in 1935 added new dimension and versatility to the picture postcard business.

The peak of early postcard production was about World War I, which resulted in an abundance of potential documents now available for at least 80 years. As a result of their popularity by collectors, many of them have been preserved. The area of postcard study has become more important in recent years as certain cards become valuable collectors items.

Whitehouse and Birches, Grand Lake Stream, Me.

Figure 10.11 Postal card of conditions many years ago. (Compare with Figure 4.1). Published in Germany for W. B. Hoar, Grand Lake Stream, ME.

Through the science of *deltiology*, postcards can be dated as to when they were produced, adding to their value as evidence concerning the image. In addition, if a card has been used, the postmark may give an indication of date, although an early card, or a card with an early image, may carry a more recent postmark since it can be mailed at any time after it was produced. A dated card will testify as to conditions prior to that date, however.

Paintings. Occasionally, a work of art may serve as a record of some aspect of the land. Landscapes, seascapes, architecture, and the like may have been done with such accuracy at the time to be reliable enough so that it may tell a story, or explain a related writing. It may document or "fix" an item or a set of conditions at a point in time. The existence of fences, buildings and roads, or the location and appearance of bodies of water, byways, or improvements may be explained through the use of paintings and other artwork.

Caution must be exercised, however, to ensure the accuracy of the artwork. Therefore, substantiation or verification is usually necessary. Most often artwork is useful as supplementary evidence to explain or support something else and will rarely stand on its own.

FOR FURTHER REFERENCE

Avery, T. Eugene. *Interpretation of Aerial Photographs.* Minneapolis: Burgess Publishing Co. 1962.

Avery, T. Eugene. *Forester's Guide to Aerial Photo Interpretation.* Agriculture Handbook 308. Washington: Forest Service, USDA. 1966.

Avery, T. Eugene. *Photointerpretation for Land Managers.* Kodak Publication M-76. Rochester: Eastman Kodak Company. 1970.

Eastman Kodak Company. *Photography From Light Planes and Helicopters.* Kodak Publication M-5. Rochester. 1971.

Eastman Kodak Company. *Photointerpretation and Its Uses.* Kodak Publication M-42. Rochester: Eastman Kodak Company.

Gilbert, Paul. *Arrow - Then . . . And Now.* Colorado Outdoors. Vol. 32(5): 44–45. Denver: Colorado Division of Wildlife.

Gillen, Larry, Ed. *Photographs and Maps Go to Court.* Falls Church: American Society of Photogrammetry and Remote Sensing. 1986.

Heckers, Jim. *Images Then and Now.* Colorado Outdoors. Vol. 32(5): 29–35. Denver: Colorado Division of Wildlife.

Hegg, Karl M. *A Photo Identification Guide for the Land and Forest Types of Interior Alaska.* U.S. Forest Service Research Paper NOR-3. Juneau: Northern Forest Experiment Station, USDA. 1967.

Sayn-Wittgenstein, L. *Recognition of Tree Species on Air Photographs by Crown Characteristics.* Forest Research Branch Tech. Note No. 95. Ottawa: Canada Department of Forestry. 1960.

Sayn-Wittgenstein, L. *Recognition of Tree Species on Aerial Photographs.* Forest Management Institute Information Report FMR-X-118. Ottawa: Canadian Forest Service. 1978.

Scott, Charles C. *Photographic Evidence.* Three Volumes. St. Paul: West Publishing Company. 1969.

Sully, Barry. *Aerial Photo Interpretation.* Agincourt: The Book Society of Canada, Ltd. 1970.

Zsilinsky, Victor G. *Photographic Interpretation of Tree Species in Ontario.* Ontario Department of Lands and Forests. 1966.

9 Am Jur Proof of
    Facts, 147.   *Photographs as Evidence.*

3 Am Jur Trials   *Preparing and Using Photographs*, XIII PHOTOGRAPHS OF PROPERTY LINES AND TERRAIN; XVII SPECIALTIES

27 ALR 913	*Admissibility of posed photograph based on recollection of position of persons or movable objects.*
77 ALR 946	*Expert testimony to interpret or explain or draw conclusion from photograph.*
9 ALR2d 899	*Authentication or verification of photograph as basis for introduction in evidence.*
19 ALR2d 877	*Admissibility of posed photograph based on recollection of position of persons or movable objects.*
53 ALR2d 1102	*Admissibility in evidence of colored photographs.*
57 ALR2d 1351	*Admissibility in evidence of aerial photographs.*
72 ALR2d 308	*Admissibility in evidence of enlarged photographs or photostatic copies.*

REFERENCES

Avery, T. Eugene. *Interpretation of Aerial Photographs.* Minneapolis: Burgess Publishing Company. 324 pp. 1968.

Klamkin, Marian. *Picture Postcards.* New York: Dodd, Mead & Company. 1974.

Staff, Frank. *The Picture Postcard and Its Origins.* New York: Frederick A. Praeger, Publishers. 1966.

Streb, Jack M. *Photography from the Air — Then and Now.* The Fifth Here's How. Rochester: Eastman Kodak Company. 1967.

CHAPTER 11

DOCUMENT EXAMINATION

The administration of justice profits by the progress of science,
and its history shows it to have been almost the earliest in
antagonism to popular delusions and superstitions. The
revelations of the microscopes are constantly resorted to,
in protection of individual and public interests. It is difficult
to conceive of any reasons why, in a court of justice, a different
rule of evidence should exist, in respect to the magnified image
presented by the lens in the photographer's camera, and
permanently delineated upon sensitive paper. Either may be
distorted or erroneous through imperfect instruments or manipulation,
but that would be apparent or easily proved. If they are relied upon
as agencies for accurate mathematical results in mensuration and
astronomy, there is no reason why they should be deemed unreliable
in matters of evidence. Wherever what they disclose can aid or
elucidate the just determination of legal controversies, there can be
no well-founded objection to resorting to them.

—Frank v. Chemical Nat.
Bank (N.Y.),
5 Jones & S. 26 (1874)

Document examination is the forensic science concerned with the study and investigation of documents, usually what is called the area of *questioned documents*. It
includes the examination and analysis of all parts of a document, including signatures

or other handwriting, typewriting, inks, papers, and anything else making up a document. Ordinarily, the examiner is concerned with possible forgeries, alterations, and dates, so he or she may also investigate possible insertions or obliterations. While most of the application of document examination is in solving crimes, and in proving guilt, many of the techniques can be used in solving routine problems likely to be encountered in the reading and interpretation of ordinary records.

Deciphering Handwriting. Many times the documents found in public repositories are not originals, they have been copied from the originals. This usually applies to deeds and related instruments, some probate records, town records, and the like. When this is the case, the key to being able to read the records is in understanding a person's handwriting. Often, the same handwriting will appear for several pages in succession, even for several successive years, so there is generally an abundance of work to compare.

Comparison is one of the methods that proves to be successful. If a word cannot be read, searching for similar words in the same document, or in other documents written by the same hand, may help to decipher it. Sometimes, it is necessary to break the word down into individual letters and apply the same technique. Comparison with known words to determine what the individual letters are often reveals the ultimate word.

Early records were transcribed by hand, so familiarity with handwriting of the period or the particular writer is important. Later records were transcribed by typewriter, and, in most cases, today's records are copied through photographic processes, either with photocopy machines or microfilm, or both. Today's processes are theoretically more accurate transcriptions, since direct copies are made. Early documents were subject to interpretation and a person copying could only write what he or she could read. The potential for a mistake was always present. Sometimes, it is necessary to view the original document, rather than the copy, if such is possible.

In genealogical work, researchers insist on the original document whenever available. Any copy, unless it is a direct reproduction, is suspect. The first time a transcription is made mistakes can occur, and they will likely be carried forward forever. One method of checking is *whenever anything appears to disagree*, or is not in complete harmony, *check the source*.

Today's records are based on earlier records, and if the early record was faulty it is probable that the current record is faulty as well. Never forget that the value of *any record is only as good as its basis.*

Constructing an Alphabet. One technique that often proves helpful is the construction of an alphabet for use with the document at hand. Taking an extra copy of the document, or other documents written in the same hand, cut out the individual letters, upper and lower case, along with particular words, especially names. Arrange them alphabetically for use as a reference tool for that particular handwriting.

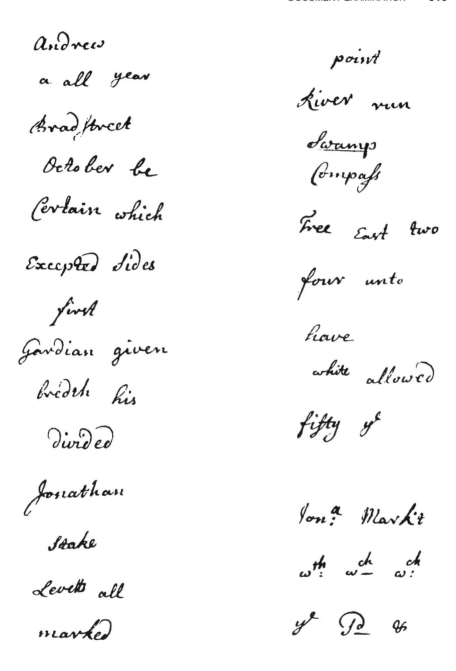

Figure 11.1 Partial alphabet made by cutting up a copy of a deed. Also included are some abbreviations and commonly used words: Jonathan, Marked, with, which, ye (the), said, and "&".

Deciphering Words. People not only wrote differently in the past, they also talked differently, and they used words not in common use today. Therefore, it is sometimes essential to consult early dictionaries and encyclopedias to locate and define early words. Colonial documents are a good example of this.

A few words of caution should be inserted here, however. First, just because you live and practice in the twenty-first century, or because your work is limited to areas outside the colonial states, this does not exempt you from these problems. Early settlers left colonial states for other areas and took their handwriting, language, and other habits with them. Second, today's records are based on earlier records, and frequently the current record is the same as the original, which may have been written a century or more ago, barring any transcription errors. Never lose sight of the fact that *the more times a writing is transcribed, the more likely it is that it* [the record] *may contain errors or mistakes*. That is why it is imperative that a description be traced back to its origin to compare with the present description for such errors or differences. If you routinely do this, you will be absolutely astounded at the frequency of errors.[1]

Early Writing and Punctuation. Knowledge of the history of writing is helpful in understanding how problems or conflicts may have originated and developed. There have been marked improvements in writing and punctuation since land records began, so it is easy to become lulled into thinking that things were always as they are now. Not only is that incorrect, but we must always keep in mind that *today's records are based on, and are sometimes the same as, those of long ago*. It is those original mistakes, or departures, that we must constantly be on guard against.

Early on, and until the middle of the sixteenth century, most public documents were written by scribes, secretaries, or priests, and at that time writing was considered very much an art form. Writing during the seventeenth century and through most of the eighteenth century was influenced by European standards. It was not until 1713 that English law provided that clerks and scribes use common English and legible handwriting. Even this law wasn't foolproof, and many old habits remained.

Latin abbreviations continued. In the English colonies some old short forms persisted that remain today, such as the ampersand for the word *and* (&), the sign for the word *at* (@), and the abbreviation of the Latin *et cetera* to *etc.*

One of the biggest problems is that in the American colonies not all scribes were English or used English words and translations. There were Dutch, French, German, Jewish, Spanish, Swedish, and many other nationalities. Each brought its own language, its own interpretations, and its own slang. In addition, each group brought its

---

[1]A trail of descriptions is sometimes like the game telephone played by children. The first person relates a story or a phrase to the second person, who retells it to a third person, and so on around the room. By the time the last person recites to the original teller, the story has changed. It may even be an entirely new story, bearing no resemblance to the original version.

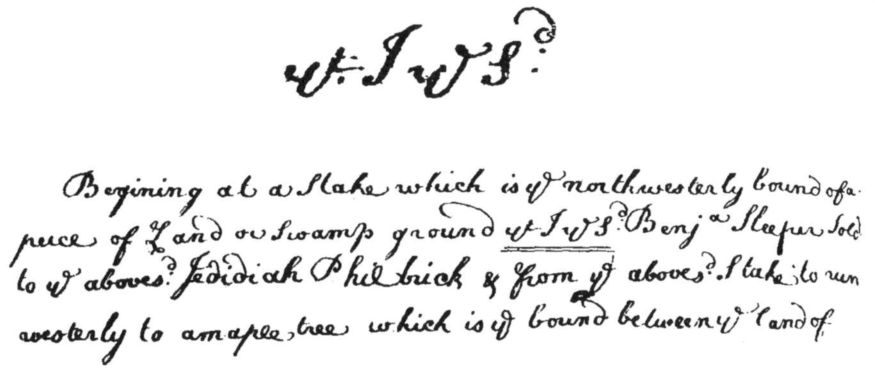

Figure 11.2 Close-up of a phrase and its accompanying text, which is abbreviated for "that I the said."

own terminology relating to land, and each group had its own ideas of land owner-ship, transfer, and tenure.

At first there were no pencils, and documents were written with quill pen. This method was popular in the American colonies until about 1830, when the modern steel nib pen was first introduced.

Document Media. Paper is another consideration, something we today take for granted. A document written before 1750 was probably written on imported paper, since paper was not made in America until the late seventeenth century, and for a long time it was cheaper to import. The first paper mill in America was established in Pittsburgh in 1816.

Punctuation. Punctuation in early documents is also different from that used today. Pauses were indicated by dots. A dot on the line indicated a brief pause, while a dot above a word indicated a full stop. A dot between words indicated phrase separation. The colon we are familiar with today and a line through part of a name both meant an abbreviation of that name.

In fact, since every stroke of the pen counted, punctuation as we know it today was mostly lacking. Page after page might be written without the single use of a period, a comma, or a capital letter. In fact, many sentences were begun with a small (lower case) letter on the first word.

Figure 11.3 Abbreviations of the words "abovesaid" and "aforesaid."

Figure 11.4   Abbreviations of "Samuel Fifield," "Benjamin Sleeper," and "Executors and administrators."

Spelling. People generally received a minimum of formal education until about 1850. Consequently, misspelled words abound, since much of the writing or spelling, especially of surnames, was phonetic. For instance, there are numerous examples where the town of Epping, New Hampshire, is written *Eppin*, just as it was pronounced.

Early Terminology. Terminology has changed significantly over the years, some of which will be found to be local or colloquial. Appendix Two contains a list of early land and measurement terms from Europe that will occasionally be encountered in early American land records.

HANDWRITING ANALYSIS

The analysis of a person's handwriting is a part of forensic document examination known as the science of *graphology*.[2] It is a behavioral science that identifies personality characteristics and expected behavior from the study of a person's handwriting.

Figure 11.5   Example of misspelled words in a sentence in a nineteenth century land description.

---

[2] The graphologist analyzes handwriting to determine the personality of the writer. The document examiner analyzes handwriting to identify the writer or detect a forgery. Dorothy V. Lehman. *Questioned Document Examination—Identification of Handwriting on Document.* 15 POF 3, p. 595.

Analysts look for the following traits in a person's handwriting:

- Slant
- Baselines
- Margin
- Spacing
- Pressure
- Size
- Speed
- Zones
- Printing vs. connected writing
- Connecting strokes
- Signatures

In addition, the following habits are studied in the handwriting itself:

- Movement, or manner of writing: line quality and alignment
- Writing habits: pen position, pen pressure, and shading
- Arrangement, size, proportion, spacing, and slant of writing
- Writing instrument
- System of writing

In so doing, graphological investigations may be made in five ways:

- Through physiological deductions
- On the basis of common sense
- By universal concepts
- By simple psychological interpretation
- Through the scientific method

ANALYSIS OF TYPEWRITING

A single typewriter makes a continuous record of its own history. The following information may be determined by careful scrutiny:

- The date of the document
- Whether the document was written continuously or at different times
- Whether the document was written on the same machine or on different machines
- Whether whole fraudulent typewritten pages have been substituted in wills
- Whether paragraphs or interlineations have been added to old letters, or to deeds and contracts
- Whether modifying conditions have been added to receipts, paid checks, and similar vouchers

Analysis of a person's typewriting examines the following habits:

- Touch
- Spacing
- Speed
- Arrangement
- Punctuation
- Incorrect use of letters, figures, or other characters
- Whether a collection of documents was produced by one or several different operators

Points for consideration are, as with handwritten letters:

- Spelling
- Punctuation
- Use of capitals
- Division of words
- Choice of words
- Construction of sentences
- Observance or nonobservance of grammatical rules
- Subject matter in general

In addition to the above, the overall style of a typewritten piece is examined for:

- Depth of indentation of paragraphs
- Spacing before or after punctuation
- Arrangement of heading
- Arrangement of conclusion

- Use of characters in an unusual way, as capital "I" for figure "1," small character "l" for capital "I", sign "&" for the word "and"
- Erroneous repetition of a letter in a word
- Striking shift-key letters in the wrong positions
- Repeated heavy impressions of certain characters
- Uniform light impressions of certain characters
- Peculiar erasures or corrections
- Uneven margin
- Balanced or unbalanced placing of matter on page
- Length of lines
- Method of writing numbers, amounts, and fractions

The phases of examination of a particular typewriting machine may be:

- Design, size, and proportion of the type faces
- Alignment
- Perpendicular position of letter in relation to the base line
- Spacing of lines, single or double
- Are added lines exactly parallel with preceding lines and/or are characters in exact vertical alignment
- Type impressions

Carbon Paper. Although not widely used today, carbon copies may be found in records and in older files. Carbon paper produces copies that, in an examination, may provide more valuable information than photocopies. The carbon cannot be easily altered.

PHOTOCOPIERS AND LASER PRINTERS

Photocopiers and laser printers use the same process as typewriters to print a page. With a photocopier, the original is placed on the glass platen and is then exposed by use of reflected light to a drum covered with photosensitive material. The image of the subject exists on the drum as an invisible positive photoelectric charge. Negatively charged toner is sprinkled onto the drum, where it sticks to only the positively charged areas, creating a visible image. Then, paper, with a positive charge, passes the drum, causing the negatively charged toner to transfer to the paper. The toner is then heat sealed to the paper, creating the printed copy. With a laser printer, the image of the original document, which is held by the computer in memory, is written to the photosensitive drum by use of a laser.

There are many ways to match a page back to its photocopier or laser printer. Since the processes are similar, the methods used to match a page back to its origin are similar.

Paper itself can yield clues: marks from the belts, pinchers, rollers, and gears that physically move the paper through machine are similar to toolmark examinations.

The toner can have unique characteristics in its chemical composition. The way the toner was placed on and fused to the paper can be characteristic. Toner can also clump on a drum, transferring blobs to the printed page.

Marks on the optics (glass platen, lenses, and mirrors) used to transfer or create an image on paper may contain unique and identifiable defects such as scratches, which will result in atypical markings on a printed page.

DOCUMENT EXAMINATION

Documents may be deliberately altered or may be injured or destroyed through carelessness or accident. Through time, any of the following may affect the readability of a document:

- Obliteration by water
- Ink faded by light
- "Bleed through" of ink from reverse side
- Staining
- Torn documents

The office copying machine can sometimes be used to enhance a document that is difficult, or impossible, to read. This is particularly true of faded lettering or anything produced with a pencil. Early blueprints, particularly if they have faded with age, frequently present a problem.

Placing a sheet of yellow acetate film between the document to be copied and the glass will usually result in a copy that is better than the original.

This technique cannot be used, however, where pages are fed *through* the machine, such as between rollers.

Excellent results have been obtained using this technique in order to be able to read and decipher early records written in ink where the ink has "bled through" and stained the reverse side of each page, which also contains writing. In addition, the author once was only able to obtain a copy of a blueprint instead of the original, and the print itself was difficult to read. Copying the copy using yellow acetate resulted in a second generation copy that was more readable than the initial blueprint (see Figure 11.6).

There are countless ways that images may be enhanced by photographic means. The use of special films, filters and developing techniques, along with combinations thereof, can sometimes produce excellent results.

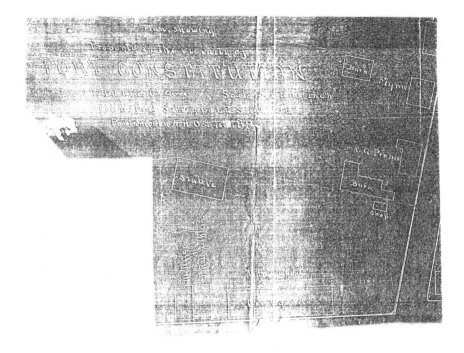

Figure 11.6   A photocopy of a faded blueprint, barely legible, enhanced by recopying using a sheet of yellow acetate.

PHOTOGRAPHIC TECHNIQUES

Use of High-Contrast Film. Photographs that are old, faded, or stained may be copied with high-contrast film, producing a photograph far superior to the original, even though there may be a slight loss of quality. The key to a satisfactory result is often the correct choice of film, filter, and developer. The yellow color can be removed through the use of a blue filter, and reddish stains removed by using a red filter. Expert assistance is frequently necessary, since in addition to the above three conditions, the selection of photographic paper may also be important. With severely faded or deteriorated images, several trials and combinations may be necessary to produce the best, or a satisfactory, result.

Ultraviolet Photography. The purpose of ultraviolet photography is to provide information about an object or material that cannot be obtained by other photographic methods. There are three types, depending upon the techniques used, (1) straight ultraviolet photography, (2) ultraviolet luminescence photography, and (3) fluorescent photography.

Straight Ultraviolet Photography. This is the method of taking a picture whereby a camera is used to record the differences in a subject's reflection, transmission, or absorption of ultraviolet rays in the same manner as visible light photography. To accomplish this, an ultraviolet light source is used and an ultraviolet filter placed on the camera lens to eliminate all visible light and transmit only ultraviolet.

Ultraviolet Luminescence Photography. This method is the process of taking a picture of a subject that is emitting invisible radiation while it is being exposed to ultraviolet rays. As an example, a document in a completely dark room can be exposed to (shortwave) ultraviolet rays and certain areas of the document, such as ink lines or paper fibers, may emit (longwave) ultraviolet radiations, which cannot be seen but can be photographed. By employing this technique the range of human vision is extended, and this enables the discovery of facts that cannot be learned any other way.

Fluorescence Photography. This technique is similar to the familiar use of ultraviolet light to "black light" in a dark room whereby objects give off a visible glow called *fluorescence*. Fluorescence is merely *visible* luminescence. Fluorescence photography is visible light photography of the image produced from exposing the subject to ultraviolet light. This is what is usually thought of or spoken of as *ultraviolet photography*.

Figure 11.7   Badly faded print (top) photographed with high-contrast film producing a legible print (bottom). *Reprinted courtesy of Eastman Kodak Company.*

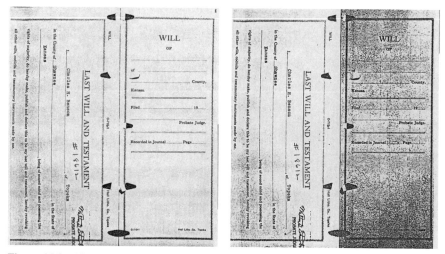

Figure 11.8 Picture on left is an ordinary photograph of a printed form appearing to be one sheet of paper which had separated at the fold. Fluorescence photograph on right shows two different pieces of paper due to different reflections because of the difference in paper composition. Reprinted from *Photographic Evidence*, second edition, Copyright © 1969 by Charles C. Scott and reproduced by permission.

Document Photography. The foregoing techniques are used routinely in document examination, perhaps more widely than in any other type of forensic examination. Fluorescence photography can demonstrate that two sheets of paper are not the same.

In this case, there were other factors that revealed a difference. One page was significantly thicker than the other as well as being wider by 1/64 of an inch, and the typewriter impressions were not the same on both pages.

Ultraviolet rays are especially helpful in bringing out erasures. In some cases, not only the fact that there was an erasure can be shown, but also the wording of the erased matter is found. Fluorescence photography will usually result in the erased writing as white lines on a dark background, while straight ultraviolet photography will show the lines more naturally, as dark lines on a lighter background.

Infrared Photography. Infrared photography deals with recording images that are invisible but which evidence themselves as heat. Recording images involving infrared rays is a complex process and includes four distinct areas, three of which are of interest here: (1) straight infrared photography, (2) indirect infrared photography, (3) infrared luminescence photography, and (4) thermography.

Straight Infrared Photography. This method involves the use of infrared film and an infrared filter over the lens. Otherwise, the procedure is the same as for ordinary black and white photography. It is useful since many subjects reflect, transmit, or absorb infrared radiation such that they produce results other than those obtained by means of ordinary visible light.

Indirect Infrared Photography. This technique involves the conversion of the infrared images on a fluorescent screen, which can be observed visually and photographed directly with fast panchromatic film. The results are very similar to those obtained by the previous procedure.

Infrared Luminescence Photography. When an object is illuminated, it not only emits ultraviolet, but also infrared radiation, not visible to the naked eye. By keeping an object from being irradiated by infrared rays, visible light will cause the object to luminesce, thereby allowing an image to be recorded on infrared film.

Figure 11.9   Alteration of "31" to "89." To the naked eye the inks appear to be the same, since they are the same in color but not in composition. To achieve this result, the suspected alteration was photographed with infrared film using an infrared filter. Reprinted from *Photographic Evidence*, second edition, Copyright © 1969 by Charles C. Scott and reproduced by permission.

Figure 11.10   Letterhead in pigment-type ink obliterated with a dye-type ink. Panchromatic photograph at top; infrared photograph at bottom. *Reprinted courtesy of Eastman Kodak Company.*

Figure 11.11   Portion of Dead Sea scroll blackened through age to be totally illegible. Ordinary photograph on left; infrared photograph on right. *Reprinted courtesy of Eastman Kodak Company.*

Document Photography. Infrared photography is one of the most important and widely used techniques for examining documents. It can demonstrate differences in inks and thereby allow deciphering of overwriting and other obscuring factors. Sometimes, an infrared photograph can make a charred or stained document readable.

Ordinarily, infrared film and a filter such as a Kodak Wratten 87 filter are employed. If stronger effect or penetration is found to be necessary, the No. 87C filter may be used.

Concerning illegible documents, the following problems are sometimes overcome through this procedure:

- Illegibility due to charring
- Deterioration because of age or dirt
- Obliteration by application of ink
- Invisible inks
- Chemical bleaching
- Mechanical erasure with subsequent overwriting

COMPARISON OF PHOTOGRAPHS TAKEN YEARS APART

A frequent use of aerial photographs is in the comparison of two or more photos of the same area taken at different times. Existing photos may show changes in conditions or changes in land use over time and are very useful in cases of boundaries affected by accretion or avulsion, erosion, timber, and gravel trespass, hazardous waste dumping, and the extent of, or change in, development.

Terrestrial photographs may be used in the same manner, if taken from the same location. Early photos may be found in archives, libraries, historical societies, private collections, and on postcards.

FOR FURTHER REFERENCE

Beyerstein, Barry L., and Dale F. Beyerstein, Eds. *The Write Stuff.* Buffalo: Prometheus Books. 1992.

Eastman Kodak Company. New York: Rochester.

Publication M-2. *Using Photographs to Preserve Evidence.* 1976.

Publication M-27. *Ultraviolet and Fluorescence Photography.* 1972.

Publication M-28. *Applied Infrared Photography.* 1977.

Publication N-9. *Basic Scientific Photography.* 1970.

Publication N-12A. *Close-Up Photography.* 1969.

Greene, James, and David Lewis. *Handwriting Analysis.* London: Treasure Press. 1990.

Kirkham, E. Kay. *The Handwriting of American Records for a Period of 300 Years.* Logan, Utah: The Everton Publishers, Inc. 1973.

McNichol, Andrea. *Handwriting Analysis. Putting It to Work for You.* Chicago: Contemporary Books. 1991.

Schwid, Bonnie L. *Forensic Document Examination.* Milwaukee: Anagraphics, Inc. 1986.

Scott, Charles C. *Photographic Evidence.* Three Volumes. St. Paul: West Publishing Company.

Stryka-Rodda, Harriet. *Understanding Colonial Handwriting.* Baltimore: Genealogical Publishing Co., Inc. 1987.

American Jurisprudence. Proof of Facts. Ancient Documents.

31 ALR 1431   *Use of photographs in examination and comparison of handwriting or typewriting.*

106 ALR 721   *Determination of date of document from inspection and examination of typewriting.*

28 ALR2d 1443   *Mutilations, alterations, and deletions as affecting admissibility in evidence of public record.*

41 ALR2d 575   *Genuineness of handwriting offered as standard or exemplar for comparison with a disputed writing or signature, or shown by age of writing.*

65 ALR2d 342   *Carbon copies of letters or other written instruments as evidence.*

11 ALR3d 1015   *Admissibility of expert evidence to decipher illegible document.*

36 ALR4th 598   *Admissibility of evidence as to linguistics or typing style (forensic linguistics) as basis of identification of typist or author.*

References

Eastman Kodak Company. New York: Rochester

> Publication M-27. *Ultraviolet and Fluorescence Photography.* 1972.

> Publication M-28. *Applied Infrared Photography.* 1977.

McNichol, Andrea. *Handwriting Analysis. Putting It to Work for You.* Chicago: Contemporary Books. 1991.

Osborn, Albert S. *Questioned Documents*, second edition. No publication data. 1929.

Schwid, Bonnie L. *Forensic Document Examination.* Milwaukee: Anagraphics, Inc. 1986.

Scott, Charles C. *Photographic Evidence.* Three Volumes. St. Paul: West Publishing Company.

Stryka-Rodda, Harriet. *Understanding Colonial Handwriting.* Baltimore: Genealogical Publishing Co., Inc. 1987.

APPENDIX ONE

DEFINITIONS OF WORDS AND PHRASES

ABOUNDING. "Abounding" means "bounding" or "bounded by." Bound is synonymous with "limit" or "border." *Barney v. City of Dayton*, 4 O.C.D. 505, 8 Cir. Ct. R. 480 (1894).

ABOUT. The word "about," where the context limits and restrains its meaning, does not materially impair the certainty of a description. *Adams v. Harrington*, 14 N.E. 603, 114 Ind. 66 (1887).

Word "about" means the same as the word "at" in a phrase fixing the time within which something should be done. *In re Heine's Estate*, Ohio Prob., 100 N.E.2d, 545 (Ohio, 1950).

"About," within deed or mineral lease describing boundary as "thence about 50 varas on J.G. Survey for corner," will be disregarded in determining distance. *Humble Oil & Refining Co. v. Luther*, 40 S.W.2d 865 (Tex. Civ. App., 1931).

"About" has no effect upon the question of boundaries, nor will it control monuments, courses, and distances where the language is clear and obvious. *Wheeler v. Randall*, 47 Mass. (6 Metc.) 529 (1843); *Whitaker v. Hall*, 4 Ky. (1 Bibb) 72 (1809); *Stephens v. Heden*, 7 Ky. (4 Bibb) 107 (1815).

The term "about" cannot be used as the equivalent of exact distances. *Picharella v. Ovens Transfer Co.*, 5 A.2d 408, 135 Pa. Super. 112 (1939).

In stating the number of acres conveyed in a conveyance, it is usual to represent it as "about" so many. Yet the word "about," although it negatives the conclusion that entire precision is intended, is without any legal operation whatever. *Purinton v. Sedgley*, 4 Me. (4 Greenl.) 283 (1826).

"More or less" are words of safety and precaution and when used in deed are intended to cover some slight inaccuracy in frontage, depth, or quantity in land conveyed, and ordinarily means "about" the same as terms "a little more than," "not quite," "not more than," or "approximately," and all are often introduced into a description practically without effect. *Harries v. Harang*, 23 So.2d 786 (La. App., 1945).

ABUT. "Abutting" means to end, to border on, to touch, and "abutting" property means any property that abuts or adjoins. *Bulen v. Moody*, 63 N.E.2d 916; 77 Ohio App. 61 (1945).

ACRE RIGHT. "The share of a citizen of a New England town in the common lands. The value of the acre right was a fixed quantity in each town, but varied in different towns. A l0-acre lot or right in a certain town was equivalent to 113 acres of upland and 12 acres of meadow, and a certain exact proportion was maintained between the acre right and salable lands." *Black's Law Dictionary.*

A.D. The capitals "A.D., as part of a date, have been adjudged to be English language by use, though in fact being the initials of two Latin words. *State v. Hodgeden*, 3 Vt. 481 (1831); *Clark v. Stoughton*, 44 Am. Dec. 361, 18 Vt. 50 (1844).

The Latin words "Anno Domini" mean, when translated, "in the year of our Lord." *Commonwealth v. Taylor*, 45 S.W. 356, 20 Ky. Law Rep. 97 (1898).

ADJACENT. "Adjacent" means lying near, close, or contiguous; neighboring; bordering on; nigh, juxtaposed, meeting, and touching. *Hall v. Gulf Ins. Co. of Dallas*, 200 S.W.2d 450 (Tex. Civ. App., 1947).

Objects are "adjacent" when they lie close to each other, but not necessarily in actual contact, but are "adjoining" when they meet at some line or point of junction. *Hauber v. Gentry*, 215 S.W.2d 754 (Mo., 1948).

ADJOINING. The word "adjoining," in the description of the premises conveyed, means "next to" or "in contact with," and excludes the idea of any intervening space. *Yard v. Ocean Beach Ass'n*, 24 A. 729, 49 N.J. Eq. 306 (1892), affirming (Ch. 1891) *Ocean Beach v. Yard*, 20 A. 763, 48 N.J. Eq. 72 (1890).

AFORESAID. "Such" is a relative adjective referring back to and identifying something previously spoken of and by grammatical usage it naturally refers to the last precedent antecedent and is equivalent to "said," "aforesaid," "aforedescribed," and "same." *In re Wallace's Estate*, 219 P.2d 910, 98 Cal. App.2d 285 (1950).

AGER. A field; land generally. A portion of land enclosed by definite boundaries. *Municipality No. 2 v. New Orleans Cotton Press*, 36 Am. Dec. 624, 18 La. 167 (1841).

ALL THE LAND BY ME OWNED. Considered to be the same as "all the land now owned by me," is equivalent to "all the land which I have not heretofore conveyed." *Fitzgerald v. Libby*, 7 N.E. 917, 142 Mass. 235 (1886).

ALNETUM. In old records, a place where alders grow or a grove of alder trees. *Black's Law Dictionary* (1944).

ALONG. When used in a description, means "by, on or over," according to the subject matter and content. *Church v. Meeker*, 34 Conn. 421 (1867).

ALSO. The word "also," in a deed, expressing what is granted thereby, means likewise, in like manner, in addition to, denoting that something is added to what precedes it. *Panton v. Tefft*, 22 Ill. 367 (1859).

AMBIT. A boundary line, as going around a place; an exterior or inclosing line or limit. *Ellicott v. Pearl*, 10 Pet. (U.S.) 412, 9 L. Ed. 475 (1836).

AND OTHERS. The Latin abbreviation "et al." means "and another" or "and others." *In re McGovern's Estate*, 250 P. 812, 77 Mont. 182 (1926).

APPROXIMATE. The word "approximate" has been defined as to be near to exactness, nearly exact, not perfectly accurate. *Fitzgerald v. Thompson*, 184 S.W.2d 198, 238 Mo. App. 546 (1944).

APPURTENANCE; APPURTENANT. "Appurtenant" means belonging to; accessory or adjacent to; adjunct, appended, or annexed to. *Appeal of Fisher*, 49 A.2d 626, 355 Pa. 364 (1946).

APPROXIMATELY. "Approximately" means very nearly but not absolutely, nearly, about, close to. *In re Searl's Estate*, 186 P.2d 913, 29 Wash.2d 230 (1947), see 173 A.L.R. 1247.

"More or less" are words of safety and precaution and when used in deed are intended to cover some slight inaccuracy in frontage, depth, or quantity in land conveyed, and ordinarily mean "about" the same as terms "a little more than," "not quite, not more than," or "approximately," and all are often introduced into a description practically without effect. Civ. Code, art. 2493, *Harries v. Harang*, 23 So.2d 786 (La. App., 1945).

AREA. The word "area" usually means tract, space, region, or a broad part of land. *Maisen v. Maxey*, Tex. Civ. App., 233 S.W.2d 309 (1950).

AT. "At" means in or near. *Chicago, L.S. & E. Ry. Co. v. McAndrews*, 124 Ill. App. 166 (1906).

"About," used to describe the location of land in an entry or grant, signifies "at unless something can be shown to evidence a contrary intention. *Simm's Lessee v., Dickson*, 22 Fed. Cas. 158, 3 Tenn. (Cooke) 137 (1812).

BACK LANDS. A term of no very definite import, but generally signifying lands lying back from (not contiguous to) a highway or a water course. See *Ryerss v. Wheeler*, 22 Wend. (N.Y.) 150 (1839).

BALDIO. In Spanish law. Waste land; land that is neither arable nor pasture. Unappropriated public domain, not set apart for the support of municipalities. *Sheldon v. Milmo*, 90 Tex. 1, 36 S.W. 413 (1896).

BANK. The "bank" of a river is that space of rising ground above low-water mark, which is usually covered by high water, and the term, when used in a grant to designate a precise line, is vague and indefinite. *Howard v. Ingersoll*, 17 Ala. 780, rev (1851) 13 How. 381, 14 L.Ed. 189 (1850).

BAY. A bending or curving of the shore of the sea or of a lake, so as to form a more or less inclosed body of water. *State v. Town of Gilmanton*, 14 N.H. 465 (1843).

An opening into the land, or an arm of the sea, where the water is shut in on all sides except at the entrance. *U.S. v. Morel*, 13 Amer. Jur. 286, Fed. Cas. No. 15, 807; *Ocean Industries v. Superior Court of California, in and for Santa Cruz County*, 252 P. 722, 200 Cal. App. 235 (1927).

BAYOU. A species of creek or stream common in Louisiana and Texas. An outlet from a swamp, pond or lagoon, to a river or the sea. *See Surgett v. Lapice*, 8 How. (U.S.) 48, 12 L.Ed. 982 (1950).

BEING THE SAME. Where two separate but adjoining parcels of real estate were deeded in one conveyance, and the grantee subsequently conveyed one of them, the deed reciting, after a detailed description, "Being the same premises" that were conveyed by the deed first mentioned, the term did not include both parcels, but would be construed to mean "being part of the same." *Fidelity Mortgage Guarantee Co. v. Babb*, 160 A. 120, 306 Pa. 411 (1932).

BEING THE SAME PROPERTY. A reference to property already adequately described in a deed as "being the same property" acquired by the vendor neither enlarges description to include lands not described therein nor is suggestive of intention to convey or acquire other land than that described. *Fortier v. Soniat*, La. App., 143 So.2d 91 (1962).

BETWEEN. When the word "between" is used with reference to a period of time bounded by two other specified periods of time, such as between two days named, the days or other periods of time named as boundaries are excluded, and "between" has a like meaning when used with reference to boundaries in space. *Winans v. Thorp*, 87 Ill. App. 297 (1899).

BLOCK. A square or portion of a city or town inclosed by streets, whether partially or wholly occupied by buildings or containing only vacant lots. *Ottawa v. Barney*, 10 Kan. 270 (1872); *Fraser v. Ott*, 30 Pac. 793, 95 Cal. 661 (1892): *State v. Deffes*, 10 So. 597, 44 La. Ann, 164 (1892); *Todd v. Railroad Co.*, 78 Ill. 530 (1875); *Harrison v. People*, 63 N.E. 191, 195 Ill. 466 (1902); *City of Mobile v. Chapman*, 79 So. 566, 202 Ala. 194 (1918); *Commerce Trust Co. v. Blakeley*, 202 S.W. 402, 274 Mo. 52 (1917). The platted portion of a city surrounded by streets. *Cravens v. Putnam*, 165 P. 801, 101 Kan.161 (1917). The term need not, however, be limited to blocks platted as such, but may mean an area bounded on all sides by streets or avenues. *St. Louis-San Francisco R. Co. v. City of Tulsa*, Okl. (C.C.A.) 15 F.(2d) 960; *Commerce Trust Co. v. Keck*, 223 S.W. 1057, 283 Mo. 209, 1061 (1920) (irregular parallelograms). Yet two blocks each, bounded by a street, do not necessarily, when thrown together by the vacation of a street, constitute a single block to be included as such within an assessment district. *Missouri, K. & T. Ry. Co. v. City of Tulsa*, 145 P. 398, 401, 45 Okl. 382 (1914).

"Block" is often synonymous with "square." *Weeks v. Hetland*, 202 N.W. 807, 52 N.D. 351 (1925). As a measure of length, "block" denotes the length of one side of such a square. *Skolnick v. Orth*, 84 Misc. 71, 145 N.Y.S. 961, 962 (1914). Sometimes it means both sides of a street measured from one intersecting street to the next. *Chamberlain v. Roberts*, 253 P. 27, 81 Colo. 23 (1927). And on occasion it may be construed not to extend between two streets that completely cross the street in question, but to stop at a street running into it though not across it. *Wise v. City of Chicago*, 183 Ill. App. 215, 216 (1913).

Under a statute providing that territory sought to be excluded from a new county must be in one block, the word "block" implies the thought of solidity or compactness, and the territory sought to be excluded must be in some regular and compact form. *State v. Moulton*, 189 P. 59, 61, 57 Mont. 414 (1920).

BLOCK OF SURVEYS. In Pennsylvania land law. Any considerable body of contiguous tracts surveyed in the name of the same warrantee, without regard to the manner in which they were originally located; a body of contiguous tracts located by exterior lines, but not separated from each other by interior lines. *Morrison v. Seaman*, 38 A. 710, 183 Pa. 74 (1897); *Ferguson v. Bloom*, 23 A. 49, 144 Pa. 549 (1891).

BLUFF. A high, steep bank, as by a river, the sea, a ravine, or a plain, or a bank or headland with a broad, steep face. *Columbia City Land Co. v. Ruhl*, 141 P. 208, 70 Or. 246 (1914).

BODILY HEIRS. Words "bodily heirs" in deed would be construed as words of limitation rather than words of purchase, since one cannot have heirs during his life. *Beasley v. Beasley*, 88 N.E. 2d 435, 404 Ill. 225 (1949).

BOUND. "Bound" is synonymous with "limit" or "border." *Barney v. Dayton*, 8 Ohio Cir. Ct. R. 480 (1894).

BRANCH OF A RIVER. "Branch," as distinguished from a channel of a river, may have two or more separate channels; "channel" meaning primarily the bed. *United States v. Hutchings* (D.C.), 252 F. 841.

BRANCH OF THE SEA. This term, as used at common law, included rivers in which the tide ebbed and flowed. *Arnold v. Mundy*, 10 Am. Dec. 356, 6 N.J. Law 86 (1821).

BUT. "But" has been judicially defined as "except," "except that," "on the contrary," "or," "and also," "yet," and "still." *State v. Marsh*, 187 N.W. 810, 108 Neb. 267 (1922).

BY. Bounding of one piece of land "by" another piece, whether such other by long or narrow, or in any other form, locates the line at the edge, and not through the middle of the adjoining premises. *Woodman v. Spencer*, 54 N.H. 507 (1874).

"By," as indicating a terminal point of time, means "not later than; as early as." *Goldman v. Broyles*, Tex. Civ. App., 141 S.W. 283 (1911).

BY LAND OF. The words "by land of" an adjoining owner, used in the description in a deed, mean along the external boundary line of that land. *Peaslee v. Gee*, 19 N.H. 273 (1848).

CAFIADA. The Spanish "cafiada," when used as a call in field notes, means valley. *Venavides v. State*, 214 S.W. 568 (1919).

CALL. A reference to, or statement of, an object, course, distance, or other matter of description, in a survey or grant, requiring or calling for a corresponding object, and so on, on the land. *King v. Watkins*, C.C.Va., 98 F. 913.

In *Bouvier's Law Dictionary*, the word "call" is defined thus: "In American land law, the designation in an entry, patent, or grant of land of visible natural objects as limits to the boundary." And *Webster* defines the word as a reference to or statement of an object, course, distance, or other matter of description in a survey or grant requiring or calling for a corresponding object, and so forth, on the land. The meaning of "calling" for a course or boundary is well understood in the law. It is necessarily applicable to written instruments, such as entries, surveys, patents, grants, and deeds. *King v. Watkins*, 98 F. 913, C.C.Va.

CANADA. Spanish, meaning valley. *Benavides v. State* (Tex. Civ. App.) 214 S.W. 568 (1919).

CATTLE PASS. As used in a statute, a narrow passage way under a railroad track high and wide enough to admit the passage of a cow, horse, or ox to and from a pasture. *True v. Maine Cent. R. Co.,* 94 A. 183, 113 Me. 375 (1915).

CATTLE RANGE. Under a statute, a range the usual and customary use of which has been for cattle. *State v. Butterfield,* 165 P. 218, 30 Idaho 415 (1917).

CEDO. I grant. The word ordinarily used in Mexican conveyances to pass title to lands. *Mulford v. Le Franc,* 26 Cal. 88 (1864).

CENTER. "Center" does not necessarily mean precise geographical or mathematical center, but in common parlance means middle or central point or portion of anything. *Bass v. Harden,* 128 S.E. 397, 160 Ga. 400 (1925).

CHAIN OF TITLE. A term applied metaphorically to the series of conveyances, or other forms of alienation, affecting a particular parcel of land, arranged consecutively, from the government or original source of title down to the present holder, each of the instruments included being called a "link." *Payne v. Markle,* 89 Ill. 66, 69 (1878); *Capper v. Poulsen,* 152 N.E. 587, 321 Ill. 480 (1926); *Maturi v. Fay,* 126 A. 170, 96 N.J. Eq. 472 (1924).

CHAMBER SURVEYS. At an early day in Pennsylvania, surveyors often made drafts on paper of pretended surveys of public lands, and returned them to the land office as duly surveyed, instead of going on the ground and establishing lines and marking corners; and these false and fraudulent pretenses of surveys never actually made were called "chamber surveys." *Schraeder Min. & Mfg. Co. v. Packer,* 129 U.S. 688, 9 S. Ct. 385, 32 L.Ed 760 (1889).

CHILD; CHILDREN. Prima facie, the word "child" or "children" when used in statute, will or deed means legitimate child or children. *McManus v. Lollar,* 235 N.Y.S.2d 61, 36 Misc.Pd 1046 (1962).

CHILDREN; HEIRS AT LAW. Where the words "Children and heirs at law" are used in a deed, the term "heirs at law" may well be construed as being used interchangeably with "children," or as meaning grandchildren or descendants, especially where, as under the statute, the issue of the person entitled takes the share of his ancestor. *Waddell v. Waddell,* 12 S.W. 349, 99 Mo. 338, 17 Am. St. Rep. 575 (1889).

CLEARED LAND. The words "cleared land" are sufficient to describe a monument and must be taken to refer to the condition of the land at the time the deed was given, and the expression is commonly used to describe land that has been cleared of timber for purposes of pasture or tillage. *Marvel v. Regienus,* 108 N.E.2d 545, 329 Mass. 414 (1952).

CLOUD ON TITLE. A "cloud on title" is an outstanding claim or encumbrance that, if valid, would affect or impair title of owner and that appears on its face to have that effect which can be shown by extrinsic evidence to be invalid. *Gary-Wheaton Bank v. Helten,* 85 N.E.2d 472, 337 Ill. App. 294 (1949).

To constitute a "cloud on title," the claim to land must be apparently valid and capable of embarrassing title, the mere verbal assertion of a claim not being such a cloud as equity will undertake to remove. *Amick v. Gauley Coal Land Co.,* 194 S.E. 268, 119 W. Va. 485 (1937).

A "cloud on title" is a semblance of title legal or equitable, or a claim of an interest in lands appearing in some legal form, but which is in fact unfounded. *Dodsworth v. Dodsworth*, 98 N.E. 279, 254 Ill. 49 (1912).

COLOR OF TITLE. "Color of title" is anything that shows the extent of the occupants' claim. *Sprott v. Sprott*, 96 S.E. 617, 110 S.C. 438(1918); *Thurmond v. Espalin*, 171 P.2d 325, 50 N.M. 109 (1946).

There must be judgement, decree, statute, or contract to give "color of title." *Griffith v. Coleman*, 135 P.2d 33, 192 Okla. 296 (1943).

"Color of title" is not title but is a void paper having the semblance of a muniment of title, to which, for certain purposes, the law attributes certain qualities of title. *Bailey v. Jarvis*, 208 S.W.2d 13, 212 Ark. 675 (1948).

The chief purpose of "color of title" is to define the limits of a claim thereunder, but it must purport to pass title and in form must be a deed, a will, or some other instrument by which title usually and ordinarily passes. *Bailey v. Jarvis*, 208 S.W.2d 13, 212 Ark. 675 (1948).

COMMON FIELD LOTS. Lots in the vicinity of the village occupied and cultivated by the inhabitants of the village in a common field. *Harrison v. Page*, 16 Mo. 182 (1852).

"Town lots, outlots, common field lots, and commons were known and recognized parts of the Spanish town or commune of St. Louis. They existed by public authority, whether by concession, custom or permission." *Vasquez v. Ewing*, 42 Mo.247 (1868).

"Common field lots," as used in Act Congress. June 13, 1812, declaring that the rights, titles and claims of town or village lots, outlots, or common field lots, and commons adjoining and belonging to the several towns and villages named in the act, including St. Louis, which lots had been inhabited, cultivated, or possessed prior to the 20th of December, 1803, were thereby confirmed to the inhabitants of the respective towns and villages according to their several rights or rights in common thereto, is a term of American invention, and adopted by Congress to designate small tracts of ground of a peculiar shape, usually from one to three arpents in front by forty in depth, used by the occupants of the French villages for the purpose of cultivation, and protected from the inroads of cattle by a common fence. The peculiar shape of the lots, its contiguity to places of similar shape, and the purpose to which it was applied, con situated it a common field lot. It could not be confounded with lots or tracts of land of any other character. *Glasgow v. Hortiz*, 66 U.S. (1 Black) 595, 17 L. Ed. 110 (1861).

COMMON FIELDS. The terms "common fields" and "commons" are of French origin and were used in several Acts of Congress relating to such lands in Louisiana Purchase territory—"common fields" to designate fields cultivated outside a village by the inhabitants, in severalty, though generally under a common fence; and "commons" to designate a large body of land held in common and used for pasturage, fuel, etc. *City of St. Louis v. St. Louis Blast Furnace Co.*, 138 S.W. 641, 235 Mo. 1 (1911).

COMMON LANDS. The phrase "common lands," in Laws 1818, c. 155, relative to the power by town trustees over common lands of the town, designates lands held in common by the proprietors and not in any technical sense, and the term is used interchangeably with the term "undivided lands." *Trustees of Freeholders, etc. of Town of Southampton v. Beets*, 47 N.Y.S. 697, 21 App. Div. 435 (1897).

CONFIRMED SWAMP LANDS. The word "confirmed" was employed in the earliest state legislation with reference to swamp lands. Gould, Dig. p. 717, § 3; Id. p. 719, § 7. In the earlier acts, the authority of state offices to make sales was confined to the confirmed lands, while the right of settlement and preemption was provided for the unconfirmed lands. So at one time the governor was authorized to issue a patent only upon being satisfied that the lands were confirmed. In determining whether the lands were "confirmed" within the meaning of the act, the court said, in *Hendry v. Willis*, 33 Ark. 833 (1878): "The selections have been made in fact by agents of the state, sent to the Secretary of the Interior through the Commissioner of the General Land Office, approved, and returned to the Governor. When those lists so approved have been transmitted to the Governor, they have been treated in our legislative and official acts as 'confirmed,' and so we must understand the word." *Chism v. Price*, 15 S.W. 883, 54 Ark. 251 (1891).

CONFIRMEE. The grantee in a deed of confirmation. 15. C.J.S.

CONFIRMOR. The grantor in a deed of confirmation. See 15 C.J.S.

CONSEDO. A term used in conveyances under Mexican law, equivalent to the English word "grant." *Mulford v. Le Franc*, 26 Cal. 103 (1864).

CONSENTABLE BOUNDARY LINE. A "consentable boundary line" connotes a new boundary line created by agreement of the parties, but not every line assented to by the parties is a consentable line. *Beals v. Allison*, 54 A.2d 84, 161 Pa.Super. 125 (1947).

CONSTRUCTION OF WILL. The object of "construction" of a will is to interpret and give effect to will if possible unless the language is such as to imperatively require the whole will to be rejected. *Cahill v. Michael*, 45 N.E.2d 657, 381 Ill. 395 (1942).

"Construction of will" is the ascertainment of the fact of intention from competent evidence, and not the application of technical rules as legal tests. *Frost v. Wingate*, 64 A. 19, 73 N.H. 535 (1906).

CONTAINING SO MANY ACRES. The words of a deed describing the length of lines and boundaries, and concluding with the words "containing so many acres," do not import a warranty of quantity. *Rickets v. Dickens*, 5 N.C. 343, 4 Am. Dec. 555 (1810).

CONTAIN 180 ACRES STRICT MEASURE. A conveyance of land described by metes and bounds, and said to contain "180 acres strict measure," is to be construed as a mere matter of description, which is subject to the description by metes and bounds; hence a deficiency of 9 acres in the quantity of land is no breach of the covenants for title. *Andrews v. Rue*, 34 N.J.L. 402 (1871).

CONTEST (claim to public land). Under Pol. Code, § 3414, West's Ann. Public Resources Code, § 7921, providing that when a contest arises concerning the approval of a survey or location of public lands, and when either party demands a trial in the courts, the surveyor general must make an order referring the contest, a "contest" arises when two persons make separate applications to purchase the same state land, which is still open to location, and it is not necessary that the contestant file a statement of the specific grounds of contest with the surveyor general. *Miller v. Engle*, 85 P. 159, 3 Cal. App. 325 (1906).

CONTIGUOUS. Two tracts of land touching at only one point are not "contiguous." *Baham v. Vernon*, 42 So.2d 141 (La. App., 1949).

"Contiguous" means in actual contact; touching; also, near, though not in contact; neighboring; adjoining; near in succession. *Ehle v. Tenney Trading Co.,* 107 P.2d 210, 56 Ariz. 241 (1940).

The word "contiguous" in its primary sense, means in actual contact or touch and when applied to tracts of land, it ordinarily conveys the idea that they border each other. *Lien v. Northwestern Engineering Co.,* 39 N.W.2d 483, 73 S.D. 84 (1949).

Two tracts of land, which touch only at common corner, are not "contiguous." *Turner v. Glass*, La. App. 195 So. 645 (1940).

Tracts that corner with one another are contiguous, "contiguous" meaning to touch. *Morris v. Gibson*, 134 S.E. 796, 35 Ga. App. 689 (1926).

CONVEY. The word "convey" is used in the law, in its technical application, to refer to deeds or other instruments passing title to real estate. *Benz v. Paulson*, 70 N.W.2d 570, 246 Iowa 1005 (1955).

CONVEYANCE. An instrument that carries from one person to another an interest in land. *State v. Sutterfield*, 176 S.W.2d 666, 237 Mo. App. 562 (1944).

The term "conveyance" as used in Real Property Law includes every instrument in writing, except a will, by which any estate or interest in real property is created, transferred, assigned, or surrendered, and every instrument creating, transferring, assigning, or surrendering an estate or interest in real property must be construed according to the intent of the parties, as far as such intent can be gathered from the whole instrument and is consistent with the rules of law. Real Property Law, § 240. *Goldberg v. Friedman*, 61 N.Y.S.2d 222, 186 Misc. 983 (1946), reversed 62 N.Y.S.2d 457, 187 Misc. 445 (1946), appeal denied 62 N.Y.S.2d 757, 270 App. Div. 939 (1946).

CONVEYANCE OF LAND. Means the land itself in fee simple absolute. *Knutson v. Hederstedt*, 264 P. 41, 125 Kan. 312 (1928).

CORNER. The intersection of two converging lines or surfaces; an angle, whether internal or external; as the "corner" of a building, the four "corners" of a square, the "corner" of two streets. A mere variation in a line does not constitute a "corner." *Christian v. Grant*, 64 S.W. 399 (Tenn. Ch., 1900).

A description of a tract as "lying in the southwest corner of a section" is sufficiently definite according to the rules of decision that a "corner" is a base point from which two sides of the land conveyed shall extend an equal distance so as to include by parallel lines the quantity conveyed. *Walsh's Lessee v. Ringer*, 15 Am. Dec. 555, 2 Ohio 328 (1826).

COUNTRY ROAD. Under the Colonial Laws, a "country road" was one which belonged to the country and was under the direct charge of the country, as distinguished from the owners of the towns and manors, and it was a necessary line of communication between sparsely settled communities, and it was for the better laying out, sparing, and preserving the public and general highways within the colony that legislation as to such roads was adopted. *Townsend v. Trustees of Freeholders & Commonality of Town of Brookhaven*, 97 App. Div. 316, 89 N.Y.S. 982 (1904).

COUNTY HIGHWAYS. Roads deeded to county before improvements were made on them were "county highways" within statutory definition. *Wine v. Boyar*, 33 Cal Rptr. 787, 220 C.A.2d 375 (1963).

COUNTY WAYS. County ways are roads leading from one town to another. *Inhabitants of Waterford v. Oxford County Com'rs*, 59 Me. 450 (1871).

COVENANT RUNNING WITH LAND. To constitute a "covenant running with the land," the covenant must have a relation to the interest in the estate conveyed, and the act to be done must concern the interest created and conveyed, but it is not necessary that privity of estate shall exist between the original grantor and a purchaser from the covenantee. *Reidville & S.E.R. Co. v. Baxter*, 795 S.E. 187, 13 Ga. App. 357 (1913).

CURVE. A curve in geometry is a line that changes its direction at every point; a line in which no three consecutive points are in the same direction or straight line. *Bishop & Babcock Mfg. Co. v. Fedders-Quigan Corp.*, D.C.N.Y., 159 F. Supp. 815.

DEMARCATION/DELIMITATION. "Demarcation" is the marking of a boundary line on the ground by physical means or a cartographic representation. "Delimitation" is the defining of a boundary line in written or verbal terms. *State ex rel. Buckson v. Pennsylvania R. Co.*, 267 A.2d 455, supplemented 273 A.2d 268 (1969).

DIRECTLY OPPOSITE. Words "directly opposite" other lots described in deed mean portion that would be included in described lots if extended in straight line. *Smith v. Chappell*, 148 So. 242, 177 La. 311 (1933).

DISTANCE. "Distance" is a straight line along a horizontal plane from point to point and is measured from the nearest point of one place to the nearest point of another. *Evans v. United States*, C.C.A.N.Y., 261 R. 902.

DOWER. "Dower" is a "life estate." *Farabow v. Perry*, 25 S.E.2d 173, 223 N.C. 21 (1943).

"Dower" and "homestead" are distinct rights. *Wallace v. King*, 170 S.W.2d 377, 205 Ark. 681 (1943).

A deed conveying certain land, except the widow's dower, meant the widow's right to the use of a third of the real estate described during her life, and did not include the reversion thereof. *Starr v. Brewer*, 3 A. 479, 58 Vt. 24 (1886).

DOWER INTEREST. A "dower interest" is only, at most, a life estate in one-third of property. *Home Ins. Co. v. Field*, 42 Ill. App. 392 (1891).

THIRDS. Where a testator bequeathed and devised all the rest, residue, and remainder of his estate, both real and personal, to his son and daughter, subject to the dower "and" thirds of his wife, the word "thirds" should be construed to mean the same thing as "dower," and the word "and" to mean "or." *O'Hara v. Dever*, 41 N.Y. 558 (1946).

DRAW. The word "draw," as used in the northwestern states, implies a depression that may run for many miles in length, in which there is not necessarily a running stream, but the water from melting snows and rains that fall on the area on either side of the draw drain into it, and thence make their way, through other channels, to the broad rivers of the West. *Lincoln & B H R Co. v. Sutherland*, 62 N.W. 859, 44 Neb. 526 (1895).

DRIFTWAY. A byroad is one that individuals who live some distance from a public road in a newly settled country make in going to the nearest public road—that is, it is a road that is used by the inhabitants, but is not laid out—and very often it is called a "driftway," and is a road of necessity in newly settled countries. *Van Blarcom v. Frike*, 29 N.J.L. (5 Dutch.) 516 (1861).

DUE. In deed description, "boundary lines to continue due west from these stakes," the phrase "due" was not ambiguous and limited the course to one that traveled directly west. *Haklits v. Oldenburg*, 201 A.2d 690, 124 Vt. 199 (1964).

"Due," as used in the description in a deed, requiring the line to be run due north, means "exactly," and adds nothing to the description. The point of a compass, if due north, is exactly north, and so is simply north. *Wells v. Jackson lron Mfg., Co.*, 47 N.H. 235, 90 Am. Dec. 575 (1866).

DWELLING. A "dwelling" as used in deed is a place of residence. *Lietz v. Pfuehler*, 215 S.2d 723, 283 Ala. 282 (1968).

E. "E," as an abbreviation, means "east." *Sibley v. Smith*, 2 Mich. 486 (1853).

EAST. Generally, the words "north," "south," "east," and "west," when used in a land description, mean, respectively, "due north," "due south," "due east," and "due west." *Plaquemines Oil & Development Co. v. State*, 23 So.2d 171, 208 La. 425 (1945).

EASTERN ONE-HALF. The term "eastern one-half" in a deed conveying one-half of a tract of land, in the absence of admissible parol evidence disclosing a different intention, would mean the eastern half, formed by a line to be run due north and south through the tract; but if it appears that before the deed was executed a division into two parts, supposedly equal in area, had been made by a line, having a different bearing, actually marked on the ground by stakes and fences, according to which possession had been held for a number of years, and the parties have since held possession according to such line, the words must be taken to mean the eastern one-half as so laid off and held in severalty. *Bank v. Catzen*, 60 S.E. 499, 63 W. Va. 535 (1908).

EAST HALF. A deed conveying the east half of certain irregularly shaped lots is presumed to mean the east half in quantity. There is no presumption that the parties intend that the tract conveyed shall be ascertained by the rule of subdivision adopted in government surveys; that is, by running a line equidistant from the opposite sides of the lot. *Cogan v. Cook*, 22 Minn. 137 (1875).

Where there is nothing to suggest the contrary, the word "half," in connection with the conveyance of a part of a tract of land is interpreted as meaning half in quantity. The words "east half" and "west half" in a deed, while naturally importing an equal division, may lose that effect when it appears that at the time some fixed line or known boundary or monument divides the premises somewhere near the center, so that the expression more properly refers to one of such parts than to a mathematical division which never has been made. *Gunn v. Brower*, 105 P. 702, 81 Kan. 242 (1909).

EAST HALF AND WEST HALF. The words "east half and west half" in a description in a deed naturally import an equal division, but they may lose that effect when it appears that at the time the deed was made some fixed boundary or monument divided the premises somewhere near the center, so that the words referred more properly to one of such parts than to a mathematical division which had never been made. *People v. Hall*, 88 N.Y.S. 276, 43 Misc. 117 (1904).

EASTWARDLY. "Eastwardly," as used in a grant, means due east, unless there be some object that can be found to control the course, in which case the course will run east, varying from that point to include the object. *Simms v. Dickson*, 3 Tenn. (1 Cooke) 137, 22 Fed. Cas. 158 (1812).

"Eastwardly," as used in a deed or grant of land to describe a call or a line to run eastwardly, is an indefinite expression, and means nothing more, necessarily, than that the land shall lie on the eastern and not on the western, side of a given line. It signifies on which side of the base of the lines marking the survey the land is to lie. *Preeble v. Vanhoozer*, 5 Ky. (2 Bibb.) 118 (1810).

ENCLOSE. The word "enclose" or "inclose" is defined as to surround, to encompass, to bound, fence, or hem in, on all sides. *White Chapel Memorial Ass'n v. Willson*, 244 N.W. 460, 260 Mich. 238 (1932).

ENCROACHMENT. An "encroachment" is a gradual entering on and taking possession by one of what is not his own; an unlawful gaining upon the rights of possession of another. *Meservey v. Gulliford*, 93 P. 780, 14 Idaho 133 (1908), citing *Chase v. City of Oshkosh*, 51 N.W. 560, 81 Wis. 313 (1892), 51 L.R.A. 553, 29 Am. St. Rep. 898.

ENCUMBRANCE. An "encumbrance" is a burden upon land, depreciative of its value, such as a lien, easement, or servitude, which, though adverse to interest of landowner, does not conflict with his conveyance of land in fee. *Moeller v. Good Hope Farms*, 215 P.2d 425, 35 Wash.2d 777 (1950).

An "encumbrance" is a right to or interest in land that may subsist in third persons to diminution of value of an estate of a tenant, but consistently with passing of fee. *Olcott v. Southworth*, 63 A.2d 189, 115 Vt. 421 (1948).

END. "End" as used in a tax judgment that is against an undivided third of the east end of each block, means the east half of the block. *Chiniquy v. People*, 78 Ill. 570 (1875).

The word "end" as used in a restriction in a deed providing that no building should be erected in the rear end of the lot conveyed within 10 feet from the line of the north side of the street, means the extremity, termination, limit. *Crofton v. St. Clement's Church*, 57 A. 570, 208 Pa . 209 (1904).

The word "end" is defined to be the extreme point of a line or anything that has more length than breadth. The end of a parallelogram is the line extending from one side line to the other at their extremities, and the width of the end is the length of such line. If the line connecting the extreme points of parallel side lines makes an angle with one greater than that made with the other, as, for instance, one being 10 and the other 170 degrees, it might not be proper to regard this line the width of the end of the figure presented, or even as the end itself. In figures having side lines irregular and not parallel with each other, a line connecting them where they terminate may be the end, and its length the width of the end, or otherwise, according to the peculiar shape of each figure. Hence the words "end" and "width of the end," are not terms of greater precision, and the meaning of parties who may use them without any words in explanation may not always be apprehended with certainty. *Kennebec Ferry Co. v. Bradstreet*, 28 Me. 374 (1848).

END LINE. The end and side lines of a lode mining claim are not necessarily those so designated by the locator, but the "end lines" are those which are crosswise of the general course of the discovery vein on the surface, although they may have been located as the side lines. *Northport Smelting & Refining Co. v. Lone Pine-Surprise Consol. Mines Co.*, D.C.Wash., 271 F. 105.

ENJOY. "Possess," used in reference to land titles and estates, means to own, have as a belonging, property; in popular usage, the word "possess" includes real and personal property to which one has title as his landed possessions. The word "enjoy" is one frequently found in instruments of transfer and means to make such use of the thing transferred as is consistent with the tenure by which it is held. "Transfer" is a word commonly used to denote a passing of title in property (usually realty) or an interest therein from one to another, the purpose of the law being to prevent ownership and legal interest in lands from passing to aliens who could never become citizens. *Exparte Okahara*, 216 P. 614, 191 Cal. 353 (1923).

The word "enjoy" means to have, possess, and use with satisfaction; to occupy or have the benefit of. A bequest of personal property to a person to hold, possess, and enjoy during her natural life gave such legatee the right to apply it in any way that would contribute to her well-being, and the fund was hers to expend for her personal benefit if she chose to do so. Her estate was a life estate with power of use. *Board of Trustees of Westminster College v. Dimmitt*, 87 S.W. 536, 113 Mo. App. 41 (1905), quoting and adopting definitions given in *Webster's Dictionary*.

Under statutes relating to possessory actions that give a right of action to a landowner who "possesses" an estate, or to one who "enjoys" a real right growing from the real estate, the word "enjoy" has the same meaning with reference to an owner of a real right that word "possess" has with reference to a landowner. The French word "*jouir*" means to enjoy, to have enjoyment of, or to possess. The word "enjoy" means to have, possess, occupy, or have the benefit of. The word "enjoyment" means the exercise of a right. The term "adverse enjoyment" means the use of an easement under a claim of right. *Allison v. Maroun*, 190 So. 408, 193 La. 286 (1939).

ELL. A measure of length, answering to the modern yard. *Black's Law Dictionary*.

ENTER; ENTRY. "Enter," when applied to the acquisition of lands from the government, means to purchase at the government land office. *Goodnow v. Wells*, 25 N.W. 864, 67 Iowa 654 (1885).

An "entry" is a description of the land entered for a survey for a patent, made in the office of the county surveyor by one having authority by virtue of a warrant or order of the county court, authorizing him to appropriate a certain quantity of vacant lands of the county. *Stephens v. Terry*, 198 S.W. 768, 178 Ky. 129 (1815).

"Entry" should be construed to mean a settlement on land with a view to purchase or homesteading. *St. Paul, M & M R Co. v. Greenhalgh*, C.C. Minn., 26 F. 563.

ESTADAL. In Spanish America, a measure of land of sixteen square varas, or yards. *Black's Law Dictionary* (1944).

ET AL. The Latin abbreviation "et al.," means "and another" or "and others." *In re McGovern's Estate*, 250 P. 812, 77 Mont. 182 (1926).

ETC. The character "etc." is used as an abbreviation of "et cetera," and et cetera means "other things." *Gray v. Central R. Co. of New Jersey, N.Y.*, 11 Hun. 70 (1877).

ETC.; &c. As commonly understood, is the abbreviation of et cetera, which means "and others," or "other things." Additional meanings have been "and so forth," "and the rest," and "and so on." Generally, the addition of this symbol is interpreted as relating to correlated matters and does not add more rights or anything more than that already described. 77 A.L.R. 879, *Content and Effect of Symbol or Abbreviation "&c." or "etc."* See also 80 Am Jur 2d Wills, § 1173.

EXACTLY. see **DUE.**

EXCEPT. The word "except" means "not including." *Austin v. Willis*, 8 So. 94, 90 Ala. 421 (1890).

EXCEPTING. Generally, a "reservation" in a deed is like an "exception," something to be deducted from the thing granted, narrowing and limiting what would otherwise pass by the general words of the grant, and the words "reserving" and "excepting," although strictly distinguishable, may be used interchangeably or indiscriminately, and the use of either term is not conclusive as to the nature of the provision. *Stephan v. Kentucky Valley Distilling Co.*, 122 S.W.2d 493, 275 Ky. 705 (1938).

EXCEPTING AND RESERVING. The phrase "excepting and reserving" is commonly used in deeds and is sometimes held to amount to an exception of part of the property that is the subject of conveyance, and sometimes to a reservation out of the estate conveyed, depending largely upon the intention of the parties, the subject matter of the grant, whether the thing excepted or reserved is a thing newly created out of the lands and tenements granted, or part of the property in existence and excepted therefrom. *Bardon v. O'Brien*, 120 N.W. 827, 140 Wis. 191, 133 Am. St. Rep. 1066 (1909).

EXTENDED LINE. The term "extended line," as used in the description in a deed, meant a produced line. *McAndres & Forbes Co. v. Camden Nat. Bank*, 94 A. 627, 87 N.J.L. 231 (1915).

EXTEND TO. That which "extends to" does not necessarily include in. *Martin v. Hunter*, 14 U.S. (1 Wheat.) 304, 4 L.Ed. 97 (1816).

FACING. In a deed containing building restrictions applicable to lots "facing" and "having a frontage" on named street, quoted words as applied to oblong lots referred to the street that buildings to be erected on the lots were intended to face. *Aller v. Berkeley Hall School Foundation*, 103 P.2d 1052, 40 Cal. App.2d 31 (1940).

FALDA. In Spanish law, the slope or skirt of a hill. *Fossat v. United States*, 2 Wall. 649, 17 L.Ed. 739 (1864).

FALSE WORDS. "False words," which may be eliminated from descriptions in wills, deeds, and so forth, are misdescriptions of property that are not applicable to any property owned or intended to be devised or conveyed. *Brown v. Ray*, 145 N.E. 676, 314 Ill. 570 (1924).

"False words" applied to descriptions and wills are misdescriptions of property devised or conveyed, not applicable to any property owned or intended to be devised, and, while oral evidence changing plain meaning of will is not admissible, it may be admitted to show latent ambiguities, put court in better position to understand will, and identify property described therein. *Armstrong v. Armstrong*, 158 N.E. 356, 327 Ill. 85 (1927).

FAMILY. The father, the mother, and the children ordinarily constitute a "family." *Daily v. Parker*, C.C.A. Ill., 152 F.2d 174.

The word "household" is synonymous with words "home" and "family." *Leteff v. Maryland Cas. Co.*, 91 So.2d 123 (La. App., 1956).

FARM. A testator, in devising his "farm at Bovingdon," meant all of the farm situated at that place, and not a part thereof. *Goodtitle v. Paul*, 2 Burrows (Eng.) 1089.

The term "farm" does not necessarily include only the land under cultivation and within a fence. It may include all the land that forms part of the tract, and may also include several connected parcels under one control, and, when one devises his "farm," he devises the whole farm, or all the land above designated. *Succession of Williams*, 61 So. 852, 132 La. 865 (1913).

FATHOM. When used in the measurement of land, the word "fathom" is to be understood as meaning a square fathom, for in its common usage it is an integral part of a unit of land measure. *Nahaolelua v. Kaaahu*, 9 Haw. 600 (1895).

FEE. The terms "fee," "fee-simple," and "fee simple absolute" are interchangeable and convey an estate the owner of which may exercise control as against all others. *Walpole v. State Board Com'rs*, 168 P. 848, 62 Colo. 554 (1917).

FENCE. A fence is a visible or tangible obstruction, which may be a hedge, ditch, wall, or frame of wood, or any line of obstacles interposed between two portions of land so as to part off and shut in the land, and set it off as private property, and such is its meaning in defining inclosed lands. *Kimball v. Carter*, 27 S.E. 823, 95 Va. 77, 38 L.R.A. 570 (1897).

FIELD. A field is cleared land for cultivation or other purposes, whether inclosed or not. *Commonwealth v. Wilson*, Va., 9 Leigh. 648 (1839).

FLAT LAND. Under allotment in partition proceeding of "5 acres of marsh meadow bounded by river," boundary of firm land by river carried with it adjacent flat land, "Meadow land" is firm and above high tide, and "flat land" is land between high and low-water mark. *Gibson v. Hoffman*, 164 A. 783, 310 Pa. 51 (1933).

FLATS. The word "flat," when used as descriptive of anything respecting an arm of the sea, means a level place over which the water stands or flows. *Church v. Meeker*, 34 Conn. 421 (1867).

FLOAT. A float, as applied to a grant of public land to be located within a certain tract or territory, whether of limited extent is a grant of quantity only within a larger tract, to be located by the consent of the government before it can attach to any specific land. *Hays v. Steiger*, 18 P. 670, 76 Cal. 555 (1888).

A grant of ten sections to the mile out of any swamp lands then belonging to, or that might thereafter belong to, the state, without any limitations or restrictions as to sections or locality, being a grant of a certain quantity out of a large quantity of land, is what is termed in land-grant law a "float." *Minneapolis & St. C.R. Co. v. Duluth & W.R. Co.,* 47 N.W. 464, 45 Minn. 104 (1890).

"Float" is the term used to designate the right of a railroad company to select lands within certain limits, to take the place of lands granted to the railroad company in its aid which have been lost by previous appropriation, which attaches to no specific tracts until the selection is actually made in the manner prescribed by law. *Elling v. Thexton*, 16 P. 931, 7 Mont. 330 (1888).

FLOWAGE RIGHT. A "flowage right," which is the right of one owner to flow the lands of another by maintenance of a dam is an "easement," which is a liberty, a privilege, or an advantage in lands without profit and distinct from an ownership. *Union Falls Power Co. v. Marinette County*, 298 N.W. 598, 238 Wis. 134 (1941), see 134 A.L.R. 958.

FOREVER. The words "forever" or "to the one and his assigns forever" add no force to a grant of, or estate in, lands. *Wollam v. Van Vleck*, 20 O.C.D. 743, 12 Cir.Ct.R., N.S. 517 (1892).

FORTY. By "forty," as used in connection with lands, is meant either the north or south half of a half of a quarter section of land. *Lente v. Clarke*, 1 So. 149, 22 Fla. 515 (1886).

In Florida, the expression "forty," as commonly used in the description of lands, means either the north or south half of a quarter section of land and usually contains about 40 acres. *Lente v. Clarke*, 1 So. 149, 22 Fla. 515 (1886).

FRACTIONAL. The words "fractional" and "part" are not synonymous. *Cates v. Cates*, 247 S.W. 780, 157 Ark. 181 (1923).

FRACTIONAL LOTS. Two lots on a plat were "fractional lots," where the irregular south boundary of the plat cuts the end line and the side line of one of the lots, leaving it roughly triangular in shape, and cuts off the corner of the other lot, leaving it in the shape of a parallelogram with one corner lopped off. *Miller v. Lavelle*, 110 N.W. 421, 130 Wis. 500 (1927).

FRACTIONAL SECTION. A section of land, as a legal subdivision under the congressional rules of survey, is a mile square, and usually contains 640 acres. When a section is not whole or regular in its contents—that is, where it does not contain approximately 640 acres—it may properly be called a "fractional section." *South Florida Farms Co. v. Goodno*, 94 So. 672, 84 Fla. 532 (1922).

FRACTIONAL TOWNSHIP. A fractional township is a township where the outer boundary lines cannot be carried out in full because of a water course or some other external interference. A township is not made fractional because, in laying it off in sections, there is an excess or deficiency to be carried into the western or northern ranges of half and quarter sections. *Goltermann v. Schiermeyer*, 19 S.W. 484, 111 Mo. 404 (1892).

FREE AND CLEAR. An agreement to convey land "free and clear" is satisfied by a conveyance passing a good title. *Meyer v. Madreperla*, 53 A. 477, 68 N.J.L. 258, 96 Am. St. Rep. 536 (1902).

FRONT. Although it is true that a corner lot does front on both streets, only that portion of the lot that is opposite the rear and faces upon the street is properly designated as the "front" of such lot. *Staley v. Mears*, 142 N.E.2d 835, 13 Ill. App. 2d 451 (1957).

FRONTAGE. "Frontage," within statute defining "business district" and "residence district," means space available for erection of buildings and does not include cross streets or space occupied by sidewalk or any ornamental spaces in plat between sidewalks and curb. Comp. Laws 1929, & 4693 (v,w). *Wallace v. Kramer*, 296 N.W. 838, 296 Mich. 680 (1941).

FRONTAGER. Frontager is a person owning or occupying land that abuts on a highway, river, seashore, or the like. See, 27 Corpus Juris.

FRONTING. Words "fronting" and "adjoining" are synonymous, and reference to lots "fronting" upon a street means that the lots touch boundary line of street. *Roach v. Soles*, D.C. Cal., 120 F.Supp. 400.

FRONTING AND ABUTTING. Very often, "fronting" signifies abutting, adjoining, or bordering on, depending largely on the context. *Rombauer v. Compton Heights Christian Church*, 40 S.W.2d, 545, 328 Mo. 1 (1931).

GENERAL DESCRIPTION. A reference to another deed for purpose of a more particular description is regarded as a "general description," within meaning of rule that a later specific description in a deed controls a prior general description. *Coffee v. Manly*, 166 S.W.2d 377 (Tex. Civ. App., 1942).

GLEBE. A "Glebe" is a grant of land belonging to or yielding revenue to a parish church. *In re Merritt's Will*, 61 N.Y.S.2d 537, 185 Misc. 979 (1945).

GOOD RECORD TITLE. A "good record title," without words of limitation, means that the proper records shall show an unincumbered, fee-simple title, the legal estate in fee, free and clear of all valid claims, liens, and incumbrances. *Riggins v. Post*, Tex., 172 S.W. 210 (1914).

GOOD TITLE. The term "good title" does not necessarily mean one perfect of record. *Block v. Ryan*, 4 App. Cas., D.C. 283.

"Good title" and "clear title" are synonymous terms, meaning that land should be free from litigation, palpable defects, and grave doubts and should consist of both legal and equitable title. *Clark v. Ray*, 96 S.W.2d 808 (Tex.Civ.App., 1936).

GORE. In old English law, a small, narrow slip of ground. In modern land law, a small triangular piece of land, such as may be left between surveys that do not close. In some of the New England states (such as Maine and Vermont) the term is applied to a subdivision of a county, having a scanty population and for that reason not organized as a town *Black's Law Dictionary*.

GORGE. A defile between hills or mountains, that is a narrow throat or outlet from a region or country. *Gibbs v. Williams*, 37 Am. Rep. 241, 25 Kan. 214 (1881).

GO TO. The words "be inherited by," in a will in which testator derives a residue of his estate during the devisee's natural life, and directs that at her decease such portion shall be inherited by her surviving issue, and so on, was constructed to be the equivalent of "go to," or "be received by." *Hill v. Giles*, 50 A. 758, 201 Pa. 215 (1902).

GRANT. The word "grant" in a deed imports a warranty in law during the lifetime of the grantor, and this warranty is not taken away by the insertion of an express warranty in the deed. *Hoxton's Lessee v. Gardiner*, Md., 1 Har. & M. 437 (1772).

A grant in its own nature amounts to an extinguishment of the right of the grantor and implies a contract not to reassert that right. A party is, therefore, always estopped by his own grant. *State of Illinois v. Illinois Cent. R. Co.,* 33 F. 721, citing *Fletcher v. Peck*, 10 U.S. 87, 6 Cranch 87, 3 L. Ed. 162 (1810).

"The words 'give, grant, sell, and convey' in a deed do not of themselves imply a warranty. Nor do they expressly and specifically convey the whole title, but are rather words of general description, susceptible of explanation or modification by other appropriate language. They are just as applicable to a conveyance of a right of redemption as to the grant of a fee." Thus their use in a deed, followed by a statement that the grantee intends to convey the same premises, and title as conveyed to him by another only, operates to pass such title, though it is only a right of redemption. *Bates v. Foster*, 8 Am.Rep. 406, 59 Me. 157 (1871).

"The term 'grant,' in a treaty, comprehends not only those which are made in form, but also any concession, warrant, order, or possession, to survey, possess, or settle, whether evidenced by writing or parol, or presumed from possession." *Bryan v. Kennett*, 5 S.Ct. 407, 113 U.S. 179, 28 L.Ed. 908 (1885); *Strother v. Lucas*, 37 U.S. 410, 12 Pet. 410, 9 L.Ed. 1137 (1838).

The word "grant" has two significations, it being often used technically to refer to lands in place, which are spoken of as granted lands in contradistinction to lands that are to be selected or indemnity lands, and then it is often used both in land legislation and opinions to refer to all lands the title to which has passed either as lands in place or by selection. *Barney v. Winona & St. P.R. Co.,* 24 F. 889.

A "grant" confirming state title to land has a technical meaning under Virginia and West Virginia law. A public "grant" means an instrument by which the state, as sovereign, passes to an individual title to land before vested in the state. A patent is only another name for a land "grant." The instrument is indifferently called a patent or "grant" in the Virginias. *State v. Harman*, 50 S.E. 828, 57 W.Va. 447 (1905).

"Sedo" is the word ordinarily used in all Mexican conveyances to pass title to lands. It is translated "I grant." *Mulford v. Le Franc*, 26 Cal. 88 (1864).

The Latin word "*concessio*," derived from the operative word in the Latin assurance, heretofore used in England, formerly was employed to designate that species of assurance, and the English word "concession," derived from the Latin word in its ordinary use, is exactly or nearly the equivalent of the word "grant," though the former is not now, in Virginia or West Virginia, generally used, as the latter is, with reference to the conveyance of land or transfer of title, right, or claim thereto. *Western Min. & Mfg. Co. v. Peytona Cannel Coal Co.,* 8 W.Va. 406 (1875).

"Survey," as used in a description in a trust deed conveying "the B survey, lying in what is known as the I. pasture, in C. and A. counties," is synonymous with the word "land," or "grant," or "location." *Clark v. Gregory*, 26 S.W. 244 (Tex., 1894).

A grant of lands by the government is equivalent to a deed in fee. *United States v. Northern Pac. R. Co.,* 12 P. 769, 6 Mont. 351 (1887).

GRANT, BARGAIN, AND SELL. Where the words, "grant, bargain, and sell," are used in a deed, without other covenants or words limiting their effect, they will amount to a covenant that the grantor has done no act and created no incumbrance whereby the estate granted can be defeated. *Finley v. Steele*, 23 Ill. 56 (1859).

GRANT TO A PUEBLO. A "grant to a pueblo," as the term was used in the Mexican law, as not a private land grant, in the sense that took title out of the state. It was the mere vesting in the pueblo of the use of the land in trust for the benefit of the inhabitants thereof, and with power as the representative of the state to make grants that should vest title in private ownership of "*solares*," or house lots, and "*suertes*," or sowing lots, to settlers, the remainder to remain vacant, to the end that gifts thereof might be made to new settlers. These "grants," as they are generally called, did not deprive the state itself, at any subsequent period, of making what are technically denominated "private land grants," vesting title in natural persons to any portion of the lands lying within the four leagues of the pueblo that had not already passed into private ownership, and this power on the part of the state was not unfrequently exercised. *United Land Ass'n v. Knight*, 24 P. 818, 85 Cal. 448 (1890).

GREAT ROADS. The term "great roads" is commonly used to designate main or principal roads or highways. *Ex parte Withers*, S.C., 3 Brev. 83 (1812); *State v. Mobley*, S.C., 1 McMul. 44 (1839).

GROUND OF RAILROAD. The phrase, "the ground of the railroad," used in describing a parcel of land in a mortgage as being "north of the ground of the railroad," evidently means its right of way. *Pence v. Armstrong*, 95 Ind. 191 (1883).

GUIA. In Spanish law, a right of way for narrow carts. *Black's Law Dictionary* (1944).

HALF. The word "half," in descriptions of premises covered by deed, is not to be understood literally if a different sense is indicated by the context, by accompanying circumstances, or by subsequent acts of the parties. *Jones v. Pashby*, 12 N.W. 884, 48 Mich. 634 (1882).

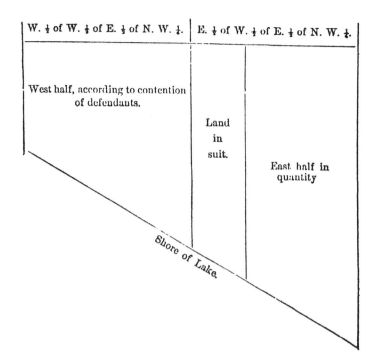

The words "the north half," used in the conveyance of a part of a platted block of land, mean the half of the block lying north of an east and west line drawn through the block, unless the surrounding facts require that these words be given a different meaning. *Lavis v. Wilcox*, 133 N.W. 563, 116 Minn. 187 (1911).

The word "half," when used in describing land, should be construed as meaning half in quantity, unless the context or surrounding facts and circumstances show a contrary intention. *Hoyne v. Schneider*, 27 P. P.2d 558, 138 Kan. 545 (1933).

"Half," as used in deeds of land according to government survey, ordinarily is used not with reference to the quantity, but with reference to a line that is equidistant from the boundary line of the parcel subdivided; but, where the government has divided a quarter of a section into what it calls "fractional 40s," the government division will govern. *Edinger v. Woodke*, 86 N.W. 397, 127 Mich. 41 (1901).

Where a deed recites that half of tract is being conveyed and there is nothing to suggest the contrary, the word "half" will be interpreted as meaning half in quantity. *McHenry v. Pence*, 212 P.2d 225, 168 Kan. 346 (1949).

"Half," when used in a conveyance that conveys the half of any particular piece of property, means an undivided half. *Baldwin v. Winslow*, 2 Minn. 213 (1858).

HALF SECTION. The general and proper acceptation of the terms "section" and "half section," as well as their construction by the general land department, denotes the land in the sections and subdivisional lines and not the exact quantity that a perfect admeasurement of an unobstructed surface would declare. *Brown v. Harden*, 21 Ark. 324 (1860).

HAND. The word "hand" as used in legal phraseology means signature or handwriting. *Kaufman v. Crist*, 124 A.2d 1, 11 Terry (Del.) 108 (1956).

HEAD OF CREEK. This term means the source of the longest branch, unless general reputation has given the appellation to another. *Davis v. Bryant*, 2 Bibb. (Ky.) 110 (1810).

HEAD OF STREAM. The highest point on the stream which furnishes a continuous stream of water, not necessarily the longest fork or prong. *Uhl v. Reynolds*, 64 S.W. 498, 23 Ky. Law Rep. 759; *State v. Coleman*, 13 N.J. Law 104 (1832).

HEADLAND. In old English law, a narrow piece of unplowed land left at the end of a plowed field for the turning of the plow. Call, also, "butt." *Black's Law Dictionary* (1944).

HEIRS. "Heirs," "heirs at law," and legal heirs are the same. *In re Fahnestock's Estate*, 50 N.E.2d 733, 384 Ill. 26 (1943).

Unless country intent appears from evidence, "heirs" will be construed as those who would take under statutes of descent. *Schaefer v. Merchants Nat. Bank of Cedar Rapids, Iowa*, 160 N.W.2d 318 (1968).

The word "heir" is used as meaning one to whom the property of a deceased person goes either by descent or by will. *Hessenmueller v. Sirilo*, Ohio, 34 O.C.D. 256, 23 Cir. Ct.R., N.S., 313 (1912).

In the word "heirs" is comprehended heirs of heirs in infinitum. *Meenan v. Meenen*, 308 P.2d 158, 180 Kan. 779 (1957).

The word "heirs" means those persons who inherit the property of a person at the time of his death. *Gorey v. Guarante*, 22 N.E.2d 99, 303 Mass. 569 (1939) .

HEIRS AND ASSIGNS. The words "heirs and assigns," used in a will devising property to a person for her natural life, "and to her heirs and assigns forever," are technical words signifying a fee. *Kendall v. Clapp*, 39 N.E. 773, 163 Mass. 69 (1895).

HEIRS APPARENT. "Heirs apparent" is a term that applies to those who will probably inherit from a live ancestor, *Ward v. Stow*, 17 N.C. 509, 27 Am. Dec. 238 (1834).

HEIRS OF THE BODY. Words "lawful issue" and "heirs of the body" mean lineal descendants by blood. *National City Bank of Cleveland v. Mitchell*, 234 N.E.2d 916, 13 Ohio App.2d 141 (1968).

HELD. The word "held" is sometimes used synonymously with "owned." *Taylor & Crate v. Asher*, 4 S.W.2d 385, 223 Ky. 574 (1928). However, the words "improved by" are not necessarily the same as "owned by" and may indicate a mere use of the land by one other than the owner. See **OWNED BY.**

IF. As used in a will, the word is usually construed as importing a condition or contingency. 80 Am Jur 2d Wills, § 1172.

IF EXTENDED. The words "if extended" as used in connection with description of a boundary, mean "if presently extended" and not "if extended some time in the future" or "in case of such extension." *Cedar Park Cemetery Ass'n v. Village of Calumet Park*, 75 N.E.2d 874, 398 Ill. 324 (1947).

INCLOSED LAND. Every man's land is, in the eye of the law, "inclosed" by a visible and material fence, or by an ideal, invisible boundary, and in every case every entry or breach constitutes a trespass carrying with it some damages for which compensation may be obtained by action. *Wood v. Snider*, 79 N.E. 858, 187 N.Y. 28, 12 L.R.A., N.S., 912 (1907).

INCLUDING SURVEY. As the term is used in reference to the public lands, it means a survey, for location purposes, of lands previously surveyed and appropriated by other persons. *Scott v. Yard*, 18 A. 359, 46 N.J. Eq. (1 Dick.) 79 (1889).

INCLUSIVE DEED. An "inclusive deed" is one which contains within the designated boundaries lands which are excepted from the operation of the deed. *Logan's Heirs v. Ward*, 52 S.E. 398, 58 W. Va. 366, 5 L.R.A., N.S., 156 (1905).

INCLUSIVE GRANT. An "inclusive grant" from the state is one that includes, within the boundaries designated in the patent, lands which are excepted from its operation. *Logan's Heirs v. Ward*, 52 S.E. 398, 58 W.Va. 366, 5 L.R.A., N.S., 156 (1905).

INCLUSIVE SURVEY. An "inclusive survey" is one that includes within its boundaries prior claims excepted from the computation of the area within such boundaries and excepted in the grant. *Stockton v. Morris*, 19 S.E. 531, 39 W.Va. 432 (1894).

INCOMPLETE TITLE. "Incomplete titles," in the sense that some French and Spanish titles required an act of confirmation by the government of the United States, were those which, while the terms of the granting act or concession sufficed to convey lands, or sufficient to evidence a present intention and purpose to convey lands, yet did not describe the particular land granted so as to identify and locate it, and no conveyance or act of location having yet (prior to the cession) been made. In other words, an incomplete title was one where the grant or concession or order of conveyance had been made, but the actual identification or location had yet to be done to distinguish and segregate the land intended to be conveyed from the mass of other land, and thus perfect the title. *Teddlie v. McNeely*, 29 So. 247, 104 La. 603 (1900).

INDEMNITY LANDS. Lands granted to railroads, in aid of their construction, being portions of the public domain, to be selected in lieu of other parcels embraced within the original grant, but which were lost to the railroad by previous disposition or by reservation for other purposes. See *Wisconsin Cent. R. Co. v. Price County*, 133 U.S. 496, 10 S. Ct. 341, 33 L. Ed. 687 (1889); *Barney v. Winona, & St. P. R.Co.*, 117 U.S. 228, 6 S. Ct. 654, 29 L. Ed. 858 (1886); *Altschul v. Clark*, 65 P. 991, 39 Or. 315 (1901). See **PLACE LANDS**.

INDIAN TITLE. The Indian title to land is the right to use, keep, and enjoy it. *Brown v. Belmarde*, 3 Kan. 41 (1864).

INHERIT. The word "inherit" is to take as an heir at law under the statutes of descent and distribution and not under a will and one who takes under a will does not inherit, strictly speaking. *Dickennson v. Buck*, 192 S.E. 748, 169 Va. 39 (1937).

INHERIT, INHERITED. While the word has been construed as meaning a taking by descent as distinguished from any other manner of acquisition, and this would appear to be its strict, technical meaning, there are instances where it is to be interpreted as meaning "take" or "go to," or may be specifically extended to include an acquisition of property by devise. 80 Am Jur 2d Wills, § 1165.

IN RANGE WITH. A line "to range" with another, from the end of which it begins, must follow the path of the other line when extended and be a continuation of it. *Lilly v. Marcum*, 283 S.W. 1059, 214 Ky. 514 (1926).

IN SQUARES TO THE CARDINAL POINTS. Where a contract calls for a survey of land "in squares to the cardinal points," without specifying whether the true or magnetic meridian shall control in determining the cardinal points, the magnetic meridian will be held to control. *Finnie v. Clay*, 5 Ky. 351, 2 Bibb. 351 (1811).

INTEREST. "Interest in land" is the legal concern of a person in the thing or property, or in the right to some of the benefits or uses from which the property is inseparable. *State Sav. Bank v. Bolton*, 273 N.W. 121, 223 Iowa 685 (1937).

The word "right" denotes among other things, "property," "interest," "power," "prerogative," "immunity," and "privilege," and in law is most frequently applied to property in its restricted sense. As an enforceable legal right, it means that which one has a legal right to do. *Shaw v. Proffitt*, 109 P. 584, 57 Or. 192 (1910).

An "interest" in land is the legal concern of a person in the thing or property, or in the right to some of the benefits or uses from which the property is inseparable. *Bahls v. Dean*, 170 N.W. 861, 222 Iowa 1291 (1937).

INTEREST IN LAND. "Interest in land" may refer to any one of a variety of estates or fractional shares in realty, including net equity or fractional property value possessed by owner over and above the sum total unpaid on outstanding incumbrances. *In re Rood's Estate*, 38 N.W.2d 70, 229 Minn. 73 (1949).

INTERPRETATION. Ascertaining testator's intent constitutes "interpretation" or will. *In re Carvalho's Will*, 57 N.Y.S.2d 307 (1944).

"Interpretation" of a statute is limited to exploration of the written text itself. *Bromley v. Mollnar*, 39 N.Y.S.2d 424, 179 Misc. 713 (1942).

The office of "interpretation" is to bring sense out of the words used and not bring sense into them. *State ex rel. McGannon v. Sayre*, 22 Ohio Dec. 234, 12 Ohio N.P., N.S., 13 (1911); *Slinguff v. Weaver*, 64 N.E. 574, 66 Ohio St. 621 (1902), 47 Wkly. Law Bull. 626.

To interpret a legal document, means first to collect the intent, to discover the writer's meaning, and then to ascertain that that meaning is expressed sufficiently. *In re Smith's Will*, 172 N.E. 499, 254 N.Y. 283 (1930).

"Interpretation" of document ascertains intention of author either from meaning of words employed, their combination, context, or even by conjecture, and serves purpose of rendering intelligible that which was ambiguous or not understood. *Hane v. Kintner*, 145 N.E. 326, 111 Ohio St. 297 (1925).

"Interpretation" is the act of making intelligible what was before not understood, ambiguous, or not obvious. It is the method by which the meaning of the language is ascertained. Resort to interpretation is never to be had where the meaning is free from doubt. *Ming v. Pratt*, 56 P. 279, 22 Mont. 262 (1899).

Interpretation is used for the purpose of ascertaining the true sense of any form of words, while construction involves the drawing of conclusions regarding subjects that are not always included in the direct expression. *Bloomer v. Todd*, 19 P. 135, 3 Wash, T. 599, 1 L.R.A. 111 (1888).

INTERSECTING METHOD. A method used in surveying for determining the center of a section of land, pursued by running straight lines from the quarter corner on the east to the quarter corner on the west, and from the quarter corner on the south to the quarter corner on the north, side of the section; the center being the points where these two lines cross. *Gerke v. Lucas*, 60 N.W. 538, 92 Iowa 79 (1864).

INTESTACY, INTESTATE. A person who dies "intestate" dies without having made a will. *Goff v. Moore*, 11 Ohio N.P., N.S., 543, 24 Ohio Dec. 552 (1911).

IN THAT CASE. The word "then" means, when used as a word of reasoning, "in that event," or "in that case," or "therefore." It also means "at that time," or "immediately afterwards." *Dudley v. Porter*, 16 Ga. 613 (1855).

IN THE DIRECTION OF THEIR LINES CONTINUED TO THE CHANNEL. As the rights granted by the riparian act of 1856 extend to "all lands covered by water lying in front of any tract of land . . . lying upon any navigable stream or bay of the sea or harbor, as far as the edge of the channel," and the right of action given the grantees to prevent encroachments extends to "all such submerged lands in the direction of their lines continued to the channel," this right of action extends to the space between lines drawn at right angles from the shore line "to the edge of the channel," where the channel runs parallel or practically so with the shore line. *Merrill-Stevens Co. v. Durkee*, 57 So. 428, 62 Fla. 549 (1912).

IN THE REAR THEREOF. "Rear" as used in a will by which the testator devised his message No. 90B. street, and the two stable lots " in the rear thereof," should be construed to apply to both the lots devised, though only one of them was directly behind No. 90; the other being behind No. 89. They are properly described as being in the rear of the testator's dwelling house lot. *Read v. Clarke*, 109 Mass. 82 (1871).

IN WHOLE TRACTS ONLY. Prior to the act of 1907, Laws 30th Leg. p. 490, c. 20, school lands could be sold in tracts of 80 acres or multiples thereof, but section 5, page 491, of such act, regulating purchases by lessees and their assignees, changed the rule, and required such sales to be made "in whole surveys only." Section 6, authorizing purchases by others, provided that no survey should be sold in any county except as a whole, though it might be leased in two or more tracts, and section 6b, page 492, provided for sales "in whole tracts only." Section 6e, page 494, declared that all surveys and unsold portions of surveys should be sold as a whole, that all unsurveyed tracts of 640 acres or less should be so sold, and that all tracts containing 100 acres or less, wheresoever situated, should be sold for cash and without condition of settlement. Held that, tracts of the public domain not yet sectionalized being referred to in the statutes as unsurveyed lands, the "surveys" should be construed to mean entire sections. *Ford v. Terrell*, 107 S.W. 40, 101 Tex. 327 (1908).

IRREGULAR HEIRS. Heirs under a testamentary succession are designated as "testamentary" or "instituted heirs," those under a legal succession, as "legal heirs" or "heirs of the blood," and those under an irregular succession, as "irregular heirs." *Succession of Wesley*, 69 So.2d 8, 224 La. 182 (1953).

ISLAND. An "island" is a body of water entirely surrounded by water. *Busch v. Wilgus*, 24 Ohio N.P., N.S. 209 (1922).

It is common knowledge that lands not in fact surrounded by water are sometimes called "islands." *Stewart v. United States*, 62 S.Ct. 1154, 316 U.S. 354, 86 L.Ed. 1529 (Cal., 1945).

ISSUE. "Issue" in commonly accepted sense means issue of the body, offspring, progeny, natural children, physically born or begotten by the person named as parent. *In re Miller's Trust*, 323 P.2d 885 (Mont., 1958).

IT IS SAID. Where a lot of land is conveyed by boundaries, and it is stated in the conveyance that the contract contains a certain number of acres, "it is said" the whole tract will pass, although it contains more than the specified number of acres. *Mann v. Pearson*, N.Y., 2 Johns, 37 (1806).

JNO. The sufficiency of an information was not affected by the abbreviating of the accused's name, "John," by use of the letters "Jno." *State v. Granger*, 102 S.W. 498, 203 Mo. 586 (1907).

JOINING. It is not necessary that property abut or have common boundary line to come within statute when words "joining" or "adjoining" are used. *Cecil v. Headly*, 373 S.W.2d 136, 237 Ark. 400 (1963).

JOINT OWNERSHIP. "Joint ownership" does not mean joint tenancy alone; partners and tenants in common are joint owners, though not joint tenants. *In re Huggin's Estate*, 125 A. 27, 96 N.J. Eq. 275 (1924).

J.P. The characters "J.P." indicate the office of Justice of the Peace. *Rowley v. Berrian,* 12 Ill. (2 Peck.) 198 (1850).

JR. The suffixes "Jr." or "Sr." are not part of a man's name and, except in a few instances, may be disregarded. *Ross v. Berry*, N.M., 124 P. 342, 17 N.M. 48 (1912).

KAHAKAI. The word "kahakai" is compounded of the words kaha and kai. "Kaha" means primarily a scratch or mark. "Kai" means the sea, or salt water. "Kahakai" then, means the mark of the sea, the junction or edge of the sea and land. *Halstead v. Gay*, 7 Haw. 587 (1884).

KULA LANDS. Where a lease, the third of a series between the same parties, conveyed all parts of "kula" lands not included in the previous leases, reserved numerous specific portions, lessor could not contend that "kula" land did not cover kuahiwi, pali, puu, mauna, auaawa or other portions of land to which various topographical terms might be applied, and hence lease would be construed as conveying the right of pasturage in kuahiwi land. *Coney v. Dowsett*, 3 Haw. 740 (1876).

KULEANA. A "kuleana" is a small Hawaiian homestead area. *U.S. v. 257.654 Acres of Land at Moanalua and Halawa, Oahu, Territory of Hawaii*, D.C. Hawaii, 72 F.Supp. 903.

The Hawaiian term "kuleana" means a small area of land, such as was awarded in fee by the Hawaiian monarch, about the year 1850, to all Hawaiians who made application therefor. *DeFries v. Scott*, C.C.A. Hawaii, 268 F. 952.

LAID OUT. "Laid out," as used in reference to a highway, has a well-known meaning, under our statutes, and plainly includes the doing of those things by the proper local officers that are essential in creating a public highway, to authorize it to be worked and traveled, and especially the surveying, marking the course or boundaries, and ordering it to be established as a highway. The affirmative action of the public authorities is indispensible in such case. *Chicago Anderson Pressed Brick Co. v. City of Chicago*, 28 N.E. 756, 138 Ill. 628 (1891).

A town site or addition is "laid out" when it is surveyed or measured and marked upon the ground. *Meachem v. City of Seattle*, 88 P. 628, 45 Wash. 380 (1907).

"Laid out," as used on the face of a map as laid out by a certain person, is equivalent to "as surveyed" by him, and embraces a reference to the monuments placed on the land by the surveyor. *Flint v. Long*, 41 P. 49, 12 Wash. 342 (1895).

LAMMAS LANDS. Lammas lands are lands that belong to a person who is absolutely the owner in fee simple, to all intents and purposes, for half the year, and the other half of the year he is still the owner in fee simple, subject to a right of pasturage over the lands by other people. See, 33 Corpus Juris.

LAND. "Land" ordinarily refers to all of the tracts conveyed by deed. *Saulsberry v. Maddix*, C.C.A. Ky., 125 F.2d 430.

The word "land" is not, in itself, an unambiguous term, and may include any estate or interest in land, either legal or equitable, easements, or incorporeal hereditaments. *Lalim v. Williams County,* 105 N.W.2d 339 (N.D., 1960).

Owner of "land" in fee has right to surface and to everything beneath and above it. *Southern Pac. Co. v. Riverside County*, 95 P.2d 688, 35 Cal.App.2d 380 (1939).

The shore is "land." *Stott v. Thompson*, 14 N.E.2d 246, 294 Ill.App. 450 (1938).

The word "lands" includes the beds of nonnavigable lakes and streams, and lands are none the less land for being covered with water. *State v. Jones*, 122 N.W. 241, 143 Iowa 398 (1909).

"Land" covered by water within the public domain of the United States is as much a part thereof as the dry land. *Kean v. Calumet Canal & Improvement Co.*, 23 S.Ct. 651, 190 U.S. 452, 47 L.Ed. 1134 (1903).

"Survey," as used in a description in a trust deed conveying "the B. survey, lying in what is known as the I. pasture, in C. and A. counties," is synonymous with the word "land," or "grant," or "location." *Clark v. Gregory*, 26 S.W. 244 (Tex., 1894).

"Lands," as used in a grant of land to colonists without including the lands "overflown by the swellings and currents of the river," meant tule or swamp lands. *The Sutter Case*, 69 U.S. (2 Wall.) 562, 17 L.Ed. 881 (1864).

"'Land,' in its most general sense, comprehends any ground, soil, or earth whatsoever, as meadow, pastures, woods, moors, waters, marshes, furzes and heaths," while "tenements" is a word of greater meaning and extent sometimes, and includes not only land, but rents, commons, and several other rights and interests issuing out of or concerning land. *Canfield & Chapman v. Ford*, N.Y., 28 Barb. 336 (1858).

LAND OF. The words "land of" in a description of land in a deed as being bounded westerly by the land of L., means land belonging to or owned by L. *Segar v. Babcock*, 26 A. 257, 18 R.I. 203 (1893).

A private right of way is an "easement" and is "land." *City of Missoula v. Mix*, Mont., 214 P.2d 212, 123 Mont. 365 (1950).

An easement of right of way is "land." *State ex rel. Polson Logging Co. v. Superior Court for Grays Harbor County*, 119 P.2d 694, 11 Wash.2d 545 (1941).

LAND LINE. The words "land line," as used in an instruction in an action of ejectment concerning a boundary, will be construed as synonymous with "boundary." *Henderson v. Dennis*, 53 N.E. 65, 177 Ill. 547 (1898).

LAND PREVIOUSLY SURVEYED. Patent excluding "all land previously surveyed" held to refer only to subsisting legal entries and surveys. *Bryant v. Meadors*, 210 S.W. 177, 183 Ky. 651 (1898).

LAND SERVICE ROAD. "Land-service road" is term used to describe ordinary road or highway that is intended primarily to enable abutting landowners to have access to outside world as distinguished from limited-access road that is a "traffic-service road" designed primarily to move through traffic. *State ex rel. State Highway Commission v. Vorhof-Duenke Co.*, 366 S.W.2d 329 (Mo., 1963).

LANDING. A place on a river or other navigable water for lading and unlading goods, or for the reception and delivery of passengers; the terminus of a road on a river or other navigable water, for the use of travelers, and the loading and unloading of goods. *State v. Randall*, 1 Strob. (S.C,) 111, 47 Am. Dec. 548 (1846).

A place for loading or unloading boats, but not a harbor for them. *Hays v. Briggs*, 74 Pa. 373 (1873).

A place laid out by a town as a common landing place and used as such, but not designated as for the particular benefit of the town, is a public landing place. *Black's Law Dictionary*.

LATENT AMBIGUITY. A "latent ambiguity" exists where face of the deed or instrument appears certain and without ambiguity, but there is some collateral matter out of the deed that breeds the ambiguity. *Weiss v. Soto*, 98 S.E.2d 727, 142 W.Va. 783 (1957).

Where deed was not ambiguous, but when its field notes were applied to ground an ambiguity arose, a "latent ambiguity" arose. *Burks v. Brinkley*, 161 S.W.2d 316 (Tex.Civ.App., 1942).

Where calls in deeds called for river as boundary, and there were two branches of river both of which carried considerable volume of water when deeds were executed, there was a "latent ambiguity." *Kuhn v. Chesapeake & O. Ry. Co.*, C.C.A.W.Va., 118 F.2d 400.

A "latent ambiguity" in description of property occurs where extrinsic evidence shows meaning to be uncertain, or where description is shown to fit different pieces of property. *People ex rel. Beedy v. Regnier*, 37 N.E.2d 186, 377 Ill. 562 (1941).

There is a latent ambiguity in the description of premises conveyed in a deed when it is necessary to resort to surrounding circumstances to determine the description intended. *Lego v. Medley*, 48 N.W. 375, 79 Wis. 211, 24 Am.St.Rep. 706 (1891).

A "latent ambiguity" arises where language used in description of land, appearing certain on its face, is shown to apply to different pieces of land, and extrinsic or parol evidence is then necessary to show which tract or parcel of land was intended. *Meinhardt v. White*, 107 S.W.2d 1061, 341 Mo. 446 (1937).

"Between a 'latent ambiguity' and a 'mistake or error in description' in a conveyance there is a manifest difference. The former may be explained, and the description aided by parol evidence in a court of law, while the other requires the jurisdiction of a court of equity for its correction. In the case of latent ambiguity in the description of land in a conveyance, the title is not thereby defeated, but parol evidence may be introduced to show the identity of the subject matter of the conveyance; in the case of an error or mistake in the conveyance, however, if such error or mistake is material to the description, no title passes, and the remedy of the purchaser is by bill in equity for a reformation of the instrument." *Donehoo v. Johnson*, 24 So. 888, 120 Ala. 438 (1898).

A "latent ambiguity" arises, not upon words of instrument as looked at in themselves, but upon those words when applied to object or to subject which they describe. *Hall v. Equity Life Assur.*, 295 N.W. 204, 295 Mich. 404 (1940).

Where there is no defect on the face of a will, but there is an uncertainty in attempting to put it into effect, the ambiguity is "latent." *Jennings v. Telbert*, 58 S.E. 420, 77 S.C. 454 (1907). see **PATENT AMBIGUITY.**

LATENT DEED. A "latent deed" is a deed kept for 20 years or more in a man's scrutoire or strong box. Such a deed, when unaccompanied with actual, distinctive, and adverse possession, is entitled to no consideration in a court of justice, and it is no ground for recovery in an action of ejectment against the actual possessor. *Wright v. Wright*, 7 N.J.L. (2 Halst.) 175, 11 Am.Dec. 546 (1824).

LATENT DEFECT. A defect that reasonably careful inspection will not reveal. *Schaff v. Ellison*, (Tex. Civ. App.) 255 S.W. 680. So, a latent defect in the title of a vendor of land is one not discoverable by inspection made with ordinary care. *Newell v. Turner*, 9 Port. (Ala.) 422 (1839).

LAWFUL HEIR. "Lawful heir" means heir capable of inheriting lands under laws of state. *Barry v. Rosenblatt*, 105 A. 609, 90 N.J.Eq. 1 (1918).

LAYING OUT ROADS. "Laying out roads" across unimproved property is the designation of boundary lines of such roads that, in such natural state, are nothing more than dirt covered by natural growth, including grass, brush, trees, and holes, within such boundary lines. *Pasco County v. Johnson*, 67 So.2d 639 (Fla., 1953).

LAY OUT. "To lay out a highway" means to locate the highway and to define its limits, which is accomplished by a survey and particular description of the highway. *Hough v. City of Bridgeport*, 18 A. 102, 57 Conn. 290 (1889).

LAYOUT A HIGHWAY. To "lay out a highway" is to locate it and define its limits. *U.S. v. Certain Parcels of Land in Riverside County,* Cal, 67 F.Supp. 780 (1946).

LEAGUE. A league of land, as used in Texas land grants, contains 4,428 acres. *Hunter v. Morses's Heirs*, 49 Tex. 219 (1878).

The jurisdiction of nations extends into the oceans to one marine league from low-water mark on the shore; a "league" in English speaking countries estimated at 3 miles. *Bolmer v. Edsall*, 106 A. 646, 90 N.J.Eq. 299 (1919).

The Mexican "league" applicable to grants of lands in the "Neutral Ground," east of the Sabine, from 1790 to 1800, being a square of 5,000 varas on each side, has always been estimated at 4,428.4 acres, the "vara" being considered 33 1/3 American inches. *U.S. v. Perot*, 98 U.S. 428, 25 L.Ed. 251 (1878).

The term "leagues" as used in a grant of four leagues of land in the province of Texas December 27, 1795, means Spanish leagues, and not American or English leagues. The old legal league, by the laws of Spain, and which was adopted in Mexico, consisted of 5,000 varas; and a vara, in Texas, is equivalent to 33 1/3 English inches; making the league equal to a little more than 2.63 miles, and the square league equal to 4,428.4 acres. *U.S. v. Perot*, 98 U.S. 428, 25 L.Ed. 251 (1878).

LEGAL NAME. A legal name under the common law consists of one Christian name and one surname, and the insertion, omission, or mistake in middle name or initial is immaterial. *Langley v. Zurich General Acc. & Liability Ins. Co.,* 275 P. 963, 97 Cal. App. 434 (1929).

LIEU LANDS. "Lieu lands" is the name applied to lands in a grant of public lands in aid of a railroad, which are to be selected by the company to take the place of lands within the limits of the grant and designated therein, which have been previously appropriated by settlers or for other purposes. The right to such lieu lands attaches to no specific tract until the selection is actually made by the company in the manner prescribed. *Elling v. Thexton*, 16 P. 931, 7 Mont. 330 (1888).

LINE. A "line" in surveying and dividing grounds, means, prima facie, a mathematical line, without breadth; yet this theoretic idea of a line may be explained, by the facts referred to and connected with the division, to mean a wall, a ditch, a crooked fence, or a hedge—a line having breadth. *Baker v. Talbott*, 22 Ky. 179, 6 T.B. Mon. 179 (1827).

A testator who, intending to devise part of his farm, begins at one of the corners of it and says, "thence as the line runs," is to be understood to mean the line of the farm, whether it be straight or crooked. A crooked line is a line just as much as a straight one. We say the line of a state, a county, a coast, or of the seashore, though it have a great many bendings, and, with equal propriety, a line of a fence, a road, or a farm. *Cubberly v. Cubberly*, 12 N.J.L. 308 (1831).

That the description of a survey called for the marked "line" of another survey between designated points which was not a straight line but consisted of four lines and three marked corners was immaterial, the word "line" being often used for the plural, and vice versa, and the singular being also sufficient to describe the exterior boundary of the survey line called for between the designated points. *Bell v. Powers*, 121 S.W. 991 (Ky., 1909).

LINEAGE. "Lineage" is defined as the line of descent from an ancestor, hence, family, race, stock. *In re Herrick's Estate*, 273 N.Y.S. 803, 152 Misc. 9 (1934).

LINE OF RAILROAD. "On the line of a railroad," as used in a deed describing the land as lying "on the line of a certain railroad," means next to or bounded by the railroad. The line of railroad must mean, as used in the contract, the boundary of the land appropriated for its use. *Burnam v. Banks*, 45 Mo. 349 (1870).

LITTLE MORE THAN. The terms "more or less," "a little more than," "not quite," "not more than," and "approximately" are terms of safety and precaution and ordinarily mean "about" when used in a deed. *Pierce v. Lefort*, 200 So. 801, 197 La. 1 (1941).

LOCATE. The words "locate" and "establish" are not synonymous. *Givens v. Woodward*, 207 S.W.2d 234 (Tex.Civ.App., 1947).

"Locate" means to designate the site or place of; to define the location or limits of, as by a survey; as, to locate a public building, a mining claim; to locate the land granted by a land warrant. *Delaware, L. & W.R.Co. v. Chiara*, C.C.A.N.J., 95 F.2d 663.

Vote, taken under G.L. 4066, by town to cause division line between towns to be "resurveyed" instead of to be "located," word "locate" being used in statute, was equivalent to vote to have line "located," and court had jurisdiction of petition to have commissioners appointed to locate line under section 4067, since "locate" means to ascertain place in which something belongs, or to locate calls in deed or survey, and word "resurvey", as used in vote taken, means to locate calls in whatever deed, charters, grants, or surveys are material and relevant to matter in dispute. *Town of Underhall v. Town of Jericho*, 140 A. 156, 101 Vt. 41 (1927).

LOCATE A ROAD. Strictly speaking, to "locate a road" is to build it, because until it is completed it is not a road. When a subscription book speaks of locating a road, everybody understands it to mean less than a finished road—a mere line surveyed and established as the locus where the road is to be built. It must be supposed and held that it was the location of the road, and not its structure, that was intended to be specified. *Warner v. Callender*, 20 Ohio St. 190 (1870).

LOCATIVE CALLS. In harmonizing conflicting calls in a deed or survey of public lands, courts will ascertain which calls are locative and which are merely directory, and conform the lines to the locative calls; "directory calls" being those that merely direct the neighborhood where the different calls may be found, whereas "locative calls" are those that serve to fix boundaries. *Cates v. Reynolds*, 228 S.W. 695, 143 Tenn. 667 (1921).

LOT. "Lot" has been defined as any portion, piece, or division of land. *Lehmann v. Revell*, 188 N.E. 531, 554 Ill. 262 (1933).

Word "lot" when used unqualifiedly, means lot in a township as duly laid out by the original proprietors. *Carney v. Dunn*, 252 S.W.2d 827, 221 Ark. 223 (1952).

"Town lots, out lots, common field lots and commons were known and recognized parts of the Spanish town or commune of St. Louis. They existed by public authority, whether by concession, custom or permission." *Vaquez v. Ewing*, 42 Mo. 247 (1868).

A "lot" is a portion of land that has been set off or allotted, whether great or small, but in common use it means simply a piece, parcel, or tract of land without regard to size and does not necessarily connect itself with buildings but may be anywhere on the earth's surface. *Schack v. Trimble*, 137 A.2d 22, 48 N.J.Super. 45 (1957).

The term "lot" applies to any portion, piece, or division of land and is not limited to parcels of land laid out into blocks and lots regularly numbered and platted. *Westbrook v. Rhodes*, 218 P. 873, 92 Okl. 149 (1923).

The words "lot," "piece," and "parcel" apply peculiarly to the land itself and are never employed to describe improvements. *Canty v. Staley*, 123 P. 252, 162 Cal. 379 (1911).

Term "lot," as used in connection with urban property, is frequently defined as subdivision of block, according to plat or survey of town or city. *Lehmann v. Revell*, 188 N.E. 531, 354 Ill. 262 (1933).

LYING IN THE SOUTHWEST CORNER OF A SECTION. A description of a tract of land as "lying in the southwest corner of a section" is sufficiently definite, according to the rules of decision that a "corner" is a base point from which two sides of the land conveyed shall extend an equal distance so as to include by parallel lines the quantity conveyed. *Walsh's Lessee v. Ringer*, 2 Ohio 327, 15 Am. Dec. 555 (1826).

MADE LAND. "Made land" is a term applied to land reclaimed from the waters of a lake by filling out into the lake. *Carli v. Stillwater St. Ry. & Transfer Co.,* 10 N.W. 205, 28 Minn. 373, 41 Am. Rep. 290 (1881).

MAP. A "map" is a picture of a survey, "field notes" constitute a description thereof, and the "survey" is the substance and consists of the actual acts of the surveyor, and, if existing established monuments are on the ground evidencing such acts, such monuments control because they are the best evidence of what surveyor actually did in making the survey and are part at least of what surveyor did. *Outlaw v. Gulf Oil Corp.,* 137 S.W.2d 787 (Tex.Civ.App., 1940).

A map is a drawing upon a plane surface representing a part of the earth's surface, and the relative position of objects thereon. It may also be so drawn as to show the geological structure and other physical facts necessary to a complete understanding of the matter at issue. *Montana Ore Purchasing Co. v. Boston & M. Consol. C. & S. Min. Co.,* 70 P. 1114, 27 Mont. 288 (1902).

A "map" that showed monuments that did not exist, and courses and distances that were erroneous was equivalent to no map. *Rowe v. Luddington*, 51 Conn. 184 (1883).

A map is but a transcript of the region that it portrays, narrowed in compass, so as to facilitate an understanding of the original. It may be said to be an abstract of the original in the only way that the subject is susceptible of being condensed and abridged. Citing *Banker v. Caldwell*, 3 Minn. 94, 103 (Gil. 46) (1859); *Jackson v. Freer*, N.Y., 17 Johns. 29 (1819). So that the making of a map of an addition implies that the addition had been surveyed, and that such survey was marked on the ground so that the streets, blocks, and lots can be identified. *Burke v. McCowen*, 47 P. 367, 115 Cal. 481 (1896).

MARINE LEAGUE. A "marine league" is equivalent to three geographical miles, or three sea miles. *Rockland, Mr. D. & S. S. Co. v. Fessenden*, 8 A. 550, 79 Me. 140 (1887).

Generally, the test applied in determining territorial rights over the sea is to fix the limit as a "marine league," or three miles from the coast. *State v. Ruvido*, 15 A.2d 293, 137 Me. 102 (1940).

MARINE MILES. "The log was invented about the same time which inaugurated measuring of the sea or marine miles, known as 'English geographical' miles. The sea mile, knot, geographical, or marine mile measures 6,086.7 feet on the sea, on the scale of 60 geographical or sea miles to a degree." In the Muscongus Grant by the council of Plymouth in Devon, England, made between 1620 and 1635, after the inauguration of the geographical or "marine mile," granting certain land, and so forth, within three miles of the main land, the three-mile limit is to be measured by the geographical or marine mile or knot, and not by the statute mile. *Lazell v. Boardman*, 69 A. 97, 103 Me. 292, 13 Ann. Cas. 673, citing *Rockland, Mt.D. & S.S.Co. v. Fessenden*, 8 A. 550, 79 Me. 140 (1887).

MARKED CORNERS. "Marked corner" (i.e., those clearly identified, and which are notorious objects) are the most satisfactory evidence of the location of a patent. *Morgan v. Renfro*, 99 S.W. 311, 124 Ky. 314 (1907).

MARKED LINE. Where bearing trees are called for at the eastern end of a northern boundary line of a survey and a stake at the western terminus, there is a presumption that the line was actually surveyed, and the corners identified by the bearing trees and the stake making the line a "marked line." *Goodson v. Fitzgerald*, 135 S.W. 696 (Tex., 1905).

MARKETABLE. Word "marketable," in an agreement for sale of land calling for a title that should be "marketable" and free and clear of all encumbrances, means saleable. *Gravino v. Gralia*, 86 A.2d 827, 18 N.J. Super. 241 (1952).

MARKETABLE TITLE. A marketable title to land is a title that is, in fact, good or one that can be made so by application of the purchase money, although it may not so appear of record. *Davis v. Buery*, 114 S.E. 773, 134 Va. 322 (1922).

Title may be "good and sufficient title," within contract of sale, and yet not "marketable title." *Texas Auto Co. v. Arbetter*, Tex., 1 S.W.2d 334 (1927).

A "marketable title" has been described as follows: It must in any event embrace the entire estate or interest sold, and that free from the lien of all burdens, charges, or incumbrances that present doubtful question of law or fact. *Frank v. Murphy*, 29 N.E.2d 41, 64 Ohio App. 501 (1940).

The terms "merchantable title," "marketable title," "clear title," "perfect title," and "good title," as used in contract for sale of realty, are synonymous. *Siedel v. Snider*, 44 N.W.2d 687, 241 Iowa 1227 (1950).

Where a contract for the sale of land warranted a fee simple, but a building on the land encroached one inch on an adjacent lot to which the vendor had no title, it was held that the conveyance of the lot owned by vendor would not give a clear title to the building and the lot on which it was situated, and that the title was not a marketable one. *Stevenson v. Fox*, 57 N.Y.S. 1094, 40 App.Div. 354 (1899).

"Marketable title" implies fee-simple title to whole of real estate, free from reservations and incumbrances. *Herman v. Engstron*, Iowa, 214 N.W. 588, 204 Iowa 341 (1927).

A "marketable title" is one free from reasonable doubt in law or fact as to its validity. *Winkler v. Neilinger*, 14 So.2d 403, 153 Fla. 288 (1943).

To be "marketable," within the meaning of that term as ordinarily understood, a title must be a clear record title, and title by adverse possession is not a marketable title, however perfect it may be. *Mays v. Blair*, 179 S.W. 331, 120 Ark. 69 (1915).

MARK OF PUNCTUATION. Punctuation marks are marks of punctuation, within the rule that the punctuation of an act or its title is not controlling in construing it for the purpose of ascertaining its real meaning. *State v. Banfield*, 72 P. 1093, 43 Or. 287 (1903).

MARY. The proper names Polly and "Mary" are sometimes used as synonyms, or as equivalents; Poll and Polly as familiar forms of "Mary." *Norton v. Jones*, 90 So. 854, 83 Fla. 81 (1922).

The proper name "Mary" is synonymous with "Mollie," the latter being a diminutive of the former. The two names are really the same and constitute only one name. *State v. Watson*, 1 P. 770, 30 Kan. 281 (1883).

MAUKA. The Hawaiian word "*mauka*" means toward the mountain, or away from the sea. *DeFries v. Scott*, C.C.A. Hawaii, 268 F. 952.

MEADOW. Term "meadows" is applied to tracts of land that lie above the shore and are overflowed by spring and extraordinary tides only, and yield grasses that are good for hay. *Church v. Meeker*, 34 Conn. 421 (1867).

The term "meadow" included salt marshes and beaches. *Sandiford v. Town of Hampstead*, 90 N.Y.S. 76, 97 App.Div. 163 (1904).

MEADOW LAND. Under allotment in partition proceeding of "5 acres of marsh meadow bounded by river," boundary of firm land by river carried with it adjacent flat land. "Meadow land" is firm land above high tide, and "flat land" is land between high and low-water mark. *Gibson v. Hoffman*, 164 A. 783, 310 Pa. 51 (1933).

MEANDER. "Meander" as used in a boundary generally relating to a river, means to follow a winding or flexuous course. Thus, when it is said "thence with the meander of the river," and so forth, it must mean a meandered line, which is a line that follows the sinuosities of the river, or in other words, that the river is the boundary of the land claimed between the points indicated. *Turner v. Parker*, 12 P. 495, 14 Or. 340 (1886).

MEANDER CORNERS. A so-called "meander corner" is not fixed point for measurements, as are section and quarter corners, but is a marker for courses. *Thunder Lake Lumber Co. v. Carpenter*, 200 N.W. 302, 184 Wis. 580 (1924).

"Meander corner" holds, as declared by the rules of the United States Land Office, the peculiar position of denoting a point on line between landowners without usually being the legal terminus or corner of the land owned. Where meander corners of a government survey are lost or obliterated, they must be restored in accordance with the circulars of the United States Land Office. *Kleven v. Gunderson*, 104 N.W. 4, 95 Minn. 246 (1905).

MEANDER LINE. A "meander line" generally contains a call for a natural object or monument that will usually control over calls for course and distance. *State v. Arnim*, 173 S.W.2d 503 (Tex.Civ.App., 1943).

A "meander line" is described by courses and distances, and the line is thus fixed by reason of difficulty of surveying a course following the sinuosities of the shore, and the impracticability of establishing a fixed boundary along shifting sands of the ocean. *Den v. Spalding*, 104 P.2d 81, 39 Cal.App.2d 623 (1940).

As a rule, "meander lines" relate to sinuosities and course of stream and do not constitute boundaries in absence of proof that clearly indicates a contrary intention. *Cox v. City & County of Dallas Levee Imp. Dist.*, 258 S.W.2d 851 (Tex.Civ.App., 1953).

The meander line run by the United States surveyors along a stream is not the boundary of the lands, but is run merely to determine the quantity of lands contained in the lots and the purchasers of the lots take to the center of the stream, and not merely to the meander line. *Jones v. Pettibone*, 2 Wis. 225 (1853).

METES. The word "metes" refers to the exact length of each line and the exact quantity of land in square feet, rods, or acres. *U.S. v. 5.324 Acres of Land*, D.C.Cal., 79 F.Supp. 748.

METES AND BOUNDS. Where land was bounded in the deed by the lands of named persons, it was described by metes and bounds, which mean the boundary lines or limits of a tract. *Moore v. Walsh*, 93 A. 355, 37 R.I. 436 (1915).

"Metes and bounds" are the boundary lines of land, with their terminal points and angles. *Lefler v. City of Dallas*, Tex.Civ.App., 177 S.W.2d 231 (1943).

MEXICAN LEAGUE. A Mexican league is not the same as the American league. The old legal league by the laws of Spain, and that was adopted in Mexico, consisted of 5,000 varas, and a vara, in Texas, has always been regarded as equivalent to 33 1/3 English inches, making the league equal to a little more than 2.65 miles, and the square league equal to 4,428.4 acres. *U.S. v. Perot*, 98 U.S. 428, 25 L.ED. 251 (1878).

MEXICAN PUEBLO. Mexican pueblo is a settlement or town under the control of the Mexican government. *City of San Francisco v. Le Roy*, 11 S.Ct. 364, 138 U.S. 656, 34 L.Ed. 1096 (1890).

MEXICAN SHORE. In a case involving the construction of a deed fixing a boundary as the shore of the bay of San Francisco, the court mentions the fact that counsel used the term "Mexican shore" to designate the shore line known to the civil law, which is the law of Mexico, which is the line of extraordinary high tides. *Valentine v. Sloss*, 37 P. 326, 103 Cal . 215 (1894).

MIDDLE. A call for a county boundary line as running west to the middle of a section calls for it to run to the center of the section measured from north to south as well as from east to west. *Alluvial Realty Co. v. Himmelbsrger-Harrlson Lumber Co.*, 229 S.W. 757, 287 Mo. 299 (1921).

The phrases "middle of the river" and "middle of the main channel" are equivalent expressions, and both mean the main line of the channel or the middle thread of the current. *Western Union Tel. Co. of Illinois v. Louisville 8 N.R. Co.,* 110 N.E. 583, 270 I11. 399 (1915), Ann.Cas. 19178, 670.

MILE. "The Arabs in the north of Africa consider it a mile when so far as not to be able to distinguish a man from a woman." *U.S. v. New Bedford Bridge*, 27 Fed.Cas. 91.

"The statute, or English, 'mile' was adopted as the standard of land measurement in the thirty-fifth year of the reign of Queen Elizabeth, 1593. The statute 'mile' measures 5,280 feet on the land." In the Muscongus Grant by the council of Plymouth in Devon, England, made between 1620 and 1635, after the inauguration of the geographical or marine mile, granting certain land, and so forth, within three miles of the mainland, the three-mile limit is to be measured by the geographical or marine mile or knot, and not by the statute "mile." *Lazell v. Boardman*, 69 A. 97, 103 Me. 292 (1908), 13 Ann.Cas. 673, citing *Rockland, M.D. & S. Steamboat Co. v. Fessenden*, 8 A. 550, 79 Me. 140 (1887).

MILITARY BOUNTY LANDS. A bounty is a grant of money or lands to persons doing military or naval services for the government. Congress, under the Constitution, has on numbers of times granted land and money to soldiers, which habitually and repeatedly have been called "bounty"—such as "bounty of three months' pay and 160 acres of land," "military bounty land," "bounty in money and land," "money bounty," "bounty of 160 acres of land," "bounty in land," "bounty right." *State of Iowa v. McFarland*, 4 S.Ct. 210, 110 U.S. 471, 28 L.Ed. 198 (1884).

MILL PRIVILEGE. The term "mill privilege" means the land and water used with the mill, and on which it and its appendages stand. *Moore v. Fletcher*, 16 Me.(4 Shep.) 63, 33 Am.Dec. 633 (1839).

The conveyance of a mill privilege will operate to convey the land occupied for the purpose, unless there be in the conveyance something indicating a different intention. *Farrar v. Cooper*, 34 Me. 394 (1852).

The right of a riparian proprietor to erect a mill on his land and to use the power furnished by the stream for the purpose of operating the mill, with due regard to the rights of other owners above and below him on the stream. *Gould v. Boston Duck Co.,* 13 Gray (Mass.) 452 (1859); *Hutchinson v. Chase*, 63 Am. Dec. 645, 39 Me. 511 (1855); *Moore v. Fletcher*, 33 Am. Dec. 633, 16 Me. 65 (1839); *Whitney v. Wheeler Cotton Mills*, 24 N.E. 774, 151 Mass. 396, 7 L.R.A. 613 (1890); *Rome Ry. & Light Co. v. Loeb*, 80 S.E. 785, 141 Ga. 202 (1914), Ann. Cas. 1915C, 1023.

MILL SITE. The grant of a "mill site," and so forth should be construed to include a water power, together with the right to maintain a dam wherever such dam would be suitable for the convenient and beneficial appropriation of the water power. *Stackpole v. Curtis*, 32 Me. 383 (1851).

An exception of a "mill site" in a grant or lease should be construed to operate as an exception of the soil of the mill site, and so much land as is necessary for the mill pond and erecting and carrying on the business of the mill. *Hasbrouch v. Vermilyea*, N.Y., 6 Cow. 677 (1827); *Burr v. Mills*, N.Y., 21 Wend. 290 (1839).

It is held that the term "mill site," in a conveyance or grant, embraces the right that the law gives the owner to erect a mill thereon, and to hold up or let out the water, at the will of the occupant, for the purpose of operating the same in a reasonable or beneficial manner. It is said that it must of necessity include a reasonable amount of fall below the mill for the purpose of letting the water flow off without obstruction, and that it is not reasonable to confine the right of the mill owner to the exact amount of the fall, to the fractional part or an inch, that will enable the water to flow away from his wheels without obstruction. *Occum Co. v. A. & W. Sprague Mfg. Co.*, 35 Conn. 496 (1868).

MINE. Mines are land, and subject to the same laws of possession and conveyance. *Byers v. Byers*, 38 A. 1027, 183 Pa. 509, 39 L.R.A. 537, 63 Am.St.Rep. 765 (1898).

The term "mine," as used in mining law, is synonymous in its meaning with the term "vein" or "lode." *Bullion, Beck, & Champion Min. Co. v. Eureka Hill Min. Co.*, 11 P. 515, 5 Utah 3 (1886).

Oil is a "mineral," and process of extracting it from rocks is "mining," and therefore oil well is a "mine." *Rice Oil Co. v. Toole County*, 284 P. 145, 86 Mont. 427 (1930).

MINERAL ACRES. It is the practice in oil fields to refer to ownership of percentage of minerals in tract of land as so many "acres of minerals" or "mineral acres." *Edwards v. Carter Oil Co.*, 288 S.W.2d 954, 226 Ark. 215 (1956).

MINERAL AND MINING RIGHTS. A reservation of "mineral and mining rights" in a deed does not include petroleum and natural gas, in the absence of a clear intent by the parties to so include them. *Preston v. South Penn Oil Co.*, 86 A. 203, 238 Pa. 301 (1913).

MINERAL DEED. A "mineral deed" is a realty conveyance involving a severance from fee of present title to minerals in place, either effecting such severance in first instance or conveying part of such mineral ownership previously severed from the fee. *Hickey v. Dirks*, 133 P.2d 107, 156 Kan. 326 (1943).

MINERAL RIGHTS. A reservation of "mineral rights" is effective to withhold oil and gas. *Missouri Pac. R. Co. v. Strohacker*, 152 S.W.2d 557, 202 Ark, 645 (1941).

A "mineral right" is right or title to all, or to certain specified, minerals in given tract. *Northwestern Imp. Co. v. Morton County*, 47 N.W.2d 543, 78 N.D. 29 (1951).

MINING CLAIM. A mining claim includes the soil, rocks, and works beneath the surface and machinery fixed thereto as well as the surface. *Mammoth Min. Co. v. Juab County*, 37 P. 348, 10 Utah 232 (1894).

A mining claim perfected under the law is property in the highest sense of that term, which may be bought, sold, and conveyed, and will pass by descent. *Sullivan v. Iron Silver Min. Co.*, 12 S.M. 555, 143 U.S. 431, 36 L.Ed. 214 (1892).

The term "mining claim" means in the mining country a portion of the public mineral lands of the United States to which a qualified person may first obtain the right of occupancy and possession by means of location, and, secondly, title by pursuing prescribed methods therefor. *Gray v. New Mexico Pumic Stone Co.*, 110 P. 603, 15 N.M. 478 (1910).

MOIETY. The half of anything. Joint tenants are said to hold by moieties. *Young v. Smithers*, 205 S.W. 949, 181 Ky. 847 (1918).

MONEY BOUNTY. See **MILITARY BOUNTY.**

MONUMENT. "Monument" may mean anything by which the memory of a person, thing, idea, art, science, or event is portrayed or perpetuated. *Odom v. Langston*, 195 S.W.2d 466, 355 Mo. 115 (1946).

"Monuments" are permanent landmarks established for the purpose of indicating boundaries. *Thompson v. Hill*, 73 S.E. 640, 137 Ga. 308 (1912).

Monuments are the visible marks or indications left on natural or other objects indicating the lines and boundaries of a survey. *Grier v. Pennsylvania Coal Co.*, 18 A. 480, 128 Pa. 79 (1889).

Where circumstances warrant, fences may be considered as "monuments." *Rodgers v. Roseville Gold Dredging Co.*, 286 P.2d 536, 135 C.A.2d 6 (1955).

Land or an adjoining proprietor is a "monument" within the rule that monuments govern measurements of land. *Di Maio v. Ranaldi*, 142 A. 145, 49 R.I. 204 (1928).

Center of section is not physical government monument, but is point capable of being mathematically ascertained, thus constituting it, in legal sense, "monument call of description." *Matthews v. Parker*, 299 P. 354, 163 Wash. 10 (1931).

MORE OR LESS. The terms "more or less," "a little more than," "not quite," "not more than," and "approximately" are terms of safety and precaution and ordinarily mean "about" when used in a deed. *Pierce v. Lefort*, 200 So. 801, 197 La. 1 (1941).

The use of the phrase "more or less" in describing a boundary line relieves a stated distance of exactness, means the parties are to risk the quantity of land conveyed, and implies waiver of the warranty as to a specified quantity. *Salyer v. Poulos*, 122 S.W.2d 996, 276 Ky. 143 (1939).

The words "more or less," used in a contract of sale in connection with an estimated quantity, when the only measure is the estimate itself, allow only a small latitude of variation. *U.S. v. Republic Bag & Paper Co.*, C.C.A.N.Y., 250 F. 79.

The words "more or less" following a statement in a deed as to the number of acres conveyed is a matter of description only. *Maffet v. Schaar*, 131 P. 589, 89 Kan. 403 (1913).

The words "more or less," used to describe a lot of land, mean merely that the lot conveyed may be in size more or less than the dimension given, but they cannot be so extended as to include a separate and distinct lot. *McCune v. Hull*, 24 Mo. 570 (1857).

The words "more or less" usually mean "about," "substantially," or "approximately," and imply that both parties assume the risk of any ordinary discrepancy. *Alexander v. Hicks*, 5 So.2d 782, 242 Ala. 243 (1942).

The words "more or less," as used in a deed declaring that the land conveyed contained a specified number of acres, more or less, did not mean as estimated, as supposed, but should be construed to mean about the specified number of acres, and are designated to cover only such small errors of surveying as usually occur in surveys. *Crislip v. Cain*, 19 W.Va. 438 (1882).

MOUNTAIN. "Mountain," in the general acceptation, is used to denote the situation, and not the quality, of land. *Lord Kildare v. Fisher*, 1 Strange (Eng.), 71.

MUNIMENT OF TITLE. "Muniments of title" is a general expression having reference to deeds and other written evidence of title, but includes all means of evidence by which an owner may defend title to property. *Lyle Cushion Co. v. McKendrick*, 87 So.2d 283, 227 Miss. 894 (1956).

N. The letter "N." in conveyances, maps, charts, and other instruments, is commonly used as an abbreviation for the word "north." *Burr v. Broadway Ins. Co.*, 16 N.Y. 267 (1857).

The Supreme Court will take judicial notice that "N." and "S." when applied to directions are properly read north and south, and that "N. to S. end of W. avenue" means north and south end of W. avenue. *Village of Bradley v. New York Cent. R. Co.*, 129 N.E. 744, 296 Ill. 383 (1921).

NAME. At common law, a legal "name" consisted of a given name and of a surname or family name. *In re Conde*, 61 A.2d 198, 137 N.J.L. 589 (1948).

N.D. N.D. is an abbreviation for "no date," and also for "northern district." See, 45 Corpus Juris.

N.E. "N.E." is an abbreviation of "northeast" in constant and universal use. *Sexton v. Appleyard*, 34 Wis. 235 (1874).

NEAR. The term "near" cannot be used as the equivalent of exact distances. *Picharella v. Ovens Transfer Co.*, 5 A.2d 408, 135 Pa. Super. 112 (1939).

NEAR RELATIVES. The term "near relatives," in a will, held to mean next of kin. *Cox v. Wills*, 22 A. 794, 49 N.J.Eq. (4 Dick.) 130 (1891).

NORTHEAST CORNER. Where description in deed was that location of square tract of land should be "in northeast corner" of M.'s league of land, description did not require that land should include extreme northeast portion of survey, where beginning point was described as northeast corner of M.'s headright league of land. *Harper v. Temple Lumber Co.*, 290 S.W. 530 (Tex., 1927).

NORTHEASTERLY. Where the terms "northerly," "northwesterly," "northeasterly," and so forth are employed to designate a line in the description of a deed, such terms will be construed as equivalent to a call to run due north, due northwest, or northeast, as the case may be, in the absence of a call for visible monuments, or any other description of a line which locates it with reasonable certainty. *Irwin v. Towne*, 42 Cal. 326 (1871).

NORTHERLY. The term "northerly" used in describing boundary implies only a general direction. *Fosburgh v. Sando*, 166 P.2d 850, 24 Wash.2d 586 (1946).

Where a course in a deed is described as running northerly, the word "northerly" must be construed as meaning due north. *Proctor v. Andover*, 42 N.H. 348 (1861).

NORTHERNLY. "The term 'northernly' in a grant, where there is no object mentioned to direct the inclination of the course toward the east or west, means due north." *State ex rel. Chandler v. Huff*, 79 S.W. 1010, 105 Mo.App. 354 (1904), quoting and adopting definition in *Brandt ex dem. Walton v. Ogden*, N.Y., 1 Johns. 156 (1806).

NORTH HALF. The words "the north half," used in the conveyance of a part of a platted block of land, mean the half of the block lying north of an east and west line drawn through the block, unless the surrounding facts require that these words be given a different meaning. *Lavis v. Wilcox*, 133 N.W. 563, 116 Minn. 187 (1911).

Under a conveyance of the "north half" of a tract of land, the east line of which is of such a shape that the north line is less than one-half the length of the south line, the grantee is entitled to one-half of the area of the tract, and not merely to one-half of the north and south length of the tract. *Robinson v. Taylor*, 123 P. 444, 68 Wash. 351 (1912), Ann.Cas. 1913E, 1011.

NORTH ONE-THIRD. The description of land as the "north one-third of lots five and six" in block 77 in itself indicates a single tract. *La Selle v. Nicholls*, 76 N.W. 870, 56 Neb. 458 (1898).

NORTH PART. A deed reciting that a tract of land conveyed the "north part" of a certain lot cannot be construed to mean the north half of the lot. *Langohr v. Smith*, 81 Ind. 495 (1882).

Under deed conveying one of two adjoining lots and reserving right of way over the "south part" of the lot conveyed, the term the "south part" constituted description in contradistinction to "north part" and did not suggest the idea of a "middle part" which is no specific part but any part that is embraced in any two lines between north and south or east and west boundaries that are parallel with and equidistant from the middle lines, while the "south part" supposes a "north part," the "south part" being all that lies south, and the "north part" being all that is north of the middle line east and west. *Roberts v. Stephens*, 40 Ill.App. 138 (1891).

NORTH SIDE. A deed describing the land conveyed as the north side of the southwest quarter of a certain block should be construed to mean the north half of such quarter. *Winslow v. Cooper*, 104 Ill. 235 (1882).

NORTHWARD. In construing a patent granting five great plains, "together with the woodland around such plains, that is to say, four English miles from the said plains eastward, four English miles northward from the said plains, four English miles westward from the said plains, and four English miles southward from the said plains," the court said: "The given object to start from is the plains; the distance to run is four miles; the courses are northward, southward, eastward, and westward; and it is a settled rule of construction that, when courses are thus given, you must run due north, south, east, and west." *Jackson v. Reeves*, N.Y., 3 Caines, 293 (1805).

NORTHWARDLY. The term "northwardly" means toward or approaching toward the north, rather than toward any of the other cardinal points. *Martt v. McBrayer*, 166 S.W.2d 823, 292 Ky. 479 (1942).

The word "north," as distinguished from the word "northwardly," conveys a definite idea, that is, indicates a particular cardinal point—while the word "northwardly" means towards or approaching towards the north, rather than towards any of the other cardinal points. *Craig v. Hawkins' Heirs*, 4 Ky. (1 Bibb) 53 (1808).

NORTHWEST. Where the base of a description in a deed was parallel of north latitude, a line called to run northwest for quantity, to adjoin a claim on the north, should not be construed to mean running north or at right angles to the base given, but the survey should be projected northwest for quantity. *Swearingen v. Smith*, 4 Ky. (1 Bibb) 92 (1809).

NORTH WITH THE HALF SECTION LINE. The words "north with the half section line" in a deed denoted nearness or in the same direction as the half section line and not necessarily upon or along such line. *Puntt v. Simmer*, 19 O.C.D. 721, 8 Cir.Ct.R., N.S., 455.

NOT MORE THAN. "More or less" are words of safety and precaution and when used in deed are intended to cover some slight inaccuracy in frontage, depth, or quantity in land conveyed, and ordinarily means "about" the same as terms "a little more than," "not quite," "not more than," or "approximately," and all are often introduced into a description practically without effect. *Harries v. Harang*, La.App., 23 So.2d 786 (1945).

NOT QUITE. See **NOT MORE THAN; MORE OR LESS.**

NOW, PRESENT, ETC. The primary question whether the term should be interpreted as to when the will was written or at the time of death, and whether the testator intended to include after-acquired property. It is necessary to examine other terms of the will and the rules in existence at the time. 80 Am Jur 2d Wills, § 1170.

OHIO. "Ohio," as used in descriptions of the surveys of the townships of land in Indiana bordering on the Ohio, means the Ohio River. *City of Madison v. Hildreth*, 2 Ind. (2 Cart.) 274 (1850).

OLD. "Old," as used in 1 Rev.St. p. 502, 9 2, authorizing commissioners of highways to lay out new roads and discontinue old ones, does not necessarily mean an ancient or long existing road. The phrase "new road" means a road newly laid out where one was not, and the words "old road" are opposite thereto, and mean one laid out and used, whether long ago or of more recent date. *People v. Griswold*, 67 N.Y. 59 (1876).

As the opposite of the term "new" is "old," the term as used in Act Cong. May 26, 1824, providing that the rights of the proprietor and occupant of certain lots should be postponed to the Spanish grantee who had obtained a "new" grant or order of survey for the same, and so forth, has no reference to time alone by reference to an epoch after which grants should be considered new, but that "new" grant was put in contradistinction to an "old" grant for the same tract of land. *Pollard's Heirs v. Kibbe*, Ala., 9 Port. 712 (1840).

OLOGRAPHIC. "Olographic" is defined as a term that signifies that an instrument is wholly written by the party. *Lovskog v. American Nat. Red Cross*, C.C.A. Alaska, 111 F.2d 88 (1940).

OLOGRAPHIC WILL. An "olographic will" is one entirely written, dated, and signed by the testator. *Succession of Vicknair*, 126 So.2d 680 (La.App., 1961).

ON. Courts will take judicial notice that surveyors generally in making field note calls such as "with a marked line" use word "with" as having same meaning as word "on." *Carter v. Texas Co.*, 87 S.W.2d 1079, 126 Tex. 388 (1935).

ON SAID WALL. The words "on said wall" in a description reading "thence southerly on said wall ten rods," show that course of line is controlled by course of wall. *Vermont Marble Co. v. Eastman*, 101 A. 151, 91 Vt. 425 (1915).

ONE-FOURTH. The conveyance of one-fourth of a tract of land, without designating by metes and bounds or otherwise locating the part conveyed, vests in the grantee, and those claiming under him, the title to one undivided fourth of the whole tract, as tenant in common with the grantor. *McCaul v. Kilpatrick*, 46 Mo. 434 (1870).

ONE LEAGUE SQUARE. The fact that a tract of land is described as one league square refers only to contents, and not to shape. "A tract of land of one square league" does not, as a term of description, suggest any boundary whatever. *Muse v. Arlington Hotel Co.,* 68 F. 637.

ONROERENDE AND VAST STAAT. The phrase "*onroerende and vast staat*" is by a statute of the colony of New York passed October, 1810, translated from English "immovable and fast estate," and therefore, in Dutch wills, deeds, and antenuptial contracts, the phrase should be construed to mean land or real estate. *Spraker v. Van Alstyne*, N.Y., 18 Wend. 200 (1837).

OPEN LOT. A lot bounded on one side by a street and on another by a court and upon a third side by a street, alley or court way, is not "an open lot" within city building ordinance, an "open lot" being one bounded upon all sides by streets. *Illinois Surety Co. v. O'Brien*, C.C.A. Ohio, 223 F. 933.

OPEN MINE. A mine lawfully leased to be opened is an "open mine." *Graham v. Smith*, 196 S.E. 600, 170 Va. 246 (1938).

OPPOSITE. Where a deed describes a boundary as ending at a point at one side of a street "opposite" a point on the other side, a straight line between the two points must cross the street at a right angle. *Bradley v. Wilson*, 58 Me. 357 (1870).

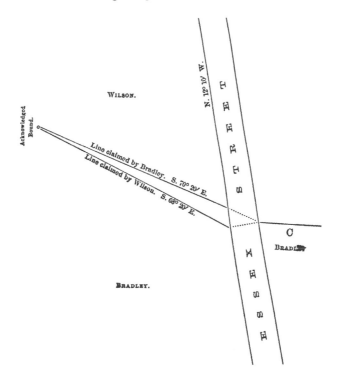

OR. Use of the word "or" in deed conveying premises to two named persons, or survivor, their heirs, executors, and administrators, did not require different construction than would have been required if word "and" had been used. *Rowerdink v. Carothers*, 54 N.W.2d 115, 366 Mich. 454 (1952).

ORAL PARTITION. A "partition" is a mere division and not a conveyance of land, and an "oral partition" merely confers upon the holder an equitable title to the specific portion allotted to him, the legal title remaining as before. *Ebner v. Nall*, 127 S.W.2d 506 (Tex.Civ.App., 1939).

ORDINARY GRANT. By an ordinary grant there is a transfer of title or estate or ownership from one to another, and the grantor, having parted with what he had, can give nothing by a second deed. *Western Elec. Co. v. Sperry Elec. Co., Ill.,* 59 F. 295, 8 C.C.A. 129.

ORIGINAL PLAT. The term "original plat" is ordinarily used to distinguish the first plat of a town from the subsequent additions, and "original town" is employed in the same way. *State v. City of Victoria*, 156 P. 705, 97 Kan. 638. (1916).

OUT-BOUNDARIES. The term "out-boundaries" was used in early Mexican land laws to designate certain boundaries within which grants of a smaller tract, which designated such out-boundaries might be located by the grantees. *U.S. v. Maxwell Land-Grant O., Colo.,* 7 S.Ct. 1015, 121 U.S. 325, 30 L.Ed. 949 (1887).

OUTLOT. In early American land law, (particularly in Missouri), a lot or parcel of land lying outside the corporate limits of a town or village but subject to its municipal jurisdiction or control. See *Kissell v. St. Louis Public Schools*, 16 Mo. 592 (1852); *St. Louis v. Toney*, 21 Mo. 243 (1855); *Eberle v. St. Louis Public Schools*, 11 Mo. 265 (1848).

"Town lots, outlots, common field lots, and commons were known and recognized parts of the Spanish town or commune of St. Louis. They existed by public authority, whether by concession, custom, or permission. It has been settled that the existence of the facts necessary to constitute an outlot is a matter of fact for the jury to determine, but that what facts will constitute an outlot is a question of law for the court." *Vasquez v. Ewing*, 42 Mo. 247 (1868).

OWN. "Owns" means to have a good title; to hold as property; to have a legal or rightful title to; to have; to possess. *Hendley v. Perry*, 47 S.E.2d 480, 229 N.C. 15 (1943).

The word "own," standing alone in deed and unqualified by reference to a lesser estate, means a complete ownership of the land in fee simple, since the word "own" is a synonym of "possess" and means to have a good legal title to. *Shewell v. Board of Ed. of Goshen Union Local School Dist.,* 96 N.E.2d 323, 88 Ohio App. 1, 43 O.O. 375 (1951).

OWNED AND OCCUPIED. Where a description in a deed calls for land "owned and occupied," the actual line of occupation is a material call to be considered in locating the lines of the land bounded therein. *Fahey v. Marsh*, 40 Mich. 239 (1879); *Cronin v. Gore*, 38 Mich. 386 (1878).

OWNED BY. The words "owned by," in a deed, mean an absolute and unqualified title, whereas the words "belonging to" do not import that whole title is meant, since a thing may belong to one who has less than an unqualified and absolute right. *Shewell v. Board of Ed. of Goshen Union Local School Dist.,* 96 N.E.2d 523, 88 Ohio App. 1, 42 O.O. 375 (1951).

OWNERSHIP IN LAND. The term "ownership in land" involves the idea of rights in some particular person or persons to use the land according to his or their pleasure, and under such concept one does not own the land but rather an estate in the land or the rights of possession, present and future. *Kristel v. Steinberg,* 69 N.Y.S.2d 476, 188 Misc. 500 (1947).

PACE. A measure of length containing two feet and a half, being the ordinary length of a step. The geometrical pace is five feet long, being the length of two steps, or the whole space passed over by the same foot from one step to another. *Black's Law Dictionary* (1944).

PAPER STREET. A street appearing on the recorded plat, but that has never been opened, nor prepared for use, nor used as a street, is known as a "paper street." *Raiolo v. Northern Pac.R.Co.,* 122 N.W. 489, 108 Minn. 431 (1909).

PARALLEL. Geometrical meaning of "parallel" is line evenly everywhere in same direction but never meeting, however far extended, and in all parts equally distant, but word also connotes with like direction or tendency, and running side by side. *Valente v. Atlantic City Elec. Co.,* 101 A.2d 106, 28 N.J.Super. 476 (1953).

By mathematical definition, parallel lines are straight lines, but, in common speech about boundaries, the words are often used to represent liens that are not straight, but photographs of each other; and courts, in passing on questions of boundaries, often use them in the latter sense. The term is used for the want of a better, and not because it in all respects affects the use to which it is applied. It is so used to avoid circumlocution, and, while such use is not exactly correct, there is no difficulty in understanding the meaning of intention. *Fratt v. Woodward,* 91 Am. Dec. 573, 32 Cal. 219 (1867).

A recital that a street runs "parallel or nearly so "with another street intimates that the streets are not absolutely parallel with one another. *Rehfuss v. Hill,* 90 N.E. 187, 243 Ill.140 (1909).

PARALLEL WITH. "Equidistant," as used in a conveyance of lots providing that the front line of all buildings thereon shall be placed equidistant from and not less than 8 feet back from the street, should be construed to mean "parallel with." *Smith v. Bradley,* 28 N.E. 14, 154 Mass. 227 (1891).

PARCEL. A part or portion of land. See *State v. Jordan,* , 17 S. 742, 36 Fla. 1(1895); *Miller v. Burke,* 6 Daly (N.Y.) 174; *Johnson v. Sirret,* 46 N.E. 1035, 153 N.Y. 51 (1897); *Chicago, M. & St. P. Ry. Co. v. Town of Churdan,* 195 N.W. 996, 196 Iowa 1057 (1923).

A part of an estate. *Martin v. Cole,* 38 Iowa 141 (1874).

It may be synonymous with lot. *Terre Haute v. Mack,* 38 N.E. 468, 139 Ind. 99 (1894).

A "parcel of land" or "parcel of real property" means a contiguous quantity of land in possession of, or owned by, or recorded as the property of, the same claimant, person, or company. *State v. Jordan,* 17 So. 742, 36 Fla. 1 (1895).

The terms "tract" and "parcel" may properly be applied to a quarter section, a half section, or a section of land. *People ex rel. Chicago General Ry. Co. v. Chase*, 70 Ill.App. 42 (1897).

The words "lot," "piece," and "parcel" apply peculiarly to the land itself, and are never employed to describe improvements. *Canty v. Staley*, 123 P. 252, 162 Cal. 379 (1911).

PARCEL OF LAND. "The terms 'tract or lot' and 'piece or parcel of real property,' or 'piece or parcel of land,' mean any contiguous quantity of land in the possession of, owned by, or recorded as the property of the same claimant, person, or company." In the connection the word "contiguous" means land that touches on the sides. Hence two quarters of the same section, which only touch at the corner, do not constitute, for the purpose of taxation, one tract or parcel of land. *Griffin v. Denison Land Co.*, 119 N.W. 1041, 18 N.D. 246 (1908).

PARTITION. The object of partition is a division of an estate between the legal owners of it, regardless of equitable claims. *Longley v. Longley*, 42 A. 798, 92 Me. 395 (1899).

PASSAGE. The word "passage" means a way, road, path, route, channel, an entrance or exit, or means of passing, and a "passageway" affords passage. *Wright v. Cherry*, 284 S.W.2d 273 (Tex. Civ. App., 1955).

PASSING CALL. Where lines of surveys of porciones running north from Rio Grande river, which was a shifting stream, began on bank of river, fact that surveys called for a designated road on both boundary lines tended to show that road was regarded as of considerable locative value, and therefore consideration would be given to calls for the road, even though they were "passing" or "incidental calls" and not strictly "locative calls." *Davenport v. Bass, Com. App.*, 153 S.W.2d 471, 137 Tex. 248 (1941).

PATENT. A "patent" is the instrument by which the government, whether state or national, passes its title. *Hayner v. Stanly*, 13 F. 217.

A "patent" is only another name for a land grant. *State v. Harman*, 50 S.E. 828, 57 W.Va. 447 (1905).

A "patent" for public land is the government deed from the premises. *Jordan v. Smith*, 73 P. 308, 12 Okl. 703 (1903).

A patent of the government to land vests in the patentee the perfect legal title, which relates back to the entry of the land. *Rankin v. Miller*, 43 Iowa 11 (1876).

A "patent" to land transfers to person in whose name it issues only the right the state possesses, and it does not prevail against any prior existing legal title. *Gray v. Gray*, 16 A.2d 166, 178 Md. 566 (1940).

PATENT AMBIGUITY. A "patent ambiguity" is one that appears upon face of the instrument, and that must be removed by construction according to settled legal principles and not by evidence. *Perigo v. Perigo*, 64 N.W.2d 789, 158 Neb. 733 (1954).

A "patent ambiguity" is an inherent uncertainty appearing on the face of the instrument, and arises at once on the reading of the instrument. *Harney v. Wirtz*, 152 N.W. 803, 30 N.D. 292 (1915).

A patent ambiguity is one that remains uncertain after all the evidence of surrounding circumstances and collateral facts admissible under the proper rules of evidence is exhausted. *McRoberts v. McArthur*, 64 N.W. 903, 62 Minn. 310 (1895); *Kretschmer v. Hard*, 32 P. 418, 18 Colo. 223 (1893).

A patent ambiguity is one that appears on the face of the instrument, and that which occurs when the expression of an instrument is so defective that a court of law which is obliged to put a construction on it, placing itself in the situation of the parties, cannot ascertain what they meant. *Grimes' Ex'rs v. Harmon*, 9 Am. Rep. 690, 35 Ind. 198 (1871).

An ambiguity resulting from blanks and omissions in document is "patent ambiguity." *Lincoln County v. Bruesch*, 254 P.2d 690, 197 Or. 571 (1953).

Ambiguity in a written agreement is "latent" where written language is apparently clear and certain but becomes doubtful in light of something extrinsic or collateral and it is "patent" when language itself is doubtful or susceptible to more than one meaning. *Stoffel v. Stoffel*, 41 N.W.2d 16, 241 Iowa 427 (1950), 14 A.L.R.2d 891.

A "patent ambiguity" is one apparent upon the face of the instrument, arising by reason of inconsistency, obscurity, or an inherent uncertainty of the language adopted, such that the effect of the words in the connection used is either to convey no definite meaning or a double one, while a "latent ambiguity" is one that arises, not upon the words of an instrument, as looked at in themselves, but upon those words when applied to the object or to the subject that they describe. *Zilwaukee Tp. v. Saginaw Bay City Ry. Co.*, 181 N.W. 37, 213 Mich. 61 (1921).

A "latent ambiguity," as distinguished from a "patent ambiguity," can only be developed by extrinsic and collateral circumstances. *Logan v. Wiley*, 55 A.2d 366, 357 Pa. 547 (1947).

A "patent ambiguity" in a will is one apparent upon the face of the instrument, while a "latent ambiguity" is discoverable only when language of will is sought to be applied to beneficiaries of property disposed of. *Hall's Adm'r v. Compton*, 281 S.W.2d 906 (Ky., 1905).

See **LATENT AMBIGUITY.**

PATH. "'Path' is constantly used in our old acts as synonymous with 'road,' and until within a few years it was the custom for large portions of the population of this state to use the term 'path' to the exclusion of 'road.'" *Singleton v. Commissioner of Roads*, S.C., 2 Nott 8 McC. 526 (1820).

PERAMBULATION. The act or custom of walking over the boundaries of a district or piece of land, either for the purpose of determining them or of preserving evidence of them. *Black's Law Dictionary* (1944).

"Perambulation," as used in Act 1872, providing for the perambulation of towns, or, as it might be more correctly called, a circumambulation, is the custom of going around the boundaries of the manor or parish with witnesses to determine and preserve recollection of its extent, and to see that no encroachment had been made upon it, and that the landmarks have not been taken away. It is a proceeding commonly regulated by the steward, who takes with him a few men and several boys, who are required to particularly observe the boundary lines traced out, and thereby qualify themselves for witnesses in the event of any dispute about the landmarks or extent of the manor at a future day. *Town of Greenville v. Town of Mason*, 57 N.H. 385 (1873).

PERCH. A perch of Paris is eighteen feet. *Sullivan v. Richardson*, 14 So. 692, 33 Fla. 1 (1884).

PERFECT AND SATISFACTORY TITLE. The term "perfect and satisfactory title," as used in a contract for the sale of land, is such a title as is free from grave and reasonable objection, and such title as would satisfy a person of ordinary prudence who is capable of passing upon the title. *Smith v. Lander*, 106 S.W. 703 (Tex., 1907).

PERFECT GRANT. There is a "perfect grant" of public lands where absolute title takes effect in praesenti. *Sena v. American Turquoise Co.*, 98 P. 170, 14 N.M. 511 (1908).

PERFECT TITLE. A "perfect title" is the possession, right of possession, and right of property. *A.D. Graham & Co. v. Pennsylvania Turnpike Commission*, 33 A.2d 22, 347 Pa. 622 (1943).

The terms "merchantable title," "marketable title," "clear title," "perfect title," and "good title," as used in contract for sale of realty, are synonymous. *Siedel v. Snider*, 44 N.W.2d 687, 241 Iowa 1227 (1950).

To constitute a perfect title there must be union of possession, right of possession, and right of property. *Donovan v. Pitcher*, 25 Am.Rep. 634, 53 Ala. 411 (1875).

PERIODICAL OVERFLOW. "Periodical overflow", as applied to lands, is not a representation that they are "swamp and overflowed lands." *Tubbs v. Wilhoit*, 14 P. 361, 73 Cal. 61 (1887).

PIECE OR PARCEL OF LAND. "The terms 'tract or lot' and 'piece or parcel of real property,' or 'piece or parcel of land,' mean any contiguous quantity of land in the possession of, owned by, or recorded as the property of the same claimant, person, or company." In the connection the word "contiguous" means land that touches on the sides. Hence two quarters of the same section, which only touch at the corner, do not constitute, for the purpose of taxation, one tract or parcel of land. *Griffin v. Denison Land Co.*, 119 N.W. 1041, 18 N.D. 246 (1908).

PITCH. By a vote of the original proprietors of the township, a committee was appointed "to run out the common lands, showing all the lands that have been pitched, and all the common land that is "left," and authorizing the committee "to lay out all metes and bounds to each person that has made his pitch, 100 acres to each right, and to make a return accordingly." "Pitch," as here used, means the selection of the general location of a claim or share by entry and occupation. *Garland v. Rollins*, 36 N.H. 349 (1858).

PLACE LANDS. Lands granted in aid of a railroad company that are within certain limits on each side of the road, and that become instantly fixed by the adoption of the line of the road. There is a well-defined difference between place lands and "indemnity lands." See *Jackson v. La Moure County*, 46 N.W. 449, 1 N.D. 238 (1890). See **INDEMNITY LANDS**.

PLACER CLAIM. The term "placer claim" means ground within definite boundaries that contains mineral or valuable mineral deposits not in place, while the term "lode" or "vein," used in the statute relating to lode claims, means lines or aggregation of mineral imbedded in quartz or other rocks in place. *United States v. Ohio Oil Co.,* D.C. Wyo., 240 F. 996.

PLAN. A "plan" is a delineation; a design; a draft or form or representation; the representation of anything drawn on a plane, as a map or chart; a scheme; a sketch; also a method of action, procedure, or arrangement. As applied to streets, a plat or survey indicating number, names, and locations of streets, their lines and courses, widths, grades, and so on, as they are or are to be laid out and opened on the land, including all particulars germane to the general subject. Of public improvements, the general plan or system of work; a profile, drawing, or picture, showing in a general way the character of the work. *Shainwald v. City of Portland,* 55 P.2d 1151, 153 Or. 167 (1936).

PLANE. "Plane" is surface in which, if any two points are taken, straight line that joins them lies wholly in that surface. *In re Vincent, Cust. & Pat.App.,* 40 F.2d 573, 17 C.C.P.A., Patents, 116.

PLANTATION. A colony; an original settlement in a new country. *Black's Law Dictionary* (1944).

A farm; a large cultivated estate. Used chiefly in the southern states. *Black's Law Dictionary* (1944).

In North Carolina, "plantation" signifies the land a man owns that he is cultivating more or less in annual crops. Strictly, it designates the place planted; but in wills it is generally used to denote more than the inclosed and cultivated fields, and to take in the necessary woodland, and, indeed, commonly all the land forming the parcel or parcels under culture as one farm, or even what is worked by one set of hands. *Stowe v. Davis,* 32 N.C. 431 (1849).

The word "plantation" in common parlance means any body of land consisting of one or several adjoining tracts, on which is a planting establishment. *State v. Blythe,* S.C., 3 McCord 363 (1825).

"Plantation," the derivative of "planting," formerly denoted, sometimes a colony, and sometimes a farm or cultivated estate. Thus, in the charter of 1662 the colony of Connecticut was called a "plantation," and its northern boundary line was the line of the "Massachusetts plantation," and in the act of 1685 the township of New Haven was empowered to manage its "plantation affairs," meaning that it might act in all the concerns of the town, *Inhabitants of Town of East Haven v. Hemingway,* 7 Conn. 186 (1828).

"The terms 'plantation,' 'town,' and 'township' seem to have been used in the early colonies of Massachusetts almost indiscriminately to indicate a cluster or body of persons inhabiting near each other, and when they became designated by name certain powers were conferred upon them by general orders and laws, such as to manage their own prudential concerns, to elect deputies, and the like, which in effect made them municipal corporations, and no formal acts of incorporation were granted until long afterwards." *Commonwealth v. City of Roxbury*, 75 Mass. (9 Gray) 451 (1857).

PLAT. A "plat" is merely a solemn written evidence of a dedication of platted lands. *Town of Kenwood Park v. Leonard*, 158 N.W. 655, 177 Iowa 337 (1916).

A "plat" is not a deed, but is a subdivision of land into lots, streets, and alleys marked upon the earth and represented upon paper. *Gannett v. Cook*, 61 N.W.2d 703, 245 Iowa 750 (1953).

A "plat" is a plan or map or a chart of a town site or a division of land. *Kane v. State*, 55 N.W.2d 333, 237 Minn. 261 (1952).

A "plat" is a map of a piece of land on which are marked courses and distances of the different lines and the quantity of land it contains. *Miller v. Lawyers Title Ins. Corp.*, D.C.Va., 112 F.Supp. 221.

A "plat" represents an ocular view of the result of a survey, constituting a visual demonstration of the work done. *Hughes v. City of Carlsbad*, 203 P.2d 995, 53 N.M. 150 (1949).

A "plat" is a representation of land on paper, appealing to the eye by means of lines and memoranda rather than by words. *Thompson v. Hill*, 73 S.E. 640, 137 Ga. 308 (1912).

A plat is a subdivision of land into lots, streets, and alleys, marked upon the earth, and represented on paper; and hence the making and platting of it implies that the land had been surveyed, and that such survey was marked on the ground, so that the streets, blocks, and lots could be identified. *McDaniel v. Mace*, 47 Iowa 509 (1877); *Burke v. McCowen*, 47 P. 367, 115 Cal. 481 (1886).

PLATTING. The word "platting," as applied to towns, is descriptive of the means of perpetuating the evidence of the creation of a town. *Mattheisen & Hegeler Zinc Co. v. City of La Salle*, 8 N.E. 81, 177 Ill. 411 (1886).

POINT. A "point" in a boundary is the extremity of a line. *Tiffany v. Town of Oyster Bay*, 126 N.Y.S. 910, 141 App. Div. 720 (1910).

"Point" as used in deed descriptions held to mean tapering end of mountain or knoll. *Staley v. Richmond*, 32 S.W.2d 546, 236 Ky. 11 (1930).

POLAR STAR RULE. The "polar star rule" is that the intent of the maker of a written document, as gathered from its four corners, shall prevail unless such intent conflicts with some statutory provision within the jurisdiction, or is against public policy. *Hanks v. McDanell*, 210 S.W.2d 784, 307 Ky. 243 (1948), see 17 A.L.R.2d 1.

POND. A "pond" is defined as a body of water naturally or artificially confined, being smaller than a lake, but of an appreciable area. *Munn v. Board of Sup'rs of Greene County*, 141 N.W. 711, 161 Iowa 26 (1913).

PORCION. In Spanish law, a part or a portion; a lot or parcel; an allotment of land. See *Downing v. Diaz*, 16 S.W. 49, 80 Tex. 436 (1891).

PRE-EMPTION. The right of pre-emption is the right to enter lands at the minimum price, in preference to any other person, if all the requirements of law are complied with. *Nix v. Allen*, Ark., 5 S.Ct. 70, 112 U.S. 129, 28 L.Ed. 675 (1884).

PREMISES. "Premises" are a piece of land or real estate. *Ford Motor Co. v. Unemployment Compensation Bd. of Review*, 79 A.2d 121, 168 Pa.Super. 446 (1951).

"Premises" is generally held to refer to that part of the deed that precedes habendum clause, including the names of the parties. *Bullock v. Porter*, 284 S.W.2d 598, 365 Mo. 572 (1955).

The word "premises" is popularly used for land, and as found in warranty in deed, means that which is conveyed. *Zandri v. Tendler*, 193 A. 598, 123 Conn. 117, 111 A.L.R. 1280 (1937).

PRIVATE LAND GRANT. A private land grant is a grant by a public authority vesting title of public land in a natural person. *United Land Ass'n v. Knight*, 24 P. 818, 85 C. 448 (1890).

PROJECTED STREET. "Projected street," as used in a deed of land by the owner while making a street, bounding the land conveyed upon the projected street, does not mean a designed, intended, or contemplated street merely, but a street already projected and then in process of construction. *Greenhood v. Carroll*, 114 Mass. 588 (1874).

PROLONGATION. The word "prolongation," as used in a deed, means a continued or extended line, though consisting of several angles, where such meaning would be consistent with the other words of description, rather than a direct line, which would render the next course in the deed inconsistent with the direction and monument by which it is described. *Chapman v. Hamblet*, 64 A. 215, 100 Me. 454 (1905).

PROPERTY. The labels "property" or "title" are used as group symbols to denote a "bundle" of rights or other relations. *Standard Oil Co. v. Clark*, C.C.A.N.Y., 163 F.2d 917.

PROPERTY AND PREMISES. The words "property and premises" when used in conveyance may be applied equally well to a conveyance of the fee or an easement. *Gulf Coast Water Co. v. Hamman Exploration Co.*, 160 S.W.2d 92 (Tex.Civ.App., 1942).

PROPERTY LINE. "Property line" is a division between two parcels of land. *Ujka v. Sturdevant*, 65 N.W.2d 292 (N.D., 1954).

PROPRIETARIOS. The word "proprietarios" in Spanish means owners but not "owners in possession." *United States v. Arredondo*, 31 U.S. 691, 6 Pet. 691, 8 L.Ed 547 (1832).

PROPRIETOR. The word "proprietor" is sufficiently broad to include owners and others with definite interest in land that is less than that of owner. *Thompson v. City of Eau Claire*, 69 N.W.2d 239, 269 Wis. 76 (1955).

The term "proprietor" had a well-defined meaning, and, according to lexicographer, means "one who has the legal right or exclusive title to anything, whether in possession or not; an owner; as the proprietor of a farm or mill." *Webster's Dictionary*. And Worcester defines the term "proprietor" as meaning a possessor in his own right; an owner; a proprietor. *Koppel v. Downing*, 11 App. D.C. 93.

PUBLIC LAND. Public lands comprise the general public domain; unappropriated lands; the lands not held back or reserved for any special governmental or public purpose. *U.S. v. Garretson*, 42 F. 22.

Land that was purchased by the federal government for prescribed uses did not fall within the category of "public land." *Rawson v. U.S.*, C.A.Or., 225 F.2d 855.

Lands owned by the province of Quebec and known as "crown lands" correspond to what is known in the country as "public lands." *Myers v. U.S.*, 140 F. 648.

PUBLIC LANDING. A "public landing" is a landing for use of public in transfer of persons and goods between land and water. *Smedberg v. Moxie Dam Co.*, 92 A.2d 606, 148 Me. 302 (1952).

The words "public landing," as used, designating a space on a plat of a town site, are at least evidence of dedication. *Village of Mankato v. Meagher*, 17 Minn. 265, 277 Gil. 243 (1871).

PUBLIC RECORD. Where, by law or regulation, a document is required to be filed in a public office, the document is a "public record". *State ex rel. Kavanaugh v. Henderson*, 169 S.W.2d 389, 350 Mo. 968 (1943).

A record that the law requires to be made is a "public record". *Mathews v. Pyle*, 251 P.2d 893, 75 Ariz. 76 (1952).

A record is not a public one unless it can be inspected by any person interested. *Keefe v. Donnell*, 42 A. 345, 92 Me. 151 (1898).

PUEBLO. "Pueblo," under Spanish law, means a small settlement or gathering of people, a steady community, and applies equally, whether settlement be a small collection of Spaniards or Indians. *Pueblo of Santa Rosa v. Fall*, D.C., 12 F.2d 332, 56 App.D.C. 259 (1927).

Under Mexican law "pueblo" answers generally to the English word "town" and may designate a collection of individuals residing at a particular place, a settlement or a village or may be applied to regular organized municipality. *Grisar v. McDowell*, 73 U.S. 363, 6 Wall. 363, 18 L.Ed. 863 (1867).

QUARTER SECTION. The general and proper acceptation of the term "quarter section," as well as its construction by the general land department, denotes the land in the subdivisional lines, and not the exact quantity that a perfect admeasurement of an unobstructed surface would declare. *Brown v. Hardin*, 21 Ark. 324 (1860).

QUARTER SECTION LINE. It is common knowledge that "quarter section lines" are lines that divide sections into four quarters. *Amundson v. Broward County Kennel Club*, 9 So.2d 793, 152 Fla. 245 (1942).

A "quarter line," as designated from a "section line," means a line running from one quarter corner to another through the center of the section. *Rud v. Pope County Com'rs*, 68 N.W. 1062, 66 Minn. 358 (1896).

"Quarter section line," as used in a description of certain lands relative to the location of a certain public highway, means the east one of three straight lines, running north and south in the section, which divide it into three equal parts. *Jackson v. Rankin*, 30 N.W. 301, 67 Wis. 285 (1886).

QUAY. A "quay" is a vacant space between the first row of buildings and the water's edge used for the reception of goods and merchandise imported or to be exported. In the Civil Code of Louisiana it is said to be common property to the use of which all the inhabitants of the city, and even strangers, are entitled in common, such as the streets and public walks. The term is well understood in all commercial countries, and, while there may be some difference of opinion as to its definition, there can be little or none in regard to its popular and commercial signification. It designates a space of ground appropriated to the public use., such use as the convenience of commerce requires. *City of New Orleans v. U.S.,* 35 U.S. 662, 10 Pet. 662, 9 L.Ed. 573 (1836).

A quay is a landing place, a place where vessels are loaded and unloaded, a wharf, usually constructed of stone, but sometimes of wood, iron, and so on, on a line of coast or the river bank, or around a harbor or a dock. *St. Anna's Asylum v. City of New Orleans*, 29 So. 117, 104 La. 392 (1900).

The word "quay" (quai) includes a levee on the bank of a river and the shore between the exterior of the levee and the water. *De Armas v. Mayor, etc., of New Orleans*, 5 La. 132 (n.d.).

QUIT. The word "quit," in a deed, is tantamount to the word "sell" or "release," and will pass the land. *Gordon v. Haywood*, 2 N.H. 402 (1821).

QUITCLAIM AND CONVEY. A deed conveying only the right, title, and interest of the grantor in land, and not the land itself, is a quitclaim deed. *Baldwin v. Drew*, Tex., 180 S.W. 614 (1915).

A deed of "all the right, title, interest, and claim which we have in and to" certain land, with undertaking to warrant and defend all such right, title, and interest, is a quitclaim. *Houston Oil Co. of Texas v. Niles*, 255 S.W. 604 (Tex., Com.App.,1923).

QUITCLAIM DEED. A "quitclaim deed," in contradistinction to conveyance of land, conveys only the vendor's chance of title. *Leal v. Leal*, 4 S.W.2d 985 (Tex., 1928).

Instrument in which warranty provisions were limited to rights, title and interest, if any, and whatsoever, in and to described lands, was a "quitclaim deed". *Simon v. Simon*, 138 So.2d 260 (La.App., 1962).

Quitclaim deeds operate as primary or original conveyances. *Smith v. Pendell*, 48 Am.Dec. 146, 19 Conn. 107 (1848).

R. "R.," as used in a deed describing land as "Secs. 22 8 23, Tp. 79, R. 13, Poweshiek County," is not uncertain or indefinite. It is a contraction in almost universal use in describing lands, and everybody understands it to mean "range." *Ottumwa, C.F. 8 St. P.R.Co. v. McWilliams*, 32 N.W. 315, 71 Iowa 164 (1887).

RADIUS. "Radius" is a straight line extending from the center of a circle to its circumference. *Sciuter v. Barile*, 70 A.2d 894, 6 N.J. Super. 595 (1950).

RAILROAD. "R.R." is a well-understood abbreviation of the word "railroad." *West Chicago St. R. Co. v. People*, 40 N.E. 599, 155 Ill. 299 (1895).

"Railroad" may mean a corporation rather than a track. *Nicolson v. Brown*, 135 F.2d 245, 77 U.S.App. D.C. 314.

The word "railroad" when used to describe a boundary of land, means the railroad right of way, and not the steel rails on the right of way, as all property essential to the operation of trains constitutes a railroad and not any individual part of railroad property. *Cross v. Bernstein*, 8 La. App. 380 (1928).

The legal signification of the term "railroad" is not only a road or way on which iron rails were laid but also a road as incident to the possession or ownership of which important franchises and rights affecting the public are attached. *Gibbs v. Drew*, 26 Am. Rep. 700, 16 Fla. 149 (1877).

RAILROAD TRACK. "Railroad track," as used in a deed bounding the land by the line of a railroad track, should be construed to mean the line of the rails, if the grantor could convey so far, and not simply to the right of way. *Reid v. Klein*, 37 N.E. 967, 138 Ind. 484 (1894).

REAL ESTATE. The phrase "real estate" in legal signification includes all interests in land, whether in possession, reversion, or remainder. *Floyd v. Carow*, 88 N.Y. 560 (1882).

RECORD TITLE. Titles granted in manner of statutes are commonly called "record" or "legal titles" as distinguished from "equitable titles," and it is policy of law to protect them, and record title is the highest evidence of ownership and is not easily defeated. *Hughes v. Cook*, 130 N.E.2d 330, 126 Ind. App. 103 (1955).

REGISTRY. The registry of a deed adds nothing to its effectiveness as a conveyance. All that it accomplishes is to impart notice. *Shirk v. Thomas*, 22 N.E. 976, 121 Ind. 147, 16 Am.St. Rep. 381 (1889).

RELEASE. A "release" is equivalent to a conveyance. *McHugh v. Duane*, D.C. Mun. App., 53 A.2d 282 (1947).

A deed of release may operate as a grant, if necessary to carry out the intention of the parties, or may be treated as a confirmation of title. *Smith v. Cantrel*, 50 S.W. 1081 (Tex., 1899).

RELICTION. "Reliction" is increase of land by retreat or recession of water. *Baumhart v. McClure*, 153 N.E. 211, 21 Ohio App. 491 (1926).

REMAINDER. Unlike a "remainder," which must be created by deed or devise, a "reversion" arises only by operation of law. *Wilson v. Phaaris*, 158 S.W.2d 274, 203 Ark. 614 (1942).

REMISE, RELEASE AND QUITCLAIM. "To remise, release and quitclaim" designated land means that grantor releases any interest he may have in land at that time, but that is all. *Williams v. Reid*, 37 S.W.2d 537 (Mo., 1931).

REMNANT RULE. The rule that width of lot, frontage of which is not specified on plat specifying frontage of all other lots in same block, is length of block, minus total width of other lots, is sometimes spoken of as "remnant rule" and is corollary to "apportionment rule," which is that any excess or deficiency, where dimensions of all lots in block are specified, will be apportioned among them according to their frontages, whether they are regular or irregular. *Routh v. Williams*, 193 So. 71, 141 Fla. 334 (1940).

RESERVATION. A "reservation" in a deed must be of some portion of granted premises that belonged to grantor and that without the reservation would be conveyed by the deed, and operates by way of a re-grant by grantee to grantor of estate or interest reserved. *Fatherree v. McCormick*, 24 So.2d 724, 199 Misc. 248 (1946).

Deed reciting that conveyance was made subject to existing right of way in favor of a third party was not a "reservation." *Ozehoski v. Scranton Spring Brook Water Service Co.*, 43 A.2d 601, 157 Pa.Super. 437 (1945).

While there is a distinct difference between "exception" and "reservation," the words are often used interchangeably, a reservation being something taken back from the grant while an exception is some part of the estate described in general terms in deed which is not granted. *Murphy v. Sunset Hills Ass'n.*, 9 N.W.2d 613, 243 Wis. 139 (1943).

Technically a "reservation" in deed is some newly created right, which grantee conveys to grantor by implication, while an "exception" is something withheld from grant that would otherwise pass as part of it. *Nelson v. Bacon*, 32 A.2d 140, 113 Vt. 161 (1943).

RESERVING ALL THE REEF IN FRONT. Expression "reserving all the reef in front" in deed from high chief to defendant's brother was inconsistent with idea that boundary line ran along outer edge of reef, since in that case there would be no reef in front of the line, and the line ran alongside the inside of coral reef, especially in view of Hawaiian version of deed reading "*Aole nae e hookomo ana i ka papa koa mawaho*" which is translated to mean "not including however, the coral reef outside." *Levi Haalelea v. Montgomery*, 2 Haw. 62 (1858).

RESERVING AND EXCEPTING. "Excluding," in a deed, is generally used as a synonym of "reserving and excepting." *Smith v. Furbish*, 44 A. 398, 68 N.H. 123 (1894), 47 L.R.A. 226.

RESERVING THE HIGHWAY. "Reserving the highway," as used in a conveyance of land containing the usual covenants of warranty, but reserving the highway, means the reservation of the highway for its use and purpose as a highway, and does not prevent the fee of the lands included in the highway from passing to the purchaser, subject only to the easement of the public. *Peck v. Smith*, 6 Am.Dec. 216, 1 Conn. 103 (1814).

RESIDUARY CLAUSE. A "residuary clause" is one that covers all of the estate not disposed of after providing for debts and particular legacies and devises. *Shannon v. Reed*, 50 A.2d 278, 355 Pa. 688 (1947).

The purpose of a "residuary clause" is to dispose of the remaining property of the testator not specifically devised or bequeathed. *Sheridan v. Perkins*, 42 S.E.2d 853, 186 Va. 465 (1947).

REST, RESIDUE, AND REMAINDER. Terms "rest," "residue," and "remainder" of estate are usually and ordinarily understood to mean part of estate left after all provisions of will have been satisfied. *In re Richter's Will*, 234 N.W. 285, 212 Iowa 38 (1931).

REVERSION. The term "reversion" has two meanings, first, as designating the estate left in the grantor during the continuance of a particular estate and also the returning of the land to the grantor or his heirs after the grant is over. *Davidson v. Davidson*, 167 S.W.2d 641, 350 Mo. 639 (1943).

REVERT. The ordinary meaning of the word "revert" is to return or to go back. *Reaney v. Wall*, 60 A.2d 505, 134 Conn. 663 (1948).

RIGHT BANK. The "right bank" of a stream as used in land description is bank on right hand as one faces downstream. *Kingsley v. Jacobs*, 149 P.2d 950, 174 Or. 514 (1944).

RIGHT OF CONQUEST. By the "right of conquest" nothing more is in fact meant than the right of discovery. It was the right of discovery as to the civilized world, and the right of conquest as to the Indians themselves. *Caldwell v. State*, Ala., 1 Stew. 8 P. 327 (1832).

RIGHT OF WAY. "Right of way" is a right of passage over another person's land. *Almada v. Superior Court in and for Napa County*, 149 P.2d 61 (Cal. App., 1944).

A "right of way" may be defined generally to be a right to pass over the land of another. It may be a private way or a public way, and it may belong to one or several persons or to the entire community. *Poole v. Greer*, 65 A. 767, 6 Pennewill (Del.), 220 (1907).

A conveyance of a "right of way" is an easement only and title does not pass, and grantee acquiring only the right to a reasonable and usual enjoyment thereof with owner of the soil retaining all rights and benefits of ownership consistent with the easement including the right to make any incidental changes consistent with the easement. *Kleih v. Van Schoyck*, 27 N.W.2d 490, 250 Wis. 413 (1947).

Term "right of way" in deed is frequently used to describe not only easement but the strip of land occupied by such use. *Moakley v. Los Angeles Pac. Ry. Co.*, 34 P.2d 218, 139 Cal. App. 421 (1934).

RIGHT OF WAY IN GROSS. A "right of way in gross" is a mere personal privilege that dies with the person who may have acquired it. *Steele v. Williams*, 28 S.E.2d 644, 204 S.C. 124 (1944).

RIGHT, TITLE, AND INTEREST. The words, "right, title, and interest," as used in deeds, have acquired a definite meaning, not importing ownership, but conveying whatever title the grantor has, and that alone. *Baker v. Davie*, 97 N.E. 1094, 211 Mass. 429 (1912), 37 L.R.A., N.S. 944.

RIPARIAN. The Century Dictionary defines "riparian" as pertaining to or situate on the bank of a river. *Mobile Transp. Co. v. City of Mobile*, 30 So. 645, 128 Ala. 335, 64 L.R.A. 333, 86 Am.St.Rep. 143 (1907).

The word "riparian" has reference to the bank and not to the bed of the stream. *Rome Ry. & Light Co. v. Loeb*, 80 S.E. 785, 141 Ga. 202 (1914), Ann.Cas. 1915C, 1023.

To be "riparian," land must ordinarily be contiguous to or abut on stream. *Rancho Santa Margarita v. Vail*, 81 P.2d 533, 11 Cal.2d 501 (1938).

A "riparian owner" is one having land bounded on a stream of water, as such owner has a qualified property in the soil to the thread of the stream. "Riparian rights" are such as grow out of the ownership of the banks of streams, and not out of the ownership of the bed of the stream. The word "riparian" has reference to the bank and not to the bed of the stream. *Rome Ry. & Light Co. v. Loeb*, 80 S.E. 785, 141 Ga. 202 (1914), Ann.Cas. 1915C, 1023.

RIPARIAN OWNER. One whose land abuts upon a river is a "riparian owner", while one whose land abuts upon a lake is a "littoral owner." *Darling v. Christensen*, 109 P.2d 585, 166 Or. 17 (1941).

RIPARIAN PROPERTY. "Riparian property" is property which has a water frontage. *Shepard's Point Land Co. v. Atlantic Hotel*, 44 S.E. 39, 132 N.C. 517, 61 L.R.A. 937 (1903).

RIPARIAN PROPRIETOR. A "riparian proprietor" is one who owns land bounded upon a water course or lake. *French-Glenn Live Stock Co. v. Springer*, 58 P. 102, 35 Or. 312 (1899).

RIVER. "Water course," "river," or "stream" consists of bed, banks, and stream of water. *Motl v. Boyd*, 286 S.W. 458, 116 Tex. 82 (1926).

RIVER BANKS. The "banks of a river" are the boundaries that confine the water to its channel throughout the entire width when stream is carrying is maximum quantity of water. *Mammoth Gold Dredging Co. v. Forbes*, 104 P.2d 131, 39 Ca.App.Pd 739 (1940).

RIVER BED. A "river bed" is that soil so usually covered by water as to be distinguishable from the banks by the character of the vegetation produced by the common presence and action of flowing water. *Howard v. Ingersoll*, 54 U.S. (13 How.) 381, 14 L.Ed. 189 (1850) .

ROAD. "Road," in its ordinary sense, is a generic term, including overland ways of every character. *Johnston v. Wortham Machinery Co.,* 151 P.2d 89, 60 Wyo. 301 (1944).

"Roads and highways" are words embracing all kinds of public ways, such a county and township roads, streets, alleys, township and plank roads, turnpike or gravel roads, tramways, ferries, canals, navigable rivers, and also railroads. *Strange v. Board of Com'rs of Grant County*, 91 N.E. 242, 172 Ind. 640 (1910).

RODEO BOUNDARIES. "Rodeo boundaries," under the customs and acknowledged usages that prevailed in California, constituted as notorious evidence of the possession of land as the cultivation or fencing in an old settled country. *Boyreau v. Campbell*, 3 Fed. Cas. 1112.

ROYAL TITLE. A "royal title" is the highest order of title known by any law, or principle, in the province of east Florida. Titles of this description were designed to convey the fee simple to the grantee. They were usually made by the acting governors of the province in the name of the king. They recited the grant to be in perpetuity, and also the specific metes and bounds of the land. This title may be said to correspond in character with that of a patent issued by our government. "Concessions without conditions" are understood to differ from a "royal title" only in this: that most of the latter recite the metes and bounds, whereas the unconditional concession, although definite in quantity and location of the land, is still subject to a survey, which, when made, was followed by maturing the concession by a royal title. There is also a peculiarity in the legal phraseology of a "royal title." In all the grants of this nature the legal right to the lands is asserted. *Florida Town Imp. Co. v. Bigalsky*, 33 So. 450, 44 Fla. 771 (1902), quoting from the report of the Land Commissioners of 1826.

R.R. "R.R." is a well-understood abbreviation of the word "railroad." *West Chicago St. R. Co. v. People ex. Rel. Kern*, 40 N.E. 599, 155 Ill. 299 (1895).

RULE OF CONSTRUCTION. Interpretation and construction of written instruments are not the same; and a "rule of construction" is one which either governs the effect of an ascertained intention, or points out what the court should do in the absence of express or implied intention, while a "rule of interpretation" is one which governs the ascertainment of the meaning of the maker of the instrument. *In re Union Trust Co.*, 151 N.Y.S. 246, 89 Misc. 69 (1915).

RULE OF INTERPRETATION. See **RULE OF CONSTRUCTION.**

RUNNING ALONG. Description of grant as "running along Smith's line" held to have made tracts contiguous to full length of named tract, leaving no vacancy between them. *Ramsay v. Butler, Purdum & Co.*, 129 A. 650, 148 Md. 438 (1925).

RUN OUT A LINE. "Run out a line," when used in connection with surveying, means that a person qualified to do such work shall go upon the ground and with proper instruments, chainmen, and necessary assistants establish and mark a line upon the surface of the earth. They clearly impose the duty of going upon the ground and actually running out the line. *Mineral County Com'rs. v. Hinsdale County Com'rs.,* 53 P. 383, 25 Colo. 95 (1898).

RUNS AND CALLS. Terms "runs and calls," "courses and distances" and "angles and distances" are synonymous; they all refer to the angles and scaled distances indicated on a plat map and must be followed in order to establish exact boundaries. *Block v. Howell*, 346 N.W.2d 441 (S.D, 1984).

S. The Supreme Court will take judicial notice that "N." and "S." when applied to directions are properly read north and south, and that "N. to S. end of W. Avenue" means north and south end of W. Avenue. *Village of Bradley v. N.Y. Cent. R. Co.,* 129 N.E. 744, 296 Ill . 383 (1921).

SAID. The word "said" is often used in wills and deeds, and so forth, to refer to some antecedent provision. *Shattuck v. Balcom*, 49 N.E. 87, 170 Mass. 245 (1898).

The word "said," when used in a document, refers to something that has been mentioned above in the document. *Com. v. Schweiters*, 93 S.W. 592, 122 Ky. 874 (1906).

The word "said" in deed ordinarily refers to next appropriate antecedent. *Ward v. Torian*, 112 So. 815, 216 Ala. 288 (1927).

SALE BY THE TRACT OR ENTIRE BODY. A deed, reciting conveyance of "lot or tract of land" constituting all of designated tract containing specified number of acres as shown by plat and having designated metes and bounds, was a "sale by tract." *Kytle v. Collins*, 19 S.E.2d 754, 67 Ga.App. 98 (1942).

SALE IN GROSS. "Sale in gross" is one wherein boundaries are specified but not quantity, or, if it is, is not material, whereas "sale by acre" is one wherein quantity is material and purchaser takes no risk of deficiency and the vendor takes no risk of excess. *Carrel v. Lux*, 420 P.2d 564, 101 Ariz. 430 (1966).

SALE OF LAND. A sale of lands "is the actual transfer of title from the grantor to the grantee by appropriate instrument of conveyance." *Ide v. Leiser*, 24 P. 695, 24 Am.St. Rep. 17, 10 Mont. 5 (1890).

SALINE LAND. Land having salt deposits. To fourteen states Congress has granted all the salt springs within them; to twelve, a limited grant of them was made. Eighteen states have received no such grant. *Montello Salt Co. v. Utah*, 221 U.S. 452, 31 S.Ct. 706, 55 L.Ed. 810 (1911), Ann. Cas. 1912D, 633.

SALT MEADOW. The term "salt meadow" is applied to the tracts of land that lie above the seashore, and that are overflowed by spring and extraordinary tides only, and yield grasses that are good for hay. *Church v. Meeker*, 34 Conn. 421 (1867).

SAME. The expression the "same" refers to something previously mentioned. *Baird v. Johnston*, 297 N.W. 315, 230 Iowa 161 (1941).

The word "same" is defined to mean "not different or other; identical." *U.S. v. East Tenn., V. & G.R.Co.*, 13 F. 642.

SATISFACTORY TITLE. A contract to convey, and to furnish a good and "satisfactory" title, is complied with by furnishing a good, marketable title, free from reasonable doubt. *Moot v. Business Men's Inv. Ass'n*, 52 N.E. 1, 157 N.Y. 201, 45 L.R.A. 666 (1898).

SAVANNA. "Savanna," as the term was used in North Carolina, meant a natural open meadow, which was not uncommon in the lower parts of the state. *Stapleford v. Brinson*, 24 N.C. 311 (1892).

S.E. The letters "S.E." mean southeast. *Bandow v. Wolven*, 107 N.W. 204, 20 S.D. 445 (1906).

S.E. 4. The abbreviation "S.E. 4," employed in the description of the property conveyed by a tax deed, will be interpreted as meaning "southeast quarter," when it is explicitly used in another part of the same instrument as the equivalent of these words. *Kennedy v. Scott*, 83 P. 971, 72 Kan. 359 (1905).

SE QR 24. In a description of property contained in a list of delinquent real property attached to a notice of tax sale describing it as "se qr 24," the number 24 being in a column headed "sec," the letters "se" clearly meant southeast, and the description given properly described the southeast quarter of section 24. *Bandow v. Wolven*, 107 N.W. 204, 20 S.D. 445 (1906).

SEA. Waters within the ebb and flow of the tides are considered the "sea." *In re Gwin's Will*, N.Y., Tuck. 44, citing Gilpin's R. 526; *Baker v. Hoag*, N.Y., 3 Barb. 203 (1848).

The word "sea" has been held to mean not only high sea, but arms of the sea, waters flowing from it into ports and havens, and as high upon rivers as the tide ebbs and flows. *Gordon v. Blackton*, 186 A. 689, 117 N.J.L. 40 (1936).

"Sea ground" is either the ground bordering on the sea or covered with the sea. In a deed, the word "ground" is sufficient to pass the soil, and the word "sea," annexed to it, only shows where it is situated. *Scratton v. Brown*, 4 Barn. & C. 485.

"Seaboard" means the country bordering on the sea, and "sea," as defined in the *Century Dictionary*, is "a more or less distinctly limited or land-locked part of the ocean, having considerable dimensions." *American Fisheries Co. v. Lennen*, 118 F. 869.

The term "high sea" is the same as the ocean or main sea. It means that part of the sea which lies without the body of the county, and is distinguished from the term "sea," which includes arms or branches of the sea that may lie within the body of the county. *De Lovio v. Boit*, 7 Fed. Cas. 418 (1815).

SEA BEACH. The "sea beach" is all that place which is covered by the waters of the sea when at its highest point during all the year. *United Land Ass'n v. Knight*, 23 P. 267, 3 C.U. 211 (1890); *United Land Ass'n v. Knight*, 24 P. 818, 85 Cal. 448 (1890).

SEAL. The purpose of a "seal" is to attest in a formal manner the execution of an instrument. *King v. Guynes*, 42 So. 959, 118 La. 344 (1907), citing Black's Law Dict.

A seal stamped upon paper of sufficient tenacity to retain the impression is a seal within the strictest rules of the common law. *Ross v. Bedell*, 12 N.Y. Super. 462, 5 Duer 462 (1856).

A mark with ink, acknowledged by the maker of the deed to be his seal, is sufficient to create a specialty, though no wax, wafer, or other similar substance be used. *U.S. v. Coffin*, 25 Fed. Cas. 485.

A piece of paper annexed to a deed by wafer, wax, gum, or any adhesive substance is a valid seal, being equivalent to the wax impression formerly used. *Maddocks v. Keene*, 96 A. 785, 114 Me. 464 (1916).

A scroll after a name attached to a deed is not a seal. *Douglas v. Oldham*, 6 N.H. 150 (1833).

The word "Seal," written opposite signature, is equivalent to a seal. *Whitley v. Davis' Lessee*, 31 Tenn. (1 Swan) 333 (1851).

SEASHORE. "Seashore" must be understood to be the margin of the sea in its usual and ordinary state. Thus when the tide is out, low-water mark is the margin of the sea; and when the tide is full, the margin is the high-water mark. The seashore is therefore all the ground between ordinary high-water mark and low-water mark. It cannot be considered as including any ground always covered by the sea; for then it would have no definite limit on the seaboard. Neither can it include any part of the upland, for the same reason. This definition of the shore seems to result necessarily from its nature and situation. *Storer v. Freeman*, 4 Am.Dec. 155, 6 Mass. 435 (1810); *Lapish v. Bangor Bank*, 8 Me. 85, 8 Greenl. 85 (1831).

SEATED LANDS. "Seated land," as used in the tax laws, means land that is used for residence or cultivation, or for the purpose of making profit in any manner. *Earley v. Euwer*, 102 Pa. 338 (1883).

SEC. The letters "sec." mean "section." *Bandow v. Wolven*, 107 N.W. 204, 20 S.D. 445 (1906).

SECTION. A "section" of a township is that which is made out on ground, and a patentee takes only such land as is included within survey of plot conveyed, and he cannot later question survey as erroneous, although in fact line in question should have been placed elsewhere. *Phelps v. Pacific Gas & Elec. Co.*, 190 P.2d 209, 94 Cal. App.2d 243 (1948).

The general and proper acceptation of the terms "section," "half section," and "quarter section" of land, as well as their construction by the general land department, denotes the land in the sectional and subdivisional lines, and not the exact quantity that a perfect admeasurement of an unobstructed surface would declare. *Brown v. Hardin*, 21 Ark. 324 (1860).

SECTION CORNER. A "quarter corner," as distinguished from a "section corner," in the government surveys means the corner or a section line midway between the section corners. *Rud v. Pope County Com'rs*, 68 N.W. 1062, 66 Minn. 358 (1896).

SECTION LINE. A "quarter line," as distinguished from a "section line," means a line running from one quarter corner to another through the center of the section. *Rud v. Pope County Com'rs*, 68 N.W. 1062, 66 Minn. 358 (1896).

SEDGE FLATS. "Sedge flats are flats which lie near the seashore, below ordinary high-water mark, and are covered by every tide, and grow a coarse or long sedge, which cattle will not eat, and which, like sea-weed, is valuable only for bedding and manure. A sedge flat, lying on the shore, which bounds an arm of the sea, is not in any popular, legal, or just sense a meadow." *Church v. Meeker*, 34 Conn. 421 (1867).

SEDO. "Sedo" is the word ordinarily used in all Mexican conveyances to pass title to lands. It is translated "I grant." *R. Mulford v. LeFranc*, 26 Cal. 88 (1864).

SEIZED. The word "seized" is used in the sense of "owned" or, sometimes, possessed. *Rouse v. Paidrick*, 49 N.E.2d 528, 221 Ind. 517 (1943).

SEIZIN. In American jurisprudence, "seizin" generally means "ownership." *McNitt v. Turner*, 83 U.S. 352, 16 Wall. 352, 21 L.Ed. 341 (1872).

SEMINARY SQUARE. The words "Seminary Square," written upon a block designated in a plat of the town filed for the purpose of laying out a town, was construed to show a dedication of the lot to the town for seminary purposes. *Miami County v. Wilgus*, 22 P. 615, 42 Kan. 457 (1889).

SENIOR. The suffixes "Jr." or "Sr." are no part of a man's name and, except in a few instances, may be disregarded. *Ross v. Berry*, 124 P. 342, 17 N.M. 48 (1912).

The word "Senior" is attached to the name of a father to distinguish him from a son of the same name. It is no part of the name. *Coit v. Starkweather*, 8 Conn. 289 (1830).

SEPARATE PROPERTY. Strictly speaking, term "separate property" applies only to property owned by married person in his or her own right during marriage. *In re Morgan's Estate*, 265 P. 241, 203 Cal. 569 (1928).

SERVITUDE. "Servitudes" are charges or incumbrances that follow the land. *Patton v. Frost Lumber Industries*, 147 So. 33, 176 La. 916 (1933).

The term "easement" or "servitude" is used to designate some privilege existing in one not the owner of real estate, constituting a privilege in reference to the land. There are two very usual methods of creating an easement or imposing a servitude on land: the owner of a tract of land may convey a portion of it, and in the deed of conveyance may retain an easement in it for the benefit of the portion that he does not dispose of; or he may in the deed convey to his grantee an easement in the land that he retains in his possession. These easements are appurtenant to the land, and pass with it to successive grantees. *Rowe v. Nally*, 32 A. 198, 81 Md. 367 (1895).

An "easement" is a right that one proprietor has to some profit, benefit, or lawful use out of or over the estate of another proprietor; while a "servitude" is the burden imposed on one tract of land for the benefit of another. *Stephenson v. St. Louis Southwestern Ry. Co. of Tex.*, 181 S.W. 568 (Tex., 1915).

Easements are property; servitudes are burdens on property; and an owner is entitled to complete dominion over and enjoyment of his property, except in so far as he voluntarily relinquishes those rights; and it is immaterial whether the easement or servitude is created by public or private grant, or by prescription. And the courts vigilantly guard a dominant tenement in its full possession of an easement, and a servient one from an increase of the servitude, regardless of whether the injury follow an abridgement of the easement or an enlargement of the servitude. *St. Louis Safe Deposit & Sav. Bank v. Kennett's Estate*, 74 S.W. 474, 101 Mo.App. 370 (1903).

SERVITUDE IN GROSS. A "servitude in gross" is a personal servitude imposed on land for the benefit of persons owning a right, irrespective of ownership of the land. *Smith v. Cooley*, 2 P. 880, 65 Cal. 46 (1884).

SET APART. In a treaty providing that certain land shall be set apart to certain Indians, the words "set apart" are sufficient to convey title in fee simple to the land reserved. *Meehan v. Jones*, 70 F. 453.

SET LOT. An oyster ground used for propagation of oysters is a "set lot." *H.J. Lewis Oyster Co. v. U.S.*, Ct.Cl., 107 F.Supp. 570 (1952).

SETTLE. The word "settle," as applied to lands, conveys the idea of a permanent inhabitance. *Burleson v. Durham*, 46 Tex. 152 (1876), citing *Webster's Dictionary*.

The word "establishment" "means nothing more, in the popular language of the French of Missouri, than is implied in the popular language of the Anglo-Americans by the word 'settlement.'" *Dent v. Bingham*, 8 Mo. 579 (1844).

The word "plant," as used a century and a half ago, had a peculiar meaning, which has since become obsolete. By the word "plant," applied to a tract of land, was meant "settle" or "establish." *Inhabitants of Town of East Haven v. Hemingway*, 7 Conn. 186 (1828).

SETTLEMENT OF AN ESTATE. Ordinarily, "settlement of an estate" means payment of taxes and debts and distribution of estate among those entitled thereto. *In re Wraught's Estate*, 32 A.2d 8, 347 Pa. 165 (1943).

SETTLER. "Settler," within the preemption acts, disposing of the public domain, is synonymous with "actual settler," or "bona fide settler," and means a purchaser who not only occupies public land, but actually resides thereon with a view to residence. *Burleson v. Durham*, 46 Tex. 152 (1876).

A "settler" in any particular locality means any one who has taken up his permanent abode in that locality. The word is ordinarily applied to those who first come to a country or section of a country, either partially or wholly inhabited, and who make their residence there. We may speak of the early settlers of a long-inhabited country, but the term is hardly applicable to any other class of residents of such a country. *Hume v. Gracy*, 27 S.W. 584, 86 Tex. 671 (1894).

SEVEN ARPENTS MORE OR LESS. Where owner of a tract specifically described as having a frontage of 5 1/2 acres and a depth of seven acres sold a portion described as having a specific depth of "seven arpents, more or less" and in other conveyances made on same day owner described the property conveyed as having a depth of seven "arpents more or less," the term "seven arpents, more or less" evidenced intention that the vendee should take the land up to actual location of the seven-acre line, whether or not, by actual measurements, its depth proved to be more or less than seven arpents. *Pierce v. Lefort*, 200 So. 801, 197 La. 1 (1941).

SHORE. The "shore" is that space of ground that is between ordinary high-water and low-water mark. *Church v. Meeker*, 34 Conn. 421 (1867).

The basic law of Spain and Mexico defines "shore" as all ground covered with water at high tide during the whole year, whether in winter or in summer. *Humble Oil & Refining Co. v. Sun Oil Co.*, C.A.Tex., 190 F.2d 191 (1951).

Flats are not "shore" unless tide ebbs and flows over them and they are not "alluvion" belonging to the proprietor of mainland unless they have been added to the shore by accretion. *Humble Oil & Refining Co. v. Sun Oil Co.*, C.A.Tex., 190 F.2d 191 (1951).

The word "shore" must be deemed to designate land washed by the sea and its waves, and to be synonymous with "beach." *Littlefield v. Littlefield*, 28 Me. 180, 15 Shep. 180 (1848).

SHORE LINE. "Meander line" is not a boundary, but water of the body that is meandered is the true boundary, regardless whether "meander line" coincides with "shore line." "Shore line" does not mean the "meander line," because the "meander line" is frequently run on the top of the bank, often many feet above the usual water line reached by the river during ordinary seasons. "Shore line" has usually been used to mean, in natural fresh-water rivers, low-water mark, and grants bounded by the shore extended to that line, making the water's edge at low water the boundary. *Schaller v. Town of Florence*, 259 N.W. 529, 193 Minn. 604 (1935).

SIDE. A "side" is a bounding line of a geometrical figure; as, the side of a field, square, river, road, etc. *Badura v. Lyons*, 23 N.W.2d 678, 147 Neb. 442 (1946).

The word "side" has many meanings. It may not always refer to the border or edge of the water of a lake; it may refer to any part or position viewed as opposite to or contrasted with another, as the south side or north side of the pond or lake. *Webster's New International Dictionary*. Whether the one meaning or the other be intended depends entirely upon the context. *White v. Knickerbocker Ice Co.*, 172 N.E. 452, 254 N.Y. 152 (1930), see 74 A.L.R. 591, reversing 241 N.Y.S. 898, 229 App.Div. 746 (1930).

SIDE LINES. The side lines of a road or railroad are the lines that include the territory covered by the road. Public roads and highways and railroads are regarded as having three lines, the side lines and the center line equidistant between the side lines. *Maynard v. Weeks*, 41 Vt. 617 (1869).

The word "front" as applied to a house is always specific, and speaking of "side line" of house as "fronting" toward street is incorrect. "Side" may be used in a generic sense so as to include the "front," but it also has a specific meaning that distinguishes it from "front." *Howland v. Andrus*, 86 A. 391, 81 N.J.Eq. 175 (1913).

SITE. A "site," according to *Webster*, is a seat or ground plot; and a mill site is the place where a mill stands. *Miller v. Alliance Ins. Co. of Boston*, 7 F. 649.

SITIO DE GRANADO MAYOR. Under Spanish or Mexican grants, a "*sitio de ganado mayor*" was a square league, and appears to have been the only unit in estimating the superficies of land. *Corrigan v. State*, 94 S.W. 95, 42 Tex.Civ.App. 171(1906).

SITO GRANADO MAYOR. "*Sito ganado mayor*," used in Mexican land grant, is to be construed as a technical Mexican and Spanish legal term well established and defined and known as the "section or township" in the survey of the United States. It was a square the "four sides of which each measured 5,000 varas," and "the distance from the center of each sito to each of its sides should be measured directly to the cardinal points of the horizon and should be 2,500 varas." *United States v. Cameron*, 21 P. 177, 3 Ariz. 100 (1889), citing *D'Aguirre's Case*, 68 U.S. 316, 1 Wall. 316, 17 L.Ed 595 (1863); *Fossat's Case*, 61 U.S. 415, 20 How. 415, 15 L.Ed 944 (1857); *Yontz's Case*, 64 U.S. 498, 23 How. 498, 16 L.Ed. 472 (1859); *United States v. Pico*, 72 U.S. 539, 5 Wall. 539, 18 L.Ed 695 (1866).

600 FEET IN WIDTH. Deed conveying "southerly 600 feet in width of lot 1," and so forth, held to contain sufficient description; the words "600 feet in width" meaning 600 feet at right angles. *Phelps v. Brevoort*, 174 N.W. 281, 207 Mich. 429 (1919).

SLOUGH. An arm of a river, flowing between islands and the mainland, and separating the islands from one another. Sloughs have not the breadth of the main river, nor does the main body of water of the stream flow through them. *Dunlieth & D. Bridge Co. v. Dubuque County*, 8 N.W. 443, 55 Iowa 565 (1881).

SO CALLED. The legal interpretation of the words, "so called," in a deed, is, not what the parties, but what the public generally, says about the premises. *Madden v. Tucker*, 46 Me. 367 (1859).

SOLE INTEREST. A "sole interest" means the same thing as an absolute interest. *Garver v. Hawkeye Ins. Co.*, 28 N.W. 555, 69 Iowa 202 (1886).

SOLE OWNERSHIP. Ownership is "sole" when no other person has any interest in property as owner. *Automobile Underwriters v. Tite*, 85 N.E.2d 365, 119 Ind.App. 251 (1949).

SO LONG AS. The words "so long as" in habendum clause of a deed are appropriate and common words of limitation. *Thypin v. Magner*, 28 N.Y.S.2d 262 (1941).

The words "so long as", "while", "until" and "during" are the usual and apt words to create a limited estate such as a determinable fee. *North Hampton School Dist. v. North Hampton Congregational Soc.*, 84 A.2d 833, 97 N.H. 219 (1951).

SOUTH. Generally, the words "north," "south," "east," and "west," when used in a land description, mean, respectively, "due north," "due south," "due east," and "due west." *Plaquemines Oil & Development Co. v. State*, 23 So.2d 171, 208 La. 425 (1945).

SOUTH COURSE. The term "south course" in deed did not mean due south, but meant a southwardly course. *Martt v. McBrayer*, 166 S.W.2d 823, 292 Ky. 479 (1942).

SOUTHERLY. In the absence of monuments and in a deed, "southerly" means due south. *Smith v. Newell*, 86 F. 56.

The words "southerly" and "westerly" used in the identifying descriptions in deeds, are not always used to indicate a direction that is due south or west. *Brown v. McCaffrey*, 60 A.2d 792, 143 Me. 221 (1948).

There are very few words in our language more indefinite and uncertain in their meaning than the words "southerly," "easterly," and "northerly." The word "southerly," as applied to the course of a proposed highway, designating the course as "thence southerly to avoid" a certain creek, and "thence easterly and northerly through" certain lands, means nearly south, but how near, and whether east or west of south, it is impossible to tell without the use of other qualifying words, and so with regard to the words "easterly" and "northerly." It is impossible to determine with any certainty the course intended thereby. *Scraper v. Pipes*, 59 Ind. 158 (1877).

SOUTH PART. Under deed conveying one of two adjoining lots and reserving right of way over the "south part" of the lot conveyed, the term the "south part" constituted description in contradistinction to "north part" and did not suggest the idea of a "middle part" which is no specific part but any part that is embraced in any two lines between north and south or east and west boundaries that are parallel with and equidistant from the middle lines, while the "south part" supposes a "north part," the "south part" being all that is north of the middle line east and west. *Roberts v. Stephens*, 40 Ill.App. 138 (1891).

SOUTHWARD. The term "southward" must be considered with reference to its subject-matter, and, as used in a conveyance describing land, may include land lying in a southwesterly direction, where that seems to be the intention. *Higgins v. Round Bottom Coal & Coke Co.*, 59 S.E. 1064, 63 W.Va. 218 (1907).

SPECIAL WARRANTY DEED. A "special warranty deed" protects the grantee against a claim under a title from his grantor, but not against a claim under a title against, or superior to, his grantor. *Central Life Assur. Soc. v. Impelmans*, 126 P.2d 757, 13 Wash.2d 632 (1912).

SQUARE. A "square" is an open area in a city or village left between streets at their intersection. *Harvey v. Mayor and Aldermen of City of Savannah*, 199 S.E. 653, 59 Ga. App.12 (1938).

Square. As used to designate a certain portion of land within the limits of a city or town, this term may be synonymous with "block," that is, the smallest subdivision that is bounded on all sides by principal streets, or it may denote a space (more or less rectangular) not built upon, and set apart for public passage, use, recreation, or ornamentation, in the nature of a "park" but smaller. See *Caldwell v. Rupert*, 10 Bush (Ky.) 179 (1873); *State v. Natal*, 7 So. 781, 42 La. Ann. 612 (1890); *Rowzee v. Pierce*, 23 So. 307, 75 Miss. 846, 40 L.R.A. 402, 65 Am. St.

Rep. 625 (1898); *Methodist Episcopal Church v. Hoboken*, 97 Am. Dec. 696, 33 N.J. Law 13 (1868); Rev. Laws Mass. 1902, p. 531, c. 52, §12 (Gen. Laws 1932, c. 85 §14).

A "block" or "square" is a portion of a city bounded on all sides by streets or avenues. *Missouri, K. & T. Ry. Co. v. City of Tulsa*, 45 382, 145 Okl. P. 398, 401; *City of Mobile v. Chapman*, 79 So. 566, 571, 202 Ala.. 194 (1918).

SQUARE BLOCK. In a technical sense the word "block" may and often does designate a territory, sometimes called a "square block," but it is also commonly used to designate that section of a square block, so called, fronting on a street between two intersecting streets. *City of Olean v. Conkling*, 283 N.Y.S. 66, 157 Misc. 63 (1835).

SQUARE LEAGUE. A "square league" is 5,000 varas square, and its area is 25,000,000 varas. *United States v. De Rodriguez*, 25 Fed.Cas. 821.

A "square league," or "sitio de ganado mayor," appears to have been the only unit in estimating the superficies of land granted by the Spanish or Mexican authorities. Eleven of these leagues was the usual extent for a rancho grant. If more or less was intended in the grant, it was carefully stated. *Corrigan v. State*, 94 S.W. 95, 42 Tex.Civ.App. 171 (1906).

SQUARES. The term "squares" is generally used synonymously with "blocks" in describing urban premises, a square or block meaning a subdivision of a city or town inclosed by streets, whether occupied by buildings or inclosures or merely comprising vacant lots. *City of Mobile v. Chapman*, 79 So. 566, 202 Ala. 194 (1918).

SR. The suffix "Sr." is not part of a man's name, and except in a few instances, may be disregarded. *Ross v. Berry*, 124 P. 342, 17 N.M. 48 (1912).

The affixes "Sr." and "Jr." do not form part of a name, but are descriptive merely; hence in legal contemplation their presence or absence is immaterial. *State v. Lewis*, 83 A. 692, 83 N.J.L. 161(1912); *Windom v. State*, 72 S.W. 193, 44 Tex.Cr.R. 514 (1903).

STAKE. It is a settled rule of construction that when "stakes" are mentioned in a deed simply, or with not other added description than that of course and distance, they are intended by the parties, and so understood, to designate imaginary points. *Massey v. Belisle*, 24 N.C. 170 (1841).

STAKED OFF. As used in an act defining mining claims and regulating title thereto, and enacting that any claim, if abandoned for 10 consecutive days after being staked off, shall be forfeited to any person who may take up the same, and that no claim shall be regarded good and valid unless staked off with the owner's name, giving the date when the same was made, the phrase "staked off" evidently refers to marked boundaries of claims by stakes, or at least with the posting of stakes along the vein or its croppings, so as to indicate to other prospectors the ground intended to be appropriated, and could hardly be intended to mean simply the erection of a single stake, with a notice somewhat similar to that required on what is known as the "discovery stake." *Becker v. Pugh*, 13 P. 906, 9 Colo. 589 (1886).

STRAND. The word "strand," when used in reference to places anywhere in the vicinity of the sea, or an arm of the sea, comprises the territory between high and low water mark, over which the tide ebbs and flows. *Doane v. Willcutt*, 66 Am.Dec. 369, 71 Mass. (5 Gray) 328 (1855).

"Strand" is synonymous with "shore," and is that portion of the land lying between ordinary high and low water mark. *Stillman v. Burfeind*, 47 N.Y.S. 280, 21 App.Div. 13 (1897), citing 3 *Kent, Comm.* 431; *Town of East Hampton v. Kirk*, N.Y., 6 Hun, 257 (1875); *Champlain & St. L. R. Railway Co. v. Valentine*, N.Y., 19 Barb. 491 (1853); *Storer v. Freeman*, 6 Mass. 435, 4 Am. Dec. 155 (1810).

STRAW MAN. "Straw man" is a person who holds naked title for the benefit of another. *Kennedy v. Innis*, 158 N.E.2d 334, 339 Mass. 195 (1959).

STREAM. "Water course," "river," or "stream" consists of bed, banks, and stream of water. *Motl v. Boyd*, 286 S.W. 458, 116 Tex. 82 (1926).

"Stream" means "a river, brook, or rivulet anything in fact that is liquid and flows in a line or course." It does not cease to be a stream because its course may be opposed by some obstruction, natural or artificial, which causes its water to be deepened or its flow diminished in velocity. *French v. Carhart*, 1 N.Y.(1 Comst.) 96 (1847).

"Stream" is water flowing in defined channel as distinguished from mere surface drainage, but size of stream is immaterial and flow need not be continuous. *Midgett v. North Carolina State Highway Commission*, 132 S.E.2d 599, 260 N.C. 241 (1963).

STREET. A "street" is a way upon land, more properly a paved way, lined, or proposed to be lined, by houses on each side. *United States v. Bain*, 24 Fed.Cas. 94G.

A "street" means more than surface, and includes so much of depth as is or can be used, not unfairly, for ordinary purposes of street. *Haven Homes, Inc. v. Raritan Tp.*, 116 A.2d 25, 19 N.J. 239 (1955).

A "street" is a road or public way in a city, town, or village. A way over land set apart for public travel in a town or city is a "street," no matter by what name it may be called; it is the purpose for which it was laid out and the use made of it that determines its character. *Greil v. Stollenwerck*, 78 So. 79, 201 Ala. 303 (1918).

A "street" is a highway in an urban district. *Fischer v. Shasta County*, 299 P.2d 222, 46 C.2d 771 (1956).

A "street" is a city road. *Muesig v. Harz*, 283 Ill.App. 115 (1935).

SUBDIVIDE. "Subdivide" means to divide a tract of land into lots to sell before developing or improving them. "Subdivision" means to divide into smaller parts the same thing or subject matter. *Cowell v. Clerk*, 99 P.2d 594, 37 Cal.App.2d 255 (1940).

SUBJECT TO. The term "subject to" as used in conveyances is a term of qualification and not of contract. *Kokernot v. Caldwell*, 231 S.W.2d 528, error refused (1950).

The words "subject to" used in their ordinary sense, mean subordinate to, "subsequent to," or limited by. *Flower v. Town of Billerica*, 87 N.E.2d 189, 324 Mass. 519 (1949).

SUBSCRIBE. "Subscription" includes a mark by or for a person who cannot write, if his name he subscribed to an instrument and witnessed by a person who writes his own name as a witness. *Terry v. Johnson*, 60 S.W. 300, 109 Ky. 589 (1901).

SUCCESSION. The word "succession" is often used synonymously with the word "descent." *Adams v. Akerlund*, 48 N.E. 454, 168 Ill. 632 (1897).

SUNK LAND. A government survey or plat of a township selected by the state under the swamp lands act, Act Cong. Sept. 28, 1850, c. 84, 9 Stat. 519, showed that a certain part of the survey was not laid out into sections and subdivisions, and that the surveyed part was separated from the unsurveyed part by a meander line, the unsurveyed part being designated as "sunk lands" and in the surveyor's field notes described as low, wet lands. The township was patented to the state according to the official plats of the survey. Held, that a "meandered line", being an ordinary line bounding a body of land, there was nothing to show that the sunk lands was a body of water, though temporarily covered with water, and under the patent the entire township passed to the state as swamp lands. *Chapman & Dewey Lumber Co. v. Board or Directors St. Francis Levee Dist.*, 139 S.W. 625, 100 Ark. 94 (1911).

SURFACE LINE. The "surface line" of a street is as essential a part of the street as its lateral lines. *Righter v. City of Philadelphia*, 28 A. 1015, 161 Pa. 73 (1894).

SURFACE RIGHTS. "S R," as used in a contract sale of surface rights of land, means the entire surface of the land, reserving the minerals to the grantor. *Keweenaw Ass'n v. Friedrich*, 70 N.W. 896, 112 Mich. 442 (1897).

SURVEY. "Survey" is the substance and consists of the actual acts of the surveyor. *Outlaw v. Gulf Oil Corp.*, Tex.Civ.App., 137 S.W.2d 787 (1940).

A "survey" is process by which parcel of land is measured and its contents ascertained and is also a statement of or a paper showing the result of the survey with the courses and distances and quantity of the land. *Overstreet v. Dixon*, 131 S.E.2d 580, 107 Ga.App. 835 (1963).

The word "survey" is commonly used in old conveyances to mean the same as "tract," "boundary," or "land." *Burke v. Owens-Illinois Glass Co.*, D.C.W.Va., 86 F.Supp. 663.

To survey land means to ascertain corners, boundaries, divisions, with distances and directions, and not necessarily to compute areas included within defined boundaries; such computation being merely a matter of mathematics. *Kerr v. Fee*, 161 N.W. 545, 179 Iowa 1097 (1917).

"Survey," as used in a description in a trust deed conveying "the B. survey, lying in what is known as the I. pasture, in C. and A. counties," is synonymous with the word "land," or "grant," or "location." *Clark v. Gregory*, 26 S.W. 244 (Tex.Civ.App., 1894).

"Laid out," as used on the face of a map as laid out by a certain person, is equivalent to "as surveyed" by him, and embraces a reference to the monuments placed on the land by the surveyor. *Flint v. Long*, 41 P. 49, 12 Wash. 342 (1895).

A boundary line fixed by agreement of adjoining landowners is an "agreed boundary," in contradistinction to the "geographical" or "surveyed boundary." *Peebles v. McDonald*, 188 S.W.2d 289, 208 Ark. 834 (1945).

SURVEY OF LANDS. A "survey of lands," under the Spanish government, as with us, meant and consisted in the actual measurement of land, ascertaining the contents by running lines and angles with compass and chain, establishing corners and boundaries, and designating the same by marking trees, fixing monuments, or referring to existing objects of notoriety on the ground, giving bearings and distances, and making descriptive field notes and plots of the work. *Winter v. U.S.,* 30 Fed. Cas. 350; *U.S. v. Hanson,* 41 U.S. 196, 16 Pet. 198, 10 L.Ed. 935 (1842).

SWAMP. A "swamp" is defined as wet, spongy land, soft, low ground saturated but not usually covered with water. *Sharpe v. Savannah River Lumber Corp.,* 87 S.E.2d 398, 211 Ga. 570 (1955).

SWAMP LAND. "Swamp lands," as distinguished from "overflowed lands," are such as require drainage to dispose of needless water or moisture on or in the lands, in order to make them fit for successful and useful cultivation. *State ex rel. Ellis v. Gerbing,* 47 So. 353, 56 Fla. 603, 22 L.R.A., N.S., 337 (1908).

SYMBOLS. Words being mere "symbols," it is necessary, in construing testator's will, to compare words with things, persons, and events, in arriving at testator's intention. *Hersh v. Rosensohn,* 3 A.2d 877, 125 N.J.Eq. 1 (1939).

TAKEN. The word "taken" implies transfer of possession, dominion, or control. *Kimbell Trust & Savings Bank v. Hartford Accident & Indemnity Co.,* 164 N.E. 661, 333 Ill. 318 (1928).

Private property is "taken for public use" when it is appropriated to the common use of the public at large. *Craighall v. Lambert,* 18 S.Ct. 217, 168 U.S. 611, 421 L.Ed. 599 (1898).

TALLY. Equal to five (Gunter's) chains. (Sorden & Vallier)[1].

TAX TITLE. "Tax title" is the title by which one holds lands purchased at a tax sale. A "tax title" is not a derivative title. If valid, it is a breaking of all other title, and is antagonistic to all other claims to the land. *Willcuts v. Rollins,* 52 N.W. 199, 85 Iowa 217 (1892).

A "tax title" is generally regarded as a definite grant from the sovereignty, which bars all other titles, of record or otherwise, and imports an absolute paramount title as against the world. It is generally accepted that a tax title is a new title, which takes its status, not from the date of the tax judgement sale, but from the date of the tax lien. *Oakland Cemetery Ass'n v. Ramsey County,* 108 N.W. 857, 98 Minn. 404, 166 Am.St.Rep. 377 (1906).

TENEMENTAL LAND. Tenemental land is that part of a manor that is granted to tenants, as distinguished from the demesne lands. *Musgrave v. Sherwood,* N.Y., 54 How. Prac. 338, 28 Hun. 674 (1878).

THEN. The word, as used in wills, may be construed as either (1) an adverb of time, its ordinary grammatical acceptation, or (2) a word of reasoning or "particle of inference" connecting a consequence with a premise and meaning "in that event," or "in that case." 80 Am Jur 2d Wills, § 1171.

THENCE. The word "thence," as used in a description of land, means "from that place." *Tracy v. Harmon,* 43 P. 500, 17 Mont. 465 (1896).

---

[1] *Lumberjack Lingo. A Dictionary of the Logging Era.* North Word, Inc. 1986.

The omission of the word "thence" between two calls of the deed is immaterial, the sense being apparent. *Johnson v. Harris*, 68 S.W. 844, 24 Ky. Law Rep. 449 (1902).

In surveying, and in descriptions of land by courses and distances, this word, preceding each course given, imports that the following course is continuous with the one before it. *Flagg v. Mason*, 6 N.E. 702, 141 Mass. 66 (1886).

THENCE DOWN THE RIVER. The expression, "thence down the river," as used in field notes of a surveyor of a patent, is construed to mean with the meanders of the river, unless there is positive evidence that the meander line as written was where the surveyor in fact ran it; for such lines are to show the general course of the stream and to be used in estimating acreage, and not necessarily boundary lines. *Burkett v. Chestnutt,* 212 S.W. 271 (Tex. Civ. App., 1919).

THIRDS. The word "thirds," as used in a devise and bequest of all testator's estate, both real and personal, subject to the dower and thirds of his wife, obviously meant the same thing as dower. *O'Hara v. Dever*, N.Y., 46 Barb. 609, 2 Abb. Prac., N.S., 418 (1866), 41 N.Y. 558, 2 Keyes 558 (1866); *O'Hara v. Dever*, N.Y., 3 Abb. Dec. 407 (1866).

THOROUGHFARE. A "thoroughfare" is a passage through, an unobstructed way open to the public, a public road, a frequented street. *Jewett v. State*, Ohio, 22 O.L.A. 37 (1936).

THREAD OF STREAM. The thread of a stream is the line midway between the banks at the ordinary stage of water, without regard to the channel, or the lowest and deepest part of the stream. *State v. Burton*, 31 So. 291, 106 La. 732 (1902).

THROUGH. The word "through" ordinarily means passage across and within interior from one boundary to another boundary, and as used in land sale contract, certainly does not mean passage on outside of something. *Desantis v. Zell*, 98 N.E.2d 68, 45 0.0. 273, 60 O.L.A. 351 (1951).

THROUGH LOT. A lot bounded on one side by a street, upon another side by a court, and upon a third side by a street, alley or court way is not a "through lot." Under building ordinance, a "through lot" A being one abutting upon a street at each end. *Illinois Surety Co. v. O'Brien*, C.C.A.Ohio, 223 F. 933.

TIDE LANDS. "Tideland" is land covered and uncovered by the daily flux and reflux of the tides; "submerged land" is land lying oceanside of the tideland. *People v. Hecker*, 4 Cal.Rptr. 334, 179 C.A.2d 823 (1960).

TIDE MILL. A "tide mill" is understood to be a mill placed upon a dam thrown across a creek or inlet from the sea into which the tide naturally ebbs and flows, and wherein the water is raised by the flow of the tide, and at high water and the turn of the tide the water is stopped by the dam and the sluice gates, and kept to that height, until the tide has so far ebbed below the dam as to create a fall, by means of which the mill is worked a few hours, until the return of the flood tide prevents it. *Murdock v. Stickney*, 62 Mass. (8 Cush.) 113 (1851).

TIDE WATER. "Tide waters" are waters, whether salt or fresh, wherever the ebb and flow of the tide of the sea is felt. *Commonwealth v. Vincent*, 108 Mass. 441 (1871).

TIDEWAY. "Tideway" is that land between high and low water mark. *In re Inwood Hill Park in Borough of Manhattan*, City of New York, 217 N.Y.S. 359, 217 App.Div. 587 (1926).

TILLAGE. Tillage means husbandry—the cultivation of the land, particularly by the plow. In its popular and ordinary meaning, "tillage" applies to land used for agricultural, not for domestic and residential, purposes. *United States v. Williams*, 18 F. 475.

TITLE. "Title" is means by which estate is acquired. *Case v. Mortgage Guarantee & Title Co.*, 158 A. 724, 52 R.I. 155 (1932).

The labels "property" and "title" are used as group symbols to devote a "bundle" of rights or other legal relations. *Standard Oil Co. v. Clark*, C.C.A.N.Y., 163 F.2d 917.

"Title" means full, independent, and fee ownership. *In re Pelis' Estate*, 271 N.Y.S. 731, 150 Misc. 918 (1934).

TO RANGE. Line "to range" with another held to follow its path extended, and to be a continuation of it. *Lilly v. Marcum*, 283 S.W. 1059, 214 Ky. 514 (1926).

TOWN. A "town," under the township organization system, is a civil subdivision of a country. *People v. Grover*, 101 N.E. 216, 258 Ill. 124, Ann. Cas. 19148, 212 (1913).

"Toon," as used by a testator in devising all his interest in real estate described to be "60 acres, Sc. 25, toon 7," and so on, means "town." *Chambers v. Watson*, 14 N.W. 339, 60 Iowa 336, 46 Am.Rep. 70 (1882).

The word "town" includes plantations, unless otherwise expressed or implied. *Parker v. Williams*, 1 A. 138, 77 Me. 418 (1885).

The word "town" is sometimes employed to designate a township, but the term "incorporated town" is seldom, if ever, employed to embrace such a body. *Harris v. Schryock*, 82 Ill. 119 (1876).

TOWN COMMONS. "Town commons" is land that is the property of the town, and remains common and for public use generally. *Cutts v. Hussey*, 15 Me. 237 (1839).

TOWN DATUM. A "town datum" is a certain monument or object, of a permanent character, which has been adopted by the municipality as a base or starting point for the grades and levels of the municipality. *Chicago Consol. Traction Co. v. Village of Oak Park*, 80 N.E. 42, 225 Ill. 9 (1906).

TOWN WAY. "Town ways" are roads within the territorial limits of a particular town. *Inhabitants of Waterford v. Oxford County*, 59 Me. 450 (1871).

TOWNSHIP. A "township" is a territorial and political subdivision of the state and county, and is organized for the convenient exercise of some of the elementary functions of government. *Powers v. Thorn*, 129 P.2d 254, 155 Kan. 758 (1942).

In the Eastern States, "townships" are small geographical units and subdivisions of a county. *Rich v. Industrial Commission*, 15 P.2d 641, 80 Utah 511 (1932).

TOWNSITE. Ordinarily, "townsite" means portion of public domain segregated by proper authority and procedure as site of a town. *Metropolitan Life Ins. Co. v. Keating*, 254 N.W. 813, 191 Minn. 520 (1934).

"Townsite," as used in the statutes of Colorado, and in the states and territories of the West generally, means that portion of the public domain that is segregated from the great body of government land by the proper authority and procedure as a site for a town, and will not be held to apply to an unincorporated town or city. *Rice v. Colorado Smelting Co.*, 66 P. 894, 28 Colo. 519 (1901).

TP. "Tp.," as used in a deed of land described as "sections 22 and 28, Tp. 79, R. 13, Poweshiek County," is not uncertain or indefinite as to the location of the land. It is a contraction in almost universal use in describing lands, and everybody understands it to mean "township." *Ottumwa, C.F. & St.P. Ry. Co. v. McWilliams*, 32 N.W. 315, 71 Iowa 164 (1887).

TR. The abbreviation "Tr.," in a description of property in an assessment as being "in Los Angeles County in Pellissier Tr.," stood for "tract." *Baird v. Monroe*, 89 P. 352, 150 Cal. 560 (1907).

TRACT. "Tract" as ordinarily understood means contiguous bodies of land embraced in one deed. *Saulsberry v. Maddix*, 125 F.2d 430 (1942), certiorari denied 63 S.Ct. 36, two cases, 317 U.S. 643, 87 L.Ed. 518 (1942).

A lot, piece, or parcel of land, of greater or less size, the term not importing, in itself, any precise dimension. See *Edwards v. Derrickson*, 28 N.J. Law, 45 (1859); *Schofield v. Harrison Land & Mining Co.*, 187 S.W. 61 (Mo. Sup., 1916); *Smith v. Heyward*, 105 S.E. 275, 115 S.C. 145 (1920).

TRACT OR LOT. "The terms 'tract or lot' and 'pieces or parcel of real property,' or 'piece or parcel of land,' means any contiguous quantity of land in the possession of, owned by, or recorded as the property of the same claimant, person or company." In this connection the word "contiguous" means land that touches on the sides. Hence two quarters of the same section that only touch at the corner do not constitute, for the purpose of taxation, one tract or parcel of land. *Griffin v. Denison Land Co.*, 119 N.W. 1041, 18 N.D. 246 (1909).

TRUE LINE. The term "a true line," used in a surveyor's field notes in describing the line between two sections, means a straight line. *Lillis v. Urrutia*, 99 P. 992, 9 Cal.App. 558 (1908).

TURNPIKE. A turnpike road is a road having tollgates or bars on it, and on which tolls are collected. *Northam Bridge Co. v. London Ry.*, 6 M. 8 W. 428.

TURNING ROW. A turning row is a strip of unplowed ground lying between plowed fields on each side, and upon which the teams used in cultivating the lands are turned around. *Langan v. Whalen*, 93 N.W. 393, 67 Neb. 299 (1903).

UNIMPROVED LAND. A statutory term that includes lands, once improved, that have reverted to a state of nature, as well as lands that have never been improved. *Moore v. Morris*, 177 S.W. 6, 118 Ark. 516 (1915).

UNLESS. The word "unless" when found in instruments relating to title or in legislative enactments means except; if not; upon any less condition; in any other case; it implies a condition the nonhappening of which prevents a right from arising; "until" means during the whole time before; up to the time of, implying cessation or reversal at that time; it is a limitation of time during which a right cannot come into being, though the right may come into being when that period is ended. *In re Wiegand*, 27 F.Supp. 725 (1939).

UNMARKETABLE TITLE. A title is "unmarketable" when for the purchaser to accept the title proffered would lay him open to a fair probability of vexatious litigation with the possibility of serious loss. *Agliata v. D'Agostino*, 124 N.Y.S.2d 212 (1953).

UP A RUN. "Up a run," in description of boundary line, does not necessarily mean following a water course, but may mean merely "in the same direction," depending on the circumstances. *Westland Realty Corporation v. Griffin*, 145 S.E. 718, 151 Va. 1005 (1928).

UPLAND. Land lying above high-water mark is customarily called "upland." *Smith Tug & Barge Co. v. Columbia-Pacific Towing Corp.,* 443 P.2d 205 (Or., 1971).

UPLANDS. Lands bordering on bodies of waters. *Martin v. Busch*, 112 So. 274, 93 Fla. 535 (1927).

UPPER. In the absence of special circumstances modifying or charging its meaning the word "upper" as used in a deed to describe a part of a survey of land, made and platted upon a map, naturally suggests the north part of such survey. *Lunn v. Scarborough*, 24 S.W. 846, 6 Tex.Civ.App. 15 (1894).

VARA. The true Mexican "vara" is slightly less than 33 American inches, but by use in California it is estimated at 33 inches, and in Texas as 33-1/3 inches. *U.S. v. Perot*, 98 U.S. 428, 25 L.Ed. 251 (1878).

A vara, in Texas, has always been regarded as equivalent to 33-1/3 inches. A standard vara is somewhat less than 33-1/3 inches. Humboldt, in 1803, found a Mexican vara to be 839.16 millimeters, or a slight fraction over 33 inches. But it seems that a vara measure of somewhat larger dimensions obtained in Texas from an early period. The standard Mexican vara is so near to 33 inches that a standard vara measure laid on an American yard would so nearly correspond with 33 inches that a difference could not be perceived by the naked eye. *U.S. v. Perot*, 98 U.S. 428, 25 L.Ed. 251 (1878).

VACANT LAND. "Vacant land" is such as is absolutely free, unclaimed, and unoccupied. *Donley v. West*, 189 P. 1052 (Cal.App., 1920).

In reference to an application for a grant of land under the Georgia headright laws, an opinion of a surveyor that the premises in question were "vacant" means that they were ungranted. *Pritchett v. Ballard,* 29 S.E. 210, 102 Ga. 20 (1897).

VALLEY. As applied to a mountainous country, lowlands, in contradistinction to mountain slopes and ridges. *Whaley v. Northern Pac. R. Co.* (C. C.) 167 F. 664.

VARIATION. "Variation" is the difference in degrees between the direction of the true pole and the magnetic pole; it is comparatively fixed for any one locality, but is not the same in different localities. *The Aakre*, C.C.A.N.Y., 122 F.2d 469.

VENDOR. The word "vendor", in its ordinary signification, means one who sells, regardless of the character of the property sold. *Lumbert v. Woodward*, 43 N.E. 302, 144 Ind. 335, 55 Am.St.Rep. 175 (1896).

VILLAS. "Villas" is the name given to villages in the Spanish law. It is distinguished from cities ("*ciudades*") and towns ("pueblos"). *Hart v. Burnett*, 15 Cal. 530 (1860).

VIZ. "Viz." is a contradiction for videlicet, to wit, namely, that is to say. See, 67 Corpus Juris.

VEIN. "Vein," "lode," and "ledge" are synonymous terms, and mean any body of mineral or mineral rock within defined boundaries in the general mass of the mountains. Nor does the fact that it was occasionally found in the general course of this vein or shoot in pockets deeper down into the earth, or higher up, affect its character as a vein, lode, or ledge. *Synnott v. Shaughnessy*, 7 P. 82, 2 Idaho 122 (1885).

"Vein," like "lode," has breadth as well as length. *Inyo Marble Co. v. Loundagin*, 7 P.2d 1067, 120 Cal.App. 298 (1932).

VIGINTILLIONTH. A tax deed described the real estate as follows: "The east vigintillionth of a vigintillionth of the east 1/64 inch of lot one in the S.W. quarter of section 25, town 38, range 7, and the east vigintillionth of a vigintillionth of the east 1 inch of lot two in northwest 1/4 of section 25, town 38, range 7, in the county of Kane and state of Illinois. Recorded in the recorder's office of said Kane county, July 10, 1888, in book 250, page 286." A tract of land described as above may perhaps be pictured in the imagination but such a tract could not be bounded. It could not be located nor could a person take possession of such a tract of land personally. Such a tract could have no existence for the purposes for which lands are acquired and held. Under such circumstances the conveyance may be regarded as void. *Glos v. Furman*, 45 N.E. 1020, 164 Ill. 585 (1887).

WAIWAI. The word "waiwai" means property. *Hapai v. Brown*, 21 Haw. 499 (1913).

WAY. A "way" implies a right of passage over another's land; as an "easement," it cannot exist without a servient as well as a dominant estate. *Abbott v. Jackson*, 24 A. 900, 84 Me. 449 (1892).

WEST. Terms "east" and "west," used in description of boundary courses without modification or variation, mean due east and due west. *E.E. McCalla Co. v. Sleeper*, 288 P. 146, 105 C.A. 562 (1930).

WESTERLY. The word "westerly," as used in an order of the county court incorporating a village, which describes the commons as "on the west side of said limits one quarter of a mile in a westerly direction," should be construed to mean due west, rendering the description definite and certain. *State ex rel. Chandler v. Huff*, 79 S.W. 1010, 105 Mo.App. 354 (1904).

WESTERLY ONE-HALF OF. The description, in a complaint in partition of the land, involved as the "westerly one-half of" a specified lot and block according to a certain recorded plat, and so on, was sufficient. *Home Security Bldg. & Loan Ass'n of Alameda County v. Western Land & Title Co.*, 78 P. 626, 145 C. 217 (1904).

WESTERN OCEAN. The term "Western Ocean," in the Warwick patent to Connecticut, of 1631, granting all the land from certain points to the Western Ocean and to the South Sea, meant the Atlantic Ocean, as is shown from the fact that a grant to Sir Henry Roswell and others, dated four years earlier than the Warwick patent, employed the words "from the Atlantic and Western Sea and Ocean on the east part, to the South Sea on the west part." *Keyser v. Coe*, 14 Fed.Cas. 442.

WEST HALF. Where there is nothing to suggest the contrary, the word "half," in connection with the conveyance of a part of a tract of land, is interpreted as meaning half in quantity. The words "east half" and "west half" in a deed, while naturally importing an equal division, may lose that effect when it appears that at the time some fixed line or known boundary or monument divides the premises somewhere near the center, so that the expression more properly refers to one of such parts than to a mathematical division that never has been made. *Gunn v. Brower*, 105 P. 702, 81 Kan. 242 (1909).

In government surveys of public lands, terms "east half" and "west half" are used, not with reference to quantity, but to a line equidistant from the boundary lines of the parcel subdivided, and those terms have the same signification in patents issued by the government. A deed of the east half of a parcel of land according to the United States survey is definite and excludes the idea of two equal quantities, and fixes the dividing line equidistant from the boundary lines of the parcel thus subdivided. *Hoyne v. Schneider*, 27 P.2d 558, 138 Kan. 545 (1933).

WEST LINE OF RAILROAD. The term "west line of railroad," when used in a deed of lands to describe a boundary beginning on the west line of a certain railroad and southeast corner of land west of said railroad, thence south on the west line of said railroad, will be construed as showing the boundary to be on the west line of the railroad and not in the center thereof. Public roads and highways, and also railroads, are regarded as having three lines—the center line, which is usually the line surveyed when the road is laid out, and on each side of which the road is laid out, and the two side lines, at equal distances from the center line, and between which lies the territory covered by the road. When, in a conveyance of real estate adjoining a highway, such highway is referred to as constituting a boundary, the center line will be held to be the boundary so referred to, unless the language used in so referring to it show clearly that a side line instead of the center was intended. If it be doubtful which is intended, the law, from considerations of public policy, will resolve the doubt in favor of the center. *Maynard v. Weeks*, 41 Vt. 617 (1869).

WESTWARDLY. Courses in a grant indicated by the term "westwardly" run due west. *Seaman v. Hogeboom*, N.Y., 21 Barb. 398 (1855).

WILD DEED. A written instrument, in the form of a deed, acknowledged and recorded, wherein named grantor, knowing he, she or it has absolutely no title of any kind to the premises described therein, nonetheless executes the instrument. *Hyland v. Kirkman*, 498 A.2d 1278 (1985).

WILD LAND. Land in a state of nature, as distinguished from improved or cultivated land. *Clark v. Phelps*, 4 Cow. (N.Y.) 203 (1825).

Land in a wilderness state, not used in connection with improved estates. *Central Maine Power Co. v. Rollins*, 138 A. 170, 126 Me. 299 (1927).

WOOD LOT. The term "lot" is sometimes used in a restricted sense, being limited at the time, as "wood lot," "house lot," or "store lot"; but where the term is used unqualifiedly, especially if said to be a lot in a certain range or right, it is almost uniformly used in a technical sense, and means lot in a township, and duly laid out by the original proprietors. *White v. Gay*, 9 N.H. 126, 31 Am.Dec. 224 (1837).

WRITTEN WORDS. The term "written words" includes printed words. *Chaffin v. Lynch*, 1 S.E. 803, 83 Va. 106 (1887).

APPENDIX TWO

DEFINITIONS OF ANCIENT LAND TERMS*

ACRE. Originally the word "acre" was not used as a measure of land, or to signify any determinate quantity of land, or to signify any open ground, wide champaign, or field. Originally a strip in the fields that was ploughed in the forenoon.
Consists of 4 roods or 6150.4 sq. yds. or 5,353,6 sq. ft. or slightly more than 5/4 of an English statute acre of 4850 sq. yards (0.405 ha.).

ACRE RIGHT. The share of a citizen of a New England town in the common lands. The value of the acre right was a fixed quantity in each town but varied in different towns. A 10-acre lot or right in a certain town was equivalent to 113 acres of upland and 12 acres of meadow, and a certain exact proportion was maintained between the acre right and salable lands.

AGER. A field; land generally. A portion of land enclosed by definite boundaries.

BUTT OF LAND. A ridge of ground between two furrows. A measure of area of land of no standardized dimensions, being a strip of land or pathway between two parallel furrows of the open field.

CANT. A method of dividing property held in common by two or more joint owners.

CHAIN. A measure of land of 892.8 inches (22.677 m.) or 100 links of 8.928 inches each. This Scottish "Gunter's or Surveyor's" chain was larger than its English counterpart of 7.92 inches (20.116 m.) or 100 links of 7.92 inches each.

CHAMBER SURVEYS. At an early day in Pennsylvania, surveyors often made drafts on paper of pretended surveys of public lands, and returned them to the land office as duly surveyed, instead of going on the ground and establishing lines and marking corners; and these false and fraudulent pretenses of surveys never actually made were called "chamber surveys."

CISTA. A box or chest for the deposit of charters, deeds, and things of value.

* From: Wilson, Donald A. "Some Known, Little-known and Unknown Terms Relating to Land"; *The Benchmark,* Vol. 2, No. 3, November 1977; "Further Peculiar Terms Relating to Land Measurement," *The Benchmark,* Vol. 6, No. 2, Summer, 1981.

CIVIL TOWNSHIP. A legal subdivision of the county for governmental purposes.

CLOSE. A portion of land, as a field, inclosed, as by a hedge, fence, or other visible inclosure. The interest of a person in any particular piece of ground, whether actually inclosed or not. The noun "close," in its legal sense, imports a portion of land inclosed, but not necessarily inclosed by actual or visible barriers. The invisible, ideal boundary, founded on limit of title, which surrounds every man's land, constitutes his close, irrespective of walls, fences, ditches, or the like.

COSS. A term used by Europeans in India to denote a road measure of about two miles, but differing in different parts.

DAVACH. A measure of land in the extreme north generally considered equal to 4 ploughgates of 104 acres each, the ploughgate being divided into 8 oxgangs of 13 acres each, or 416 acres (c. 212.16 ha.) in all. The actual number of acres would vary, however, depending on the quality of the soil and other agricultural or topographic features.

DAYWERE. A term applied to land and signifying as much arable land as could be plowed up in one day's work.

DENARIATE. As much land as is worth one penny per annum.

DOMESDAY, DOMESDAY BOOK. An ancient record made in the time of William the Conqueror, and now remaining in the English exchequer, consisting of two volumes of unequal size, containing minute and accurate surveys of the lands in England.

DOWLE STONES. Stones dividing lands and so on.

ESTADAL. In Spanish America, a measure of land of 16 square varas, or yards.

FALL. Contains 6 ells (5.669 m.) or 6.2 English yards and also a measure of area consisting of 36 sq. ells (32.140 sq. m.) or 38.44 English sq. yds.

FANAGA. A measure of land varying in different provinces but in the Spanish settlements in America consisting of 6,400 sq. varas or yards.

FARDEL OF LAND. In old English law, the fourth part of a yard-land. It has been said an eighth only, because, two fardels make a nook, and four nooks a yard-land.

FARDING-DEAL. The fourth part of an acre of land.

FARTHING OF LAND. A great quantity of land, differing much from farding-deal.

FORSCHEL. A strip of land lying next to the highway.

GENERAL FIELD. Several distinct lots or pieces of land inclosed and fenced in as one common field.

GLEBE. The land possessed as part of the endowment or revenue of a church or ecclesiastical benefice.

HEADLAND. In old English law, a narrow piece of unplowed land left at the end of a plowed field for the turning of the plow.

HOLOGRAPH. A will or deed written entirely by the testator or grantor with his own hand.

HUEBRAS. In Spanish law, a measure of land equal to as much as a yoke of oxen can plow in one day.

HYD. In old English law, a measure of land containing, according to some, a hundred acres.

INBOUND COMMON. An uninclosed common, marked out, however, by boundaries.

INHOC. In old records, a nook or corner of a common or fallow field, inclosed and cultivated.

INNINGS. In old records, lands recovered from the sea by draining and banking.

JORNALE. In old English law, as much land as could be plowed in one day.

JUGERUM. An acre. As much as a yoke (jugum) of oxen could plow in one day.

LACE. A measure of land equal to one pole.

LINK. A measure of length of 8.928 inches (0.227 m.) equal to 1/100 of a chain. This Scottish "Gunter's or Surveyor's" link was larger than its English counterpart of 7.92 inches (0.2012 m.)

MERE-STONE. In old English law, a stone for bounding or dividing lands.

MERK. A measure of area for land in Shetland (c. 1800) varying from 1/2 to 2 acres (c. 0.20 to c. 0.81 ha.).

MET. A Scottish variant of "measure."

MORGEN. In old New York law, a measure of land, equal to about two acres.

NOOK OF LAND. In English law, twelve acres and a half.

OUT-BOUNDARIES. A term used in early Mexican land laws to designate certain boundaries within which grants of a smaller tract, which designated such out-boundaries, might be located by the grantee.

OUTLOT. In early American land law, a lot or parcel of land lying outside the corporate limits of a town or village but subject to its municipal jurisdiction of control.

OXGANG. A measure of area for land. Like the acreage of other superficial measures, the total acreage of the oxgang depended on local soil conditions, and so forth, but dimensions varying from 4 to 50 acres (c. 2.04 to c. 25.50 ha.) were the most common. The only time it was ever given legal definition was in the Assize of David I where it consisted of 13 acres.

PARTICULA. A small piece of land.

PASCUA. A particular meadow or pasture land set apart to feed cattle.

PECIA. A piece or small quantity of ground.

PERCA. A perch of land; 16 1/2 feet.

PICK OF LAND. A narrow slip of land running into a corner.

PLOUGHLAND. Generally considered equal to 8 oxgangs of varying acreage.

PLOW-LAND. A quantity of land "not of any certain content, but as much as a plow can, by course of husbandry, plow in a year."

POUND OF LAND. An uncertain quantity of land, said to be about 52 acres.

ROOD. Consists of 40 sq. falls (1,285.587 sq. m.) or 1,440 sq. ells equal to 13,838.4 English sq. feet or 0.3177 English acres. The English rood consisted of 40 sq. perches (0.101 ha.) equal to 1/4 statute acre of 160 sq. perches.

SELION OF LAND. In old English law, a ridge of ground rising between two furrows, containing no certain quantity, but sometimes more and sometimes less.

SHEEP COMMON. As much land as a sheep can graze in a day.

SLADE. In old records, a long, flat, and narrow piece or strip of ground.

STRAND. A shore or band of the sea or a river.

SWOLING OF LAND. So much land as one's plow can till in a year; a hide of land.

VERGE, or VIRGE. An uncertain quantity of land from 15 to 30 acres.

VIRGATA. A quarter of an acre of land. It might also be used to express a quarter of a hide of land.

WISTA. Half a hide of land, or 60 acres.

ZYGOCEPHALUM. A measure of land, as much land as a yoke of oxen could plow in a day.

INDEX